METHODS IN MOLECULAR BIOL

Series Editor
John M. Walker
School of Life and Medical Sciences
University of Hertfordshire
Hatfield, Hertfordshire, UK

For further volumes:
http://www.springer.com/series/7651

Statistical Genomics

Methods and Protocols

Edited by

Ewy Mathé

Biomedical Informatics, College of Medicine, Ohio State University, Columbus, OH, USA

Sean Davis

National Institutes of Health, National Cancer Institute, Bethesda, MD, USA

Humana Press

Editors
Ewy Mathé
Biomedical Informatics, College of Medicine
Ohio State University
Columbus, OH, USA

Sean Davis
National Institutes of Health
National Cancer Institute
Bethesda, MD, USA

ISSN 1064-3745 ISSN 1940-6029 (electronic)
Methods in Molecular Biology
ISBN 978-1-4939-3576-5 ISBN 978-1-4939-3578-9 (eBook)
DOI 10.1007/978-1-4939-3578-9

Library of Congress Control Number: 2016933669

Printed on acid-free paper

This Humana Press imprint is published by Springer Nature
The registered company is Springer Science+Business Media LLC New York

Preface

Statistical Analysis of Genomic Data is, indeed, a very broad topic. We have attempted in this volume to provide chapters with cross-cutting groundwork materials, public data repositories, common applications of statistical analysis in genomics, and some representative toolsets for operating on genomic data. While we cannot be comprehensive in a single volume, we have tried to provide a breadth of both applications and tools. The authors of the individual chapters have largely focused on practical aspects of their topics, as we feel that application is an integral part of learning about statistical analysis of genomic data.

More specifically, the volume is divided into four parts. In the first part, we have included overview material and resources that can be applied across topics later in the book. In the second part, a couple of prominent public repositories for genomic data are covered in some depth. In the third part, several different biological applications of statistical genomics are presented. In the fourth and last part, software tools that can be used to facilitate ad hoc analysis and data integration are highlighted. Finally, we thank the chapter authors for the generosity of their time and insight in preparing their excellent contributions.

Columbus, OH, USA *Ewy Mathé*
Bethesda, MD, USA *Sean Davis*

Contents

PART IV TOOLS

Contributors

NAOMI S. ALTMAN • *Department of Statistics and Huck Institutes of Life Sciences, The Pennsylvania State University, PA, USA*
SONALI ARORA • *Computational Biology Program, Fred Hutchinson Cancer Research Center, Seattle, WA, USA*
SONGJOON BAEK • *Laboratory of Receptor Biology and Gene Expression, National Cancer Institute, National Institutes of Health, Bethesda, MD, USA*
TANYA BARRETT • *National Center for Biotechnology Information, National Library of Medicine, National Institutes of Health, Bethesda, MD, USA*
GABRIEL BECKER • *Genentech, South San Francisco, CA, USA*
MATTHIAS BLUM • *Equipe Labellisée Ligue Contre le Cancer, Department of Functional Genomics and Cancer, Institut de Génétique et de Biologie Moléculaire et Cellulaire (IGBMC)/CNRS/INSERM/Université de Strasbourg, Illkirch Cedex, France*
MATTHEW BRAUER • *Genentech, South San Francisco, CA, USA*
MARC R.J. CARLSON • *Computational Biology Program, Fred Hutchinson Cancer Research Center, Seattle, WA, USA; Seattle Children's Research Institute, Seattle, WA, USA*
MARK A. CARTY • *Institute for Computational Biomedicine, Weill Cornell Medical College, New York, NY, USA; Memorial Sloan Kettering Cancer Center, New York, NY, USA*
BENILTON S. CARVALHO • *Brazilian Institute of Neuroscience and Neurotechnology (BRAINN) and Department of Statistics, University of Campinas, Campinas, São Paulo, Brazil*
LUKAS CHAVEZ • *German Cancer Research Center, Heidelberg, Germany*
YUNSHUN CHEN • *Walter and Eliza Hall Institute of Medical Research, Parkville, VIC, Australia*
PIERRE-ETIENNE CHOLLEY • *Equipe Labellisée Ligue Contre le Cancer, Department of Functional Genomics and Cancer, Institut de Génétique et de Biologie Moléculaire et Cellulaire (IGBMC)/CNRS/INSERM/Université de Strasbourg, Illkirch Cedex, France*
EMILY CLOUGH • *National Center for Biotechnology Information, National Library of Medicine, National Institutes of Health, Bethesda, MD, USA*
AEDIN CULHANE • *Department of Biostatistics and Computational Biology, Dana-Farber Cancer Institute, Boston, MA, USA; Department of Biostatistics, Harvard T.H. Chan School of Public Health, Boston, MA, USA*
OLIVIER ELEMENTO • *Institute for Computational Biomedicine, Weill Cornell Medical College, New York, NY, USA*
EIBE FRANK • *Department of Computer Science, University of Waikato, Hamilton, New Zealand*
HINRICH GRONEMEYER • *Equipe Labellisée Ligue Contre le Cancer, Department of Functional Genomics and Cancer, Institut de Génétique et de Biologie Moléculaire et Cellulaire (IGBMC)/CNRS/INSERM/Université de Strasbourg, Illkirch Cedex, France*
FLORIAN HAHNE • *Novartis Institute for Biomedical Research, Basel, Switzerland*
NANCY F. HANSEN • *National Human Genome Research Institute, Rockville, MD, USA*
LOREN A. HONAAS • *Department of Biology, The Pennsylvania State University, Wenatchee, WA, USA*

ROBERT IVANEK • *Department of Biomedicine, University of Basel, Basel, Switzerland*
MARK A. JENSEN • *Research Administration Directorate, Leidos Biomedical Research Inc., Frederick National Laboratory for Cancer Research, Frederick, MD, USA*
MARTIN KRZYWINSKI • *Canada's Michael Smith Genome Sciences Centre, Vancouver, BC, Canada*
M. SCOTT KUEHN • *Opower Inc., San Francisco, CA, USA*
MICHAEL LAWRENCE • *Genentech, South San Francisco, CA, USA*
MATTHIAS LIENHARD • *Max Planck Institute for Molecular Genetics, Berlin, Germany*
AARON T. L. LUN • *Walter and Eliza Hall Institute of Medical Research, Parkville, VIC, Australia*
MARCO ANTONIO MENDOZA-PARRA • *Equipe Labellisée Ligue Contre le Cancer, Department of Functional Genomics and Cancer, Institut de Génétique et de Biologie Moléculaire et Cellulaire (IGBMC)/CNRS/INSERM/Université de Strasbourg, Illkirch Cedex, France*
CHEN MENG • *Chair of Proteomics and Bioanalytics, Technische Universität Mnchen, Freising, Germany*
MARTIN MORGAN • *Computational Biology Program, Fred Hutchinson Cancer Research Center, Seattle, WA, USA*
SHANE NEPH • *Department of Genome Sciences, Altius Institute for Biomedical Sciences, Seattle, WA, USA*
VALERIE OBENCHAIN • *Computational Biology Program, Fred Hutchinson Cancer Research Center, Seattle, WA, USA*
HERVÉ PAGÈS • *Computational Biology Program, Fred Hutchinson Cancer Research Center, Seattle, WA, USA*
JENS REEDER • *Genentech, South San Francisco, CA, USA*
ALEX P. REYNOLDS • *Department of Genome Sciences, Altius Institute for Biomedical Sciences, Seattle, WA, USA*
MARKUS RIESTER • *Novartis Institutes for BioMedical Research (NIBR), Cambridge, MA, USA*
MOHAMED-ASHICK M. SALEEM • *Equipe Labellisée Ligue Contre le Cancer, Department of Functional Genomics and Cancer, Institut de Génétique et de Biologie Moléculaire et Cellulaire (IGBMC)/CNRS/INSERM/Université de Strasbourg, Illkirch Cedex, France*
TONY C. SMITH • *Department of Computer Science, University of Waikato, Hamilton, New Zealand*
GORDON K. SMYTH • *Walter and Eliza Hall Institute of Medical Research, Parkville, VIC, Australia*
JOHN A. STAMATOYANNOPOULOS • *Altius Institute for Biomedical Sciences, Seattle, WA, USA; Department of Medicine, University of Washington, Seattle, WA, USA; Department of Genome Sciences, University of Washington, Seattle, WA, USA*
MYONG-HEE SUNG • *Laboratory of Receptor Biology and Gene Expression, National Cancer Institute, National Institutes of Health, Bethesda, MD, USA; Laboratory of Molecular Biology and Immunology, National Institute on Aging, National Institutes of Health, Baltimore, MD, USA*
LEVI WALDRON • *Department of Epidemiology and Biostatistics, City University of New York, School of Public Health, New York, NY, USA*
ZHINING WANG • *Center for Cancer Genomics, National Cancer Institute, National Institutes of Health, Bethesda, MD, USA*

THOMAS D. WU • *Genentech, South San Francisco, CA, USA*
ZHIJIN WU • *Department of Biostatistics, Brown University, Providence, RI, USA*
HAO WU • *Department of Biostatistics and Bioinformatics, Rollins School of Public Health, Emory University, Atlanta, GA, USA*
JEAN CLAUDE ZENKLUSEN • *Center for Cancer Genomics, National Cancer Institute, National Institutes of Health, Bethesda, MD, USA*
HONGEN ZHANG • *Center for Cancer Research, National Institutes of Health, National Cancer Institute, Bethesda, MD, USA*

Part I

Groundwork

Chapter 1

Overview of Sequence Data Formats

Hongen Zhang

Abstract

Next-generation sequencing experiment can generate billions of short reads for each sample and processing of the raw reads will add more information. Various file formats have been introduced/developed in order to store and manipulate this information. This chapter presents an overview of the file formats including FASTQ, FASTA, SAM/BAM, GFF/GTF, BED, and VCF that are commonly used in analysis of next-generation sequencing data.

Key words Sequencing data file format, Next-generation sequencing, Sequencing data, FASTQ, FASTA, SAM/BAM, GFF/GTF, BED, VCF

1 Introduction

Next-generation sequencing (NGS) refers to high-throughput technologies for large scale DNA sequencing (such as whole genome sequencing, whole-exome sequencing, RNA-seq, miRNA-seq, ChIP-seq, and DNA Methylation) and could be conducted with different platforms, for example, Illumina(Solexa), Roche 454, Ion Torrent, and SOliD [1–5]. The main output of a next-generation sequencing experiment is short reads (short DNA sequences of <200 bases). In general, one NGS experiment can generate billions of short reads for each sample and stores the short reads in one or multiple files. While these raw sequencing data may contain some meta-data, genomic position information is not included in each short read. Therefore a series of procedures must be applied to parse and manipulate the raw reads in downstream analysis in order to uncovering interested genomic structures and variations [6–11]. In addition to short sequences, downstream analysis will add more information such as description and annotations, which may increase the file size from few gigabytes to hundreds gigabytes. To efficiently store, view, and manipulate such information, various file formats have been introduced/developed [12–16].

Ewy Mathé and Sean Davis (eds.), *Statistical Genomics: Methods and Protocols*, Methods in Molecular Biology, vol. 1418, DOI 10.1007/978-1-4939-3578-9_1, © Springer Science+Business Media New York 2016

This chapter gives an overview of file formats commonly used in storage and analysis of next-generation sequencing data.

2 Results

2.1 FASTQ: Format of Raw Short Read Files

FASTQ format was originally invented by Jim Mullikin at the Wellcome Trust Sanger Institute to storing both nucleotide sequence and its corresponding Phred quality scores [17, 18] and currently it has became a common file format for sharing sequencing read data [13–16].

FASTQ files are plain text file with extension ".fq" or ".fastq" and could be viewed directly from command line on computers with Unix/Linux operating system. In FASTQ file, each sequence (short read of NGS) is defined by four lines of text:

The first line starts with a "@" character and is followed by a sequence identifier and an *optional* description.

The second line is the raw sequence letters: A, T, G, C, and N (unknown).

The third line begins with a "+" symbol and is optionally followed by the same sequence identifier (and any description) again. The "+" sign serves as a marker indicating the end of sequence.

The fourth line is the quality values for sequence in the second line, and must contain the same number of symbols as letters in the sequence.

Following is an example of a single sequence from Illumina sequence system:

```
@HWI-ST193:542:C2H0GACXX:8:1101:4404:2179 1:Y:0:ACACGA
ATGCNTTTTATAATCAAAAGCGAAGACCTAGCAGGAGGTTAAAAACCTTT
+
<<<<#2<@@5:9@44:@@?4(-8@(<9@<<658.==5=0<>??????9??
```

The first line is sequencer identifier and description items separated by ":" or " ". The descriptions may vary for different sequencers or processing pipelines. In this example, each item identifies as following:

HWI-ST193: sequencer name (should be unique, usually from manufacturer).

542: run id.

C2H0GACXX: flowcell id.

8: flowcell lane.

1101: tile number within the flowcell lane.

4404: *x*-coordinate of the cluster within the tile.

2179: *y*-coordinate of the cluster within the tile.

1: the member of a pair (1 for single read, and 2 for *paired-end and mate-pair reads*).

Y: *filter status* (Y if the read is filtered, N otherwise).

0: status of control bits (0 for none of the control bits are on, otherwise it is an even number).

ACACGA: index sequence.

The second line of above example is the sequence contents and the fifth base is unable to determine so an N is used. The third line has only a "+" sign and has no more information added.

The fourth line is the quality scores (Phred quality scores) for corresponding bases in the second line.

The Phred quality scores stand for sequencing quality for each base and the higher the Phred score the more accurate the base to be determined. Phred scores range from 0 to 93 and are encoded with following characters in the order from left to right:

```
!"#$%&'()*+,-./0123456789:;<=>?@ABCDEFGHIJKLMNOPQRSTU
VWXYZ[\]^_`abcdefghijklmnopqrstuvwxyz{|}~
```

The "!" represents the lowest Phred score and the "~" is for the highest one.

Phred score encoding varies for different sequencers and cannot always be directly compared among them until they have been converted to probabilities through a transformation defined by the sequencer manufacturer.

FASTQ format has no restriction on total number of single sequences in a FASTQ file but since the number of short reads from next-generation sequencing experiment is very huge, for convenience, each sample will use several FASTQ files to hold all reads. In most case, the FASTQ files are compressed with software to GNU zip format (with .gz file extension) to reduce file size.

Due to the high volume of sequence reads in a FASTQ file and no genomic position information for each reads, it is not practical to view or check a single reads manually. During the analysis, the quality analysis of short reads is usually performed with computer software. The most commonly used one is FASTQC developed by Simon Andrews at Babraham Institute [19] which takes FASTQ files as input and generates summary graphs and tables for a quick overview of the raw sequence read quality.

2.2 FASTA: Format of Reference Genome Files

FASTA format is also text based format and commonly uses file extension ".fa" or ".fasta". FASTA has been standard format for nucleotide sequence since the first generation sequencing [20, 21]. For next-generation sequencing, it is the standard format for reference genome sequences used by mapping/alignment software tools [22–25].

FASTA format is very simple. It starts with a header line followed by lines of sequence. The header line begins with a ">" sign for sequence identifier and may contain optional descriptive information. Sequence lines consist of characters representing nucleotide bases in the sequence and are usually no more than 80 characters per line but have no limitation for total number of lines. Same as sequence in FASTQ files, each nucleotide base is encoded as a single character (one of A, T, G, C, and N if undetermined, with case insensitive).

FASTA format can be used to display sequence of a single gene, a single chromosome, or multiple ones. Following is the header line and first two lines of human p53 gene (NC_000017.11) sequence in FASTA format (http://www.ncbi.nlm.nih.gov/nuccore/NC_000017.11?report=fasta&from=7668402&to=7687550&strand=true):

```
>gi|568815581:c7687550-7668402 Homo sapiens chromosome
17, GRCh38 Primary Assembly
GATGGGATTGGGGTTTTCCCCTCCCATGTGCTCAAGACTGGCGCTAAAAGT TTT
GAGCTTCTCAAAAG TC
TAGAGCCACCGTCCAGGGAGCAGGTAGCTGCTGGGCTCCGGGGACACTTTG CGT
TCGGGCTGGGAGCGTG
```

For a single chromosome, the sequence identifier can be only the chromosome name, for example, below is partial sequence of human chromosome 1 in FASTA format:

```
>chr1
NNNNNNNNNNNNNNNNNNNNNNNNNNNNNNNNNNN NNNNNNNNNNNNNNNNNN
TAACCCTAACCCTAACCCTAACCCTAACCCTAACCCTAACCCTAACCCTA
ACCCTAACCCTAACCCTAACCCTAACCCTAACCCTAACCCTAACCCTAAC
CCTAACCCAACCCTAACCCTAACCCTAACCCTAACCCTAACCCTAACCCC
```

A small FASTA file could be viewed with text editing tools and command line of UNIX/LINUX platforms. For big FASTA files such as a reference genome in FASTA format, Java based stand along software, Integrative Genomics Viewer (IGV) developed by Broad Institute in Boston, is a commonly used tool to navigate genome sequence [26, 27] and accompanying features aligned to it.

2.3 BAM/SAM: Format of Alignment (Mapping) Data Files

BAM/SAM file has import role in analysis of next-generation sequencing data because the raw sequence reads from sequencers have no genomic position information associated and they must be mapped/aligned to known reference genome. The mapping/alignment outputs are BAM/SAM files that will serve as inputs for various downstream analysis such as feature counts and variant calling.

SAM stands for *S*equence *A*lignment/*M*ap format, and is a generic alignment format for storing read alignments against reference sequences, supporting short and long reads (up to 128 Mbp) produced by different sequencing platforms [14]. A SAM file often uses file extension ".sam" and is a tab-delimited text file that contains sequence alignment data. BAM is the binary version of a SAM file and usually has a file extension ".bam". Both BAM file and SAM file contain same information and with SAMtools software, a BAM file can be easily converted to SAM file.

A BAM/SAM file has two sections: header section and alignment section. Details of BAM/SAM format is described on the SAMtools website (http://samtools.sourceforge.net) and here is a short description of header section and alignment section.

2.3.1 Header Section

The header section is optional for BAM/SAM file. If present, it provides generic information for BAM/SAM file. Header section can have multiple lines and each line must start with a "@" symbol. The information can be divided into four types: @HD, @SQ, @RG, and @PG and contents of each line are tab-delimited TAG:Value field(s). An optional comment line starting with @CO may exist at the end of header section.

@HD Line

The header line. If header section exists in a BAM/SAM file, the first line must start with @HD followed by required and optional field(s) including:

- VN: format version. This field is required.
- SO: sorting order of alignments. Valid values include: unknown (default), unsorted, queryname, and coordinate.
- GO: group order (full sorting is not imposed in a group). Valid values are: none, query, or reference.

@SQ Lines

@SQ line serves as reference sequence dictionary. A BAM/SAM file usually has multiple @SQ lines and the order of @SQ lines defines the alignment sorting order. @SQ line has following fields:

- SN: reference sequence name. This field is required and each @SQ line must have a unique SN tag.
- LN: reference sequence length with range of 1 to $2^{31} - 1$. This is a required field.
- AS: genome assembly identifier.
- M5: MD5 checksum of the sequence in the uppercase, excluding spaces but including pads (as "*"s).
- SP: species.
- UR: URI of the sequence. This value may start with one of the standard protocols, e.g., http: or ftp:. If it does not start with one of these protocols, it is assumed to be a file-system path.

@RG Lines

Read group information is listed in @RG line(s). Unordered multiple @RG lines are allowed. @RG line has also required and optional fields:

- ID: read group identifier. This is a required field and each @RG line must have a unique ID. The value of ID is used in the RG tags of alignment records. It must be unique among all read groups in header section. Read group IDs may be modified when merging SAM files in order to handle collisions.
- CN: name of sequencing center producing the read.
- DS: description.
- DT: the date the run was produced (using ISO8601 date or date/time format).
- FO: flow order.
- KS: the array of nucleotide bases that correspond to the key sequence of each read.
- LB: library.
- PG: programs used for processing the read group.
- PI: predicted median insert size.
- PL: platform/technology used to produce the reads. Valid values: CAPILLARY, LS454, ILLUMINA, SOLID, HELICOS, IONTORRENT, and PACBIO.
- PM: Platform model. Free-form text providing further details of the platform/technology used.
- PU: platform unit (e.g., flowcell-barcode lane for Illumina or slide for SOLiD). It must be a unique identifier.
- SM: sample name. Use pool name where a pool is being sequenced.

@PG Lines

These lines describe the program used to generate BAM/SAM file with following fields:

- ID: program record identifier. Each @PG line must have a unique ID. The value of ID is used in the alignment PG tag and PP tags of other @PG lines.
- PN: program name.
- CL: command line.
- PP: previous @PG-ID. It must match another @PG header's ID tag.
- DS: description of the program.
- VN: program version.

@CO Line

One-line text comment. Unordered multiple @CO lines are allowed.

Following is an example of a BAM file header viewed with SAMtools software:

```
@HD VN:1.4 GO:none SO:coordinate
@SQ SN:chr1 LN:249250621
```

```
@SQ SN:chr2 LN:243199373
@SQ SN:chr3 LN:198022430
@RG ID:1048 PL:Illumina LB:4949 SM:PAVECB_Tumor
@RG ID:1054 PL:Illumina LB:4949 SM:PAVECB_Tumor
@RG ID:26343 PL:Illumina LB:5144 SM:PAVECB_Tumor
@PG ID:GATK IndelRealigner VN:3.1-1-g07a4bf8
CL:knownAlleles=[(RodBinding name=knownAlleles
source=/data/CCRBioinfo/public/GATK/bundle/2.3/hg19/Mills_and_1000G_gold_
standard.indels.hg19.vcf)]
targetIntervals=bam/SAMPLE/DNA/SAMPLE.intervals
LODThresholdForCleaning=5.0 consensusDeterminationModel=USE_READS entropy
Threshold=0.15
maxReadsInMemory=150000maxIsizeForMovement=3000
maxPositionalMoveAllowed=200maxConsensuses=30 maxReadsForConsensuses=120
maxReadsForRealignment=20000 noOriginalAlignmentTags=false nWayOut=null
generate_nWayOut_md5s=false check_early=false noPGTag= false
keepPGTags=false indelsFileForDebugging=null
statisticsFileForDebugging=null SNPsFileForDebugging=null
```

This example shows the BAM file version is 1.4 and alignments are sorted by genomic coordinates. Reference sequences used are chromosome 1 ~ 3 and sequence length are listed for each chromosome. The platform is Illumina sequencer and sample name is PAVECB_Tumor. Software used to generate this BAM file is GATK IndelRealigner, version 3.1-1-g07a4bf8, and a copy of the command line text is included.

2.3.2 Alignment Section

Alignment section in SAM/BAM file contains sequences with genomic position and other descriptive information. In BAM/SAM format, each single sequence (short reads from FASTQ file) and its associated information are presented as one line text and each line consists of multiple tab-delimited text fields. There are 11 mandatory fields for each line and these fields are always in the same order. Mandatory fields provide:

- QNAME: query name, same as the sequence identifier in FASTQ file.
- FLAG: bitwise flag of two byte length to indicate the read property such as the segment mapped or unmapped, passed or not passed quality controls. An easy way to convert the flag to human readable meaning can be found at https://broadinstitute.github.io/picard/explain-flags.html.
- RNAME: reference sequence name, one from SN field defined in @SQ line of header section.
- POS: the leftmost position of the first matching base in reference sequence.

- MAPQ: mapping quality.
- CIGAR: a character string indicating the match status of bases in the short read. For example, a CIGAR string "3M1I3M1D5M" indicates that, in a short read with 13 base long, the first 3 bases match to reference sequence, the fourth base is an insertion (not found in reference sequence), following 3 bases match to reference sequence and 1 base is missing in the read comparing to reference sequence, and the last 5 bases match to reference sequence.
- RNEXT: reference name of next aligned read.
- PNEXT: position of the primary alignment of the next read.
- TLEN: signed observed template length.
- SEQ: sequence of the short read. Same as from FASTQ file.
- QUAL: quality scores for each base in the short read, same as the one in FASTQ file.

One or more optional fields may follow after the mandatory fields. Details can be found in SAM file documentation.

Below is a single line from BAM file viewed with SAMtools software:

```
HWI-ST1108:4:1214:9025:65227#1 1187 chr1 12111 0 101M =
12121 111 CATTGTTCATCTTCT
GGCCCCTGTTGTCTGCATGTAACTTAATACCACAACCAGGCATAGGGGAAA
GATTGGAGGAAAGATGAGTGACAGCATCAACTTCT
CCCFFFFFHHHHHJJJJJIJJJJJJJJJJJJJJIJJJJJIJJJIJJJIJJJIJ
JJIJIJJJJFHJJIHHHHHFFFFDEEEEDDCDDDDDDDDDDDDDDA      MC:
Z:101M RG:Z:26316 NM:i:1 MQ:i:0 AS:i:96 XS:i:96
```

In addition to the SAMtools, the software of Integrative Genomics Viewer [26, 27] is commonly used to view BAM/SAM file contents against a reference genome.

2.4 GFF/GTF and BED: Formats for Genomic Feature Annotation Files

Uncovering the genetic structures/variations and their biological meaning is one of final goals of next-generation sequencing data analysis. To meet this goal, next-generation sequencing data is often summarized at different annotation levels. Currently, the GFF/GTF and BED format are among of most commonly used data formats to provide genomic annotation information for next-generation sequencing data analysis.

The GFF stands for *G*eneral *F*eature *F*ormat. GFF format was initially proposed by Richard Durbin and David Haussler in Wellcome Trust Sanger Institute in England and its version 2 was introduced in 1998. The GTF (*G*eneral *T*ransfer *F*ormat) format is identical to GFF version 2 [28, 29]. GFF file uses file extension ".gff" and GTF files have extension ".gtf".

In GFF/GTF file, each feature is represented with one line text and each line has nine columns (fields) of data values, plus optional track definition lines. Data fields must be tab-separated and each field must contain a value. "empty" columns should be denoted with a ".". Data fields in each line must be in the same order as <seqname> <source> <feature> <start> <end> <score> <strand> <frame> [attributes] [comments] (<> denotes mandatory fields and [ ] is optional).

Following are field names and descriptions of values for relevant fields:

- seqname: The name of the sequence. Having an explicit sequence name allows a feature file to be prepared for a data set of multiple sequences. Normally the seqname will be the identifier of the sequence in an accompanying fasta format file. An alternative is that <seqname> is the identifier for a sequence in a public database, such as an EMBL/Genbank/DDBJ accession number. Which is the case, and which file or database to use, should be explained in accompanying information.
- source: name of the program that generated this feature, or the data source (database or project name), the source of this feature. This field will normally be used to indicate the program making the prediction, or if it comes from public database annotation, or is experimentally verified, etc.
- feature: feature type name, e.g., gene, exon, transcript.
- start: Start position of the feature, with sequence numbering starting at 1.
- end: End position of the feature, with sequence numbering starting at 1.
- score: A floating point value.
- strand: defined as + (forward) or – (reverse).
- frame: One of "0", "1", "2", or ".". "0" indicates that the specified region is in frame, i.e., that its first base corresponds to the first base of a codon. "1" indicates that there is one extra base, i.e., that the second base of the region corresponds to the first base of a codon, and "2" means that the third base of the region is the first base of a codon. If the strand is "–", then the first base of the region is value of <end>, because the corresponding coding region will run from <end> to <start> on the reverse strand. As with <strand>, if the frame is not relevant then set <frame> to ".". It has been pointed out that "phase" might be a better descriptor than "frame" for this field.
- attribute: A semicolon-separated list of tag-value pairs, providing additional information about each feature. From version 2 onwards, the attribute field must have an tag value structure following the syntax used within objects in a .ace file,

flattened onto one line by semicolon separators. Tags must be standard identifiers ([A-Za-z][A-Za-z0-9_]*). Free text values must be quoted with double quotes.

GFF/GTF files may contains comment lines at the beginning and all comment lines must start with "#". For example, the first line of a GFF file may be

```
##gff-version 2
```

which indicates the file uses version 2 of GFF format.

Since GFF/GTF files are text based they can be viewed from command line of Unix/Linux computers. Microsoft Excel and other text edit software can also be used to access small GFF/GTF files. Below is partial content of a GFF/GTF file.

```
#!genome-build GRCh38
#!genome-date 2013-12
#!genome-build-accession NCBI:GCA_000001405.15
#!genebuild-last-updated 2014-08
1 havana    gene  11869  14409  .  +  .  gene_id "ENSG00000223972"
1 havana    exon  11869  14409  .  +  .  gene_id "ENSG00000223972"
1  havana   exon  11869  12227  .  +  .  gene_id "ENSG00000223972"
1  havana   exon  12613  12721  .  +  .  gene_id "ENSG00000223972"
```

In this example, the first four lines are comments showing the genome version is GRCh38 generated in December of 2013 and accession number is NCBI:GCA_000001405.15. All gene information was updated in August, 2014. The following four lines display the information of Ensembl gene ENSG00000223972 in the order of chromosome name, data source name, start and end position, score, strand, frame information, and attribute (here gene id).

BED format is another flexible way to define the features of genome annotation [30, 31]. BED format is also tab-delimited text file and each line represents one feature. A basic BED file has only three required fields in each line, i.e., chrom, chromStart, and chromEnd to define chromosome name, start, and end position of the feature on the chromosome. A detailed BED file will have extra optional fields to describe how the feature will be displayed (specifically, in Web browser). The optional fields are:

- name: defines the name of the BED line. This label is displayed to the left of the BED line in a genome browser window when the track is open to full display mode or directly to the left of the item in pack mode.
- score: a score between 0 and 1000. If the track line *useScore* attribute is set to 1 for this annotation data set, the *score* value will determine the level of gray in which this feature is displayed (higher numbers = darker gray).
- strand: defined as + (forward) or – (reverse).

- thickStart: The starting position at which the feature is drawn thickly (for example, the start codon in gene displays). When there is no thick part, thickStart and thickEnd are usually set to the chromStart position.
- thickEnd: The ending position at which the feature is drawn thickly (for example, the stop codon in gene displays).
- itemRgb: An RGB value of the form R,G,B (e.g., 255,0,0). If the track line *itemRgb* attribute is set to "On", this RBG value will determine the display color of the data contained in this BED line.
- blockCount—The number of blocks (exons) in the BED line.
- blockSizes—A comma-separated list of the block sizes. The number of items in this list should correspond to blockCount.
- blockStarts—A comma-separated list of block starts. All of the blockStart positions should be calculated relative to chromStart. The number of items in this list should correspond to blockCount.

BED file can also contain track line definition to configure the display further, e.g., by grouping features into separate tracks. Track lines should be placed at the beginning of the list of features they are to affect. And they should consist of the word "track" followed by space-separated key=value pairs which include name, description, priority, and useScore of the track.

An example of BED file with track line definition looks like below:

```
track name="ItemRGBDemo" description="Item RGB demonstration" itemRgb="On"
chr7 127471196 127472363 Pos1 0 + 127471196 127472363 255,0,0
chr7 127472363 127473530 Pos2 0 + 127472363 127473530 255,0,0
chr7 127473530 127474697 Pos3 0 + 127473530 127474697 255,0,0
```

2.5 VCF: A Special File Format for Genomic Variations

VCF (*V*ariant *C*all *F*ormat) format was initially introduced by the 1000 Genomes Project to store the most prevalent types of genomic sequence variation, such as SNPs (single nucleotide polymorphism) and small INDELS (insertions/deletions), enriched by annotations [16, 32, 33]. VCF file is text file and contains meta-information (header) lines and data lines. Most VCF files contain both header lines and data lines. The header lines start with "##" and have a field in the format of "ID=value". The first header line is always the VCF format version followed by lines starting with ##INFO=, #FILTER=, and ##FORMAT, which define the name, length, value type, and description of each item in relevant fields (columns) of each data line.

Below is an example of VCF file header:

```
##fileformat=VCFv4.1
##source=VarScan2
##INFO=<ID=ADP,Number=1,Type=Integer,Description="Average per-sample depth of bases with Phred score >= 15">
##INFO=<ID=WT,Number=1,Type=Integer,Description="Number of samples called reference (wild-type)">
##INFO=<ID=HET,Number=1,Type=Integer,Description="Number of samples called heterozygous-variant">
##INFO=<ID=HOM,Number=1,Type=Integer,Description="Number of samples called homozygous-variant">
##INFO=<ID=NC,Number=1,Type=Integer,Description="Number of samples not called">
##FILTER=<ID=str10, Description="Less than 10 % or more than 90 % of variant supporting reads on one strand">
##FILTER=<ID=indelError,Description="Likely artifact due to indel reads at this position">
##FORMAT=<ID=GT,Number=1,Type=String,Description="Genotype">
##FORMAT=<ID=GQ,Number=1,Type=Integer,Description="Genotype Quality">
##FORMAT=<ID=SDP,Number=1,Type=Integer,Description="Raw Read Depth as reported by SAMtools">
##FORMAT=<ID=DP,Number=1,Type=Integer,Description="Quality Read Depth of bases with Phred score >= 15">
##FORMAT=<ID=RD,Number=1,Type=Integer,Description="Depth of reference-supporting bases (reads1)">
##FORMAT=<ID=AD,Number=1,Type=Integer,Description="Depth of variant-supporting bases (reads2)">
##FORMAT=<ID=FREQ,Number=1,Type=String,Description="Variant allele frequency">
##FORMAT=<ID=PVAL,Number=1,Type=String,Description="P-value from Fisher's Exact Test">
##FORMAT=<ID=RBQ,Number=1,Type=Integer,Description="Average quality of reference-supporting bases (qual1)">
##FORMAT=<ID=ABQ,Number=1,Type=Integer,Description="Average quality of variant-supporting bases (qual2)">
##FORMAT=<ID=RDF,Number=1,Type=Integer,Description="Depth of reference-supporting bases on forward strand (reads1plus)">
##FORMAT=<ID=RDR,Number=1,Type=Integer,Description="Depth of reference-supporting bases on reverse strand (reads1minus)">
##FORMAT=<ID=ADF,Number=1,Type=Integer,Description="Depth of variant-supporting bases on forward strand (reads2plus)">
##FORMAT=<ID=ADR,Number=1,Type=Integer,Description="Depth of variant-supporting bases on reverse strand (reads2minus)">
```

In this example, VCF file version is v4.1 and the file is generated with VarScan2 software. The ##INFO lines listed five IDs which will be used in the INFO column of data line to describe, for the variant in each line, the average read depth of bases with Phred score ≥ 15, how many of samples are wild type, heterozygous, homozygous, or no call. The two ##FILTER lines listed how a variant being filtered. The nine ##FORMAT lines show what contents will and how be arranged in the FORMAT column of each data line.

Data lines in VCF file are tab-delimited text lines and each line contains nine fixed fields (columns) followed by one or more sample column(s). The first nine fields are:

- CHROM: chromosome name.
- POS: the leftmost position of the variant in the sequence.
- ID: variant identifier such as SNP id. "." denotes unavailable.
- REF: reference base (for SNP) or sequence (for INDEL).
- ALT: alternate base or bases (the observed variant).
- QUAL: Phred-scaled quality score.
- FILTER: fiter status. PASS for passed all filters or a semicolon-separated list of code for filters that fail.
- INFO: additional information in <key>=<data> format.
- FORMAT: colon separated key and value.
- Sample field(s): one or more sample columns may exist in a VCF file. This field has the values for a sample arranged in the format defined in previous FORMAT field.

The first seven columns in each VCF data line are information of the variant itself, for example, the line below shows variant is on chromosome 1 at position of 880639 base, no ID (snp) since it is a deletion, reference bases are TC and C is missing in sample sequence. The call of this variant has no quality score but has passed the filtering.

```
#CHROM  POS  ID  REF  ALT  QUAL  FILTER
1  880639  .  TC  T  .  PASS
```

The INFO and FORMAT columns in VCF data line list information of the variant in sample(s). To understand the meaning of contents in these two columns, one must refer to the information given by header lines. For example, the contents of "ADP=10; WT=0;HET=1;HOM=0;NC=0" in INFO column indicate that average depth (ADP) for the variant in the samples is 10, one heterozygous (HET) is called and no other calls. The FORMAT column lists all tags (type of information) and their order for each sample. If a FORMAT column has following content: "GT:GQ: SDP:DP:RD:AD:FREQ:PVAL:RBQ:ABQ:RDF:RDR:ADF:ADR" and values in a sample column is as "0/1:22:10:10:4:6:60 %:5.418E-3:36:38:3:1:0:6", the information listed for the sample will be genotype (0/1), genotype quality (22), raw read depth from SAMtools (10), quality read depth with Phred score>=15 (10), depth of reference-supporting bases (4), depth of variant-supporting bases (6), variant allele frequency (60 %), *P*-value from Fisher's Exact Test (0.005418), average quality of reference-supporting bases (36), average quality of variant-supporting bases (38), depth of reference-supporting bases on forward strand (3), depth of reference-supporting bases on reverse strand (1), depth of variant-supporting bases on forward strand (0), and depth of variant-supporting bases on reverse strand (6).

Detailed specification of VCF format can be found at https://github.com/samtools/hts-specs. VCFtools [16] is a software equipped with numerous functionalities to process VCF files. A small VCF file can also be viewed from command line of Unix/Linux computer or Microsoft Excel.

3 Conclusions

Next-generation sequencing technologies have brought new file formats and resurrected old formats. Understanding the basics of these formats is important to knowing what information they contain and how they can be used in various steps from raw data generation to interpretable biological results.

References

1. Shendure J, Ji H (2008) Next-generation DNA sequencing. Nat Biotechnol 26:1135–1145
2. Metzker ML (2010) Sequencing technologies—the next generation. Nat Rev Genet 11:31–46
3. Quail MA, Smith M, Cooupland P et al (2012) A tale of three next generation sequencing platforms: comparison of Ion Torrent, Pacific Biosciences and Illumina MiSeq sequencers. BMC Genomics 13:341
4. Mardis ER (2008) Next-generation DNA sequencing methods. Annu Rev Genomics Hum Genet 9:387–402
5. Mardis ER (2013) Next-generation sequencing platforms. Annu Rev Anal Chem 6:287–303
6. Flicek P, Birney E (2009) Sense from sequence reads: methods for alignment and assembly. Nat Methods 6(Suppl 11):S6–S12
7. Medvedev P, Stanciu M, Brudno M (2009) Computational methods for discovering structural variation with next-generation sequencing. Nat Methods 6(Suppl 11):S13–S20
8. Pepke S, Wold B, Mortazavi A (2009) Computation for ChIP-seq and RNA-seq studies. Nat Methods 6(Suppl 11):S22–S32
9. van Dijk EL, Auger H, Jaszczyszyn Y et al (2014) Ten years of next-generation sequencing technology. Trends Genet 30:418–426
10. Voelkerding KV, Dames SA, Durtschi JD (2009) Next-generation sequencing: from basic research to diagnostics. Clin Chem 55:641–658
11. Pavlopoulos GA, Oulas A, Lacucci E et al (2013) Unraveling genomic variation from next generation sequencing data. BioData Min 6:13
12. Allcock RJN (2014) Production and analytic bioinformatics for next-generation DNA sequencing. In: Trent R (ed) Clinical bioinformatics, 2nd edn. Humana, New York, pp 17–30
13. Cock PJ, Fields CJ, Goto N et al (2010) The Sanger FASTQ file format for sequences with quality scores, and the solexa/illumina FASTQ variants. Nucleic Acids Res 38:1767–1771
14. Li H, Handsaker B, Wysoker A et al (2009) The sequence alignment/Map format and SAMtools. Bioinformatics 25:2078–2079
15. The SAM/BAM Format Specification Working Group (2014) Sequence alignment/map format specification. http://samtools.github.io/hts-specs/SAMv1.pdf

16. Danecek P, Auton A, Abecasis G et al (2011) The variant call format and VCFtools. Bioinformatics 27:2156–2158
17. Ewing B, Hillier L, Wendl MC et al (1998) Base-calling of automated sequencer traces using Phred. I. Accuracy assessment. Genome Res 8:175–185
18. Ewing B, Green P (1998) Base-calling of automated sequencer traces using Phred. II. Error probabilities. Genome Res 8:186–194
19. Andrews S (2010) FastQC: a quality control tool for high throughput sequence data., Available online at http://www.bioinformatics.babraham.ac.uk/projects/fastqc
20. Lipman D, Pearson W (1985) Rapid and sensitive protein similarity searches. Science 227:1435–1441
21. Pearson WR, Lipman DJ (1988) Improved tools for biological sequence comparison. Proc Natl Acad Sci 85:2444–2448
22. Wu TD, Watanabe CK (2005) GMAP: a genomic mapping and alignment program for mRNA and EST sequences. Bioinformatics 21:1859–1875
23. Langmead B, Trapnell C, Pop M et al (2009) Ultrafast and memory-efficient alignment of short DNA sequences to the human genome. Genome Biol 10:R25
24. Li H, Durbin R (2010) Fast and accurate long-read alignment with Burrows-Wheeler transform. Bioinformatics 26:589–595
25. Dobin A, Davis CA, Schlesinger F et al (2013) STAR: ultrafast universal RNA-seq aligner. Bioinformatics 29:15–21
26. Robinson JT, Thorvaldsdóttir H, Winckler W et al (2011) Integrative Genomics Viewer. Nat Biotechnol 29:24–26
27. Thorvaldsdóttir H, Robinson JT, Mesirov JP (2013) Integrative Genomics Viewer (IGV): high-performance genomics data visualization and exploration. Brief Bioinform 14:178–192
28. Generic Feature Format (GFF). http://www.sanger.ac.uk/resources/software/gff/spec.html
29. GFF/GTF File Format—Definition and supported options. http://www.ensembl.org/info/website/upload/gff.html
30. BED File Format. Definition and supported options. http://useast.ensembl.org/info/website/upload/bed.html
31. BED format. http://genome.ucsc.edu/FAQ/FAQformat.html#format1
32. The 1000 Genomes Project Consortium (2010) A map of human genome variation from population-scale sequencing. Nature 467:1061–1073
33. McVean GA, Abecasis DM, Auton R et al (2012) An integrated map of genetic variation from 1,092 human genomes. Nature 491:56–65

Chapter 2

Integrative Exploratory Analysis of Two or More Genomic Datasets

Chen Meng and Aedin Culhane

Abstract

Exploratory analysis is an essential step in the analysis of high throughput data. Multivariate approaches such as correspondence analysis (CA), principal component analysis, and multidimensional scaling are widely used in the exploratory analysis of single dataset. Modern biological studies often assay multiple types of biological molecules (e.g., mRNA, protein, phosphoproteins) on a same set of biological samples, thereby creating multiple different types of omics data or multiassay data. Integrative exploratory analysis of these multiple omics data is required to leverage the potential of multiple omics studies. In this chapter, we describe the application of co-inertia analysis (CIA; for analyzing two datasets) and multiple co-inertia analysis (MCIA; for three or more datasets) to address this problem. These methods are powerful yet simple multivariate approaches that represent samples using a lower number of variables, allowing a more easily identification of the correlated structure in and between multiple high dimensional datasets. Graphical representations can be employed to this purpose. In addition, the methods simultaneously project samples and variables (genes, proteins) onto the same lower dimensional space, so the most variant variables from each dataset can be selected and associated with samples, which can be further used to facilitate biological interpretation and pathway analysis. We applied CIA to explore the concordance between mRNA and protein expression in a panel of 60 tumor cell lines from the National Cancer Institute. In the same 60 cell lines, we used MCIA to perform a cross-platform comparison of mRNA gene expression profiles obtained on four different microarray platforms. Last, as an example of integrative analysis of multiassay or multi-omics data we analyzed transcriptomic, proteomic, and phosphoproteomic data from pluripotent (iPS) and embryonic stem (ES) cell lines.

Key words Multivariate, Dimension reduction, Multi-omics, Multiassay, Data integration

1 Introduction

High throughput technologies including microarray, sequencing, mass spectrometry based proteomics which assay biological molecules have developed rapidly in the past decades. These technologies generate vast amounts of data that describe biological samples at genomic scale and are often called omics data. The capacity and performance of these technologies have improved concurrently with dramatic decreases in cost, and therefore modern

Ewy Mathé and Sean Davis (eds.), *Statistical Genomics: Methods and Protocols*, Methods in Molecular Biology, vol. 1418, DOI 10.1007/978-1-4939-3578-9_2, © Springer Science+Business Media New York 2016

omics studies frequently apply multiple omics techniques to describe the same set of biological observations, such studies include The Cancer Genome Atlas (TCGA), Cancer Cell Line Encyclopedia (CCLE), and ENCyclopedia of DNA Elements (ENCODE). These projects systematically profile large numbers of biological samples resulting in multiple levels of qualitative or quantitative omics data. Whilst the systematically measuring large number of biological molecules (genes, proteins) can reveal novel knowledge that cannot be discovered by traditional methods, the accumulation of multiple omics data presents new challenges for data integration and interpretation.

Several exploratory data analysis (EDA) methods including correspondence analysis (CA), principal component analysis (PCA) have been widely applied to study single omics data [1, 2]. These EDA methods are frequently performed in the early stage of analysis for quality control, detecting batch effect or exploring basic cluster structure in a dataset. In the analysis of multiple omics data, EDA also needs to identify correlations and associations between each of the high dimensional datasets. In this chapter, we describe the following EDA methods that enable researchers to identify relationships between two or more high dimensional datasets:

- Co-inertia analysis (CIA) can be used to explore relationships between two datasets [3];
- Multiple co-inertia analysis (MCIA) can be applied to analyze multiple datasets [4].

Both methods project observations (samples) and variables onto a lower dimensional space, but constrain the dimension reduction such that the new variables represent covariant structure among datasets. Variables from each dataset are transformed onto the same scale. The association between variables and samples can be visualized in this new space, which greatly facilitates the detection of global variance structure and identification of the most informative variables across datasets.

2 Methods

2.1 Analysis of Two Datasets Using Co-inertia Analysis

2.1.1 Co-inertia Analysis

Co-inertia analysis is a multivariate exploratory approach used to identify the covariance between two datasets that have the same set of observations [5]. In the field of omics data analysis, CIA was first introduced to cross-platform comparison of microarray data [3]. With increasing availability of other omics data, it has also been applied to integration of different types of omics data [6].

Two omics datasets may be represented by two matrices, X and Y. In this chapter, we assume the rows of a matrix are samples (observations) and columns are variables, such as genes, proteins, other small molecules, etc. Similar to PCA, CIA is a dimension reduction technique but it considers two datasets simultaneously. For the *i*th dimension, CIA finds a pair of new variables, designated as co-inertia components or dimensions, using a linear combination of the original variables in X and Y, so as to maximize the squared covariance between them.

$$argmax_{u^i, v^i} cov^2(Xu^i, Yv^i) \quad (1)$$

Xu^i and Yv^i are the co-inertia components for matrix X and Y, respectively; v^i and u^i are the linear combination coefficients, which is comparable to the loading vectors in the PCA. Due to the optimized criteria, the co-inertia components capture the most important covariance structure between the two datasets. The co-structure between the two datasets may be visualized by the co-inertia components in a lower dimensional space.

2.2 Case Study I: Integration of NCI-60 Cell Line Transcriptomic and Proteomic Data

The NCI-60 panel is a collection of 60 cancer cell lines from nine different tissues of origin. It includes leukemia, melanoma, ovarian, renal, breast, prostate, colon, lung, and central nervous system (CNS). These cell lines are widely used for in vitro screening of anti-cancer compounds. In attempts to discover gene–drug interactions, several genome wide data profiling approaches have been applied to these cell lines, including DNA copy number variation, DNA mutation, gene expression, protein expression, drug sensitivity, etc. In this case study, we will examine the mRNA expression measure by Agilent GE 4x44K microarray platform (downloaded from [7]) and the proteome data (mass spectrometry based proteomics) [8]. We will use CIA to explore the similarity between datasets and cell lines.

We load the required package and data using:

```
library(omicade4)
library(made4)
load("../data/NCI60_rnaprotein.RDA")
```

`NCI60_rnaprotein` is an object of class *list*, which consists of two numerical matrices, `mRNA` and `protein`. These two matrices have the following dimensions:

```
summary(NCI60_rnaprotein)

##         Length Class  Mode
## mRNA    640958 -none- numeric
## protein 431288 -none- numeric

sapply(NCI60_rnaprotein, dim)

##       mRNA protein
## [1,] 11051    7436
## [2,]    58      58
```

Each of the matrices has 58 cell lines in columns. Due to the problem of data quality, we removed two cell lines, resulting in 58 cell lines included in this analysis. CIA requires that the columns in the matrices are correctly matched, to verify this:

```
identical(colnames(NCI60_rnaprotein$mRNA),colnames(NCI60_rnaprotein$protein))

## [1] TRUE
```

However, the number of rows in different matrices may be different. To facilitate the visualization later, we first create some auxiliary variables to indicate the names of cell lines, tissues of origin of cell lines, and the color for each.

```
names <- strsplit(colnames(NCI60_rnaprotein$mRNA), "\\.")
tumorType <- sapply(names, "[", 1)
cellline <- sapply(names, "[", 2)

# color vector
tumorColor <- as.factor(tumorType)
levels(tumorColor) <- c("red", "green", "blue", "cyan", "pink",
                       "brown", "gray25", "orange", "gray75")
tumorColor <- as.character(tumorColor)

phenoData<-cbind(tumorType=tumorType,cellline=cellline,tumorColor=tumorColor)
rownames(phenoData) = colnames(NCI60_rnaprotein$mRNA)
phenoData[1:4,]

##               tumorType cellline     tumorColor
## BR.MCF7       "BR"      "MCF7"       "red"
## BR.MDA_MB_231 "BR"      "MDA_MB_231" "red"
## BR.HS578T     "BR"      "HS578T"     "red"
## BR.BT_549     "BR"      "BT_549"     "red"
```

Note that Bioconductor is developing a "multi-assay" data object class https://github.com/vjcitn/biocMultiAssay which should be helpful and will be recommended when analyzing multi-assay data.

2.2.1 PCA of Individual Datasets

Before performing the integrative analysis, we first perform basic exploratory analysis and PCA on each individual dataset. For example, exploring the distribution of datasets:

```
layout(matrix(1:2, 1, 2))
boxplot(NCI60_rnaprotein$mRNA, main="mRNA", col=tumorColor)
boxplot(NCI60_rnaprotein$protein, main="Protein", col=tumorColor)
```

The plot is not shown here. But the lower boundary of the boxes in proteomic data reaches 0. This is because the missing values in the proteomics data are replaced with 0. Before the integrative analysis of two datasets, we analyze each of single datasets using PCA:

```
pca_mrna <- prcomp(t(NCI60_rnaprotein$mRNA))
pca_protein <- prcomp(t(NCI60_rnaprotein$protein))
```

The variance of principal components (PCs) and cell lines in the first two PCs may be visualized by:

```
layout(matrix(1:4, 2, 2))
par(mar=c(3, 3, 1.5, 0.5))
plot(pca_mrna, main="mRNA")
legend("topright", col = unique(tumorColor), pch=20, legend = unique(tumorType))
plot(pca_mrna$x[, 1:2], col=tumorColor, pch=20)
abline(v=0, h=0)
plot(pca_protein, main="Protein")
plot(pca_protein$x[, 1:2], col=tumorColor, pch=20)
abline(v=0, h=0)
```

The output is shown in Fig. 1. We observe that the first two PCs in transcriptomic data explain a larger proportion of variance than those in proteomic data. We see cell lines with different anatomical tissue of origin are better separated in the transcriptomic data. But this analysis does not evaluate the co-structure between the two datasets.

2.2.2 CIA of Both Datasets

To visualize the correlated structure between the datasets, we perform CIA using R function `cia`,

```
mRNA <- NCI60_rnaprotein$mRNA
protein <- NCI60_rnaprotein$protein
coin <- cia(mRNA, protein, cia.nf = 5)
```

The output of the `cia` function is an object of class `cia`, which can be easily visualized using the `plot` function

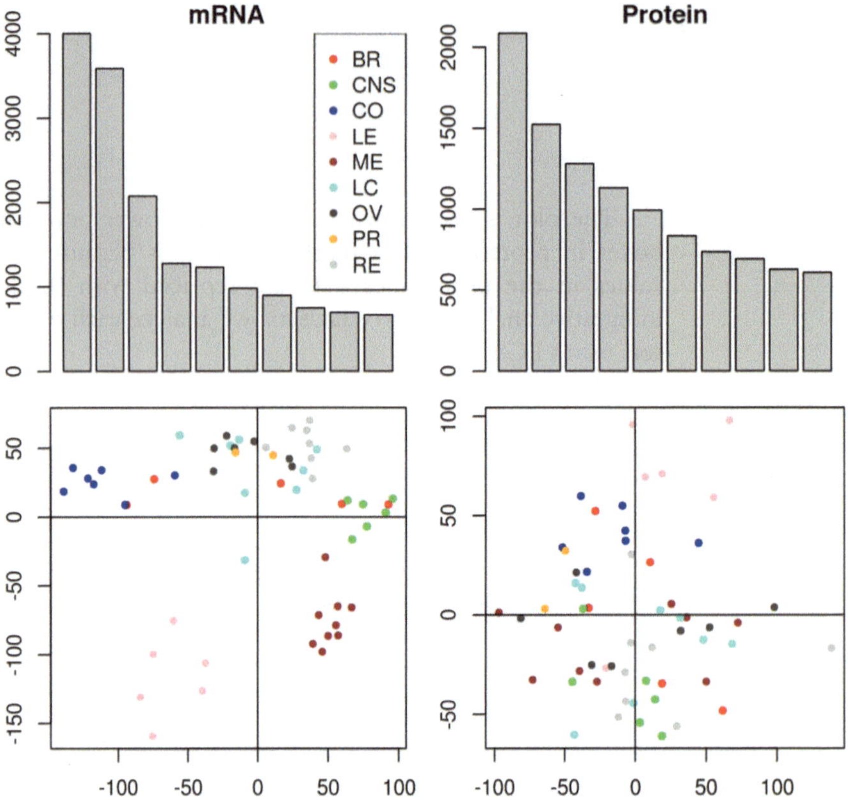

Fig. 1 PCA of individual datasets of mRNA gene expression and protein expression profiles of NCI60 cell lines. The *upper panels* show the variance associated with each principal component of the PCA. The *lower panels* show a plot of the first (horizontal axis) and second (vertical axis) PC for the mRNA and proteins data, respectively

```
barplot(coin$coinertia$eig[1:20])
plot(coin, col=tumorColor)
```

These commands generate Fig. 2. The scree plot in panel a shows the variance associated with each of the co-inertia components, which may be interpreted similarly to variance of PCs in PCA. Fig. 2a shows that the first three components have significantly higher variance than the others. Therefore, these three components should be carefully interpreted and compared with biological or batch factors. The interpretation of components often involves the visualization of samples and variables. Panel b shows the first two co-inertia components. In this plot, samples from the transcriptomic data are shown as the head of arrows. The corresponding cell lines in proteomic data are the ends of arrows. Therefore, this plot is also denoted as sample space. We can see that the leukemia cell lines and melanoma cell lines are weighted on the negative end of the second component, indicating these

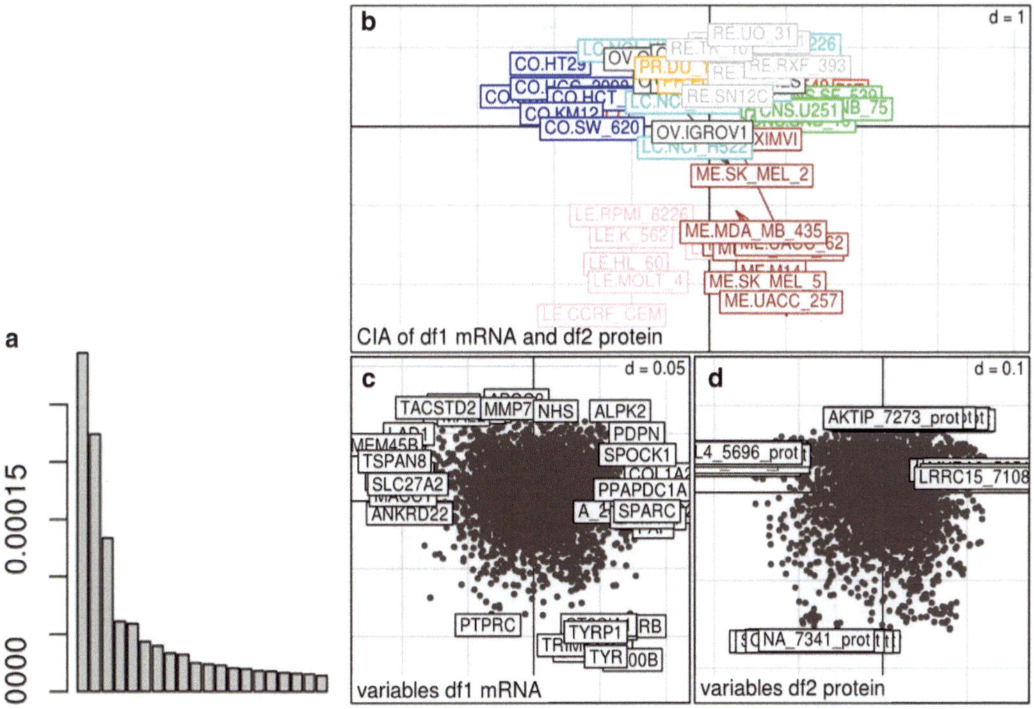

Fig. 2 A plot showing results of a CIA of NCI60 transcriptomic and proteomics data (the same data as in Fig. 1). (**a**): the variance associated with each co-inertia components; (**b**): sample space; (**c**): variables space for mRNA data; (**d**): variable space for protein data

two cell lines are most different with others. The lengths of the arrows describe the similarity between the samples from the two different datasets. Highly correlated pairs of samples will be projected close to each other and, therefore, are linked by a short arrow. In practice, we often need to extract the co-inertia components and customize the plot. As an example, we remove some labels of cell lines:

```
ax1=1
ax2=2
par(mar=c(0.1, 0.1, 0.1, 0.1))
plot(coin$coinertia$mX[, c(ax1, ax2)], col=tumorColor, pch=20,
     xlim=c(-2.5, 2.5), axes=FALSE, frame.plot = TRUE)
abline(v=0, h=0)
arrows(coin$coinertia$mX[, ax1], coin$coinertia$mX[, ax2],
       coin$coinertia$mY[, ax1], coin$coinertia$mY[, ax2],
       angle = 15, length = 0.1, col=tumorColor)
legend("bottomleft", col = unique(tumorColor), pch=20, legend = unique(tumorType))
# text(coin$coinertia$mX[, c(ax1, ax2)], labels = cellline, cex=0.7)
```

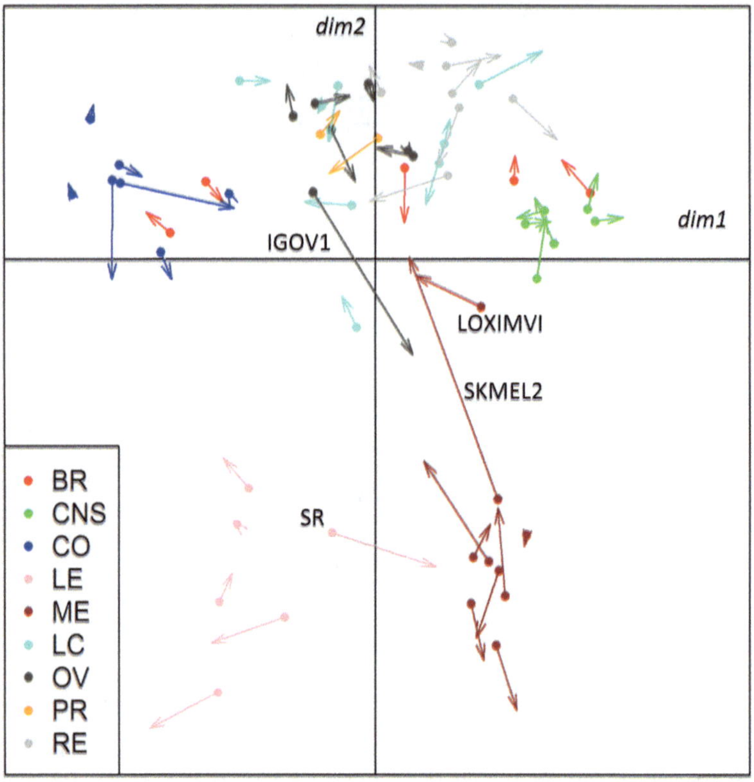

Fig. 3 A customized plot of the sample space shows the project of NCI60 cells when a CIA is performed on NCI60 transcriptomic and proteomic data analysis. Each cell line is colored by its anatomical tissue of origin

These commands generate a plot similar to Fig. 3. In the plot, we see most of the arrows are short in length, which indicates high overall similarity or considerable correlated structure between the datasets. This can be confirmed using the RV coefficient, which is also included in the `cia` object and may be extracted by

```
coin$coinertia$RV

## [1] 0.7464801
```

The RV coefficient is a multivariate extension of the Pearson correlation coefficient to measure the overall similarity between two matrices. It ranges from 0 to 1. A high RV coefficient indicates high degree of similarity. In this case, the RV coefficients of 0.75 suggest that a relative high correlation exists between the two datasets.

Despite a good overall similarity, some cell lines have lower correlation between their mRNA and protein profiles. These include lung cancer cell line IGOV1, leukemia cell line SR, and melanoma cell line SKMEL2. For example, whilst the SKMEL2 is projected close to other melanoma cell lines in transcriptomic data, in the proteomics data this cell line is closer to the plot origin and is far from most other melanoma cell lines. Similarly, the proteome of leukemia cell line SR is closer to the melanoma cell lines in comparison with other leukemia cell lines. This discrepancy may reflect biological variance, a batch effect, or a technical artifact, such as sample mis-labeling.

2.2.3 CIA: Exploring the Variables

In CIA, the projection of each sample is determined by its variable measurements. The variables from both datasets are transformed onto the same scale and projected into the same space, thereby enabling exploration of relationships between variables, and between samples and variables. The loadings of the mRNA and protein variables are shown in panel c and d of Fig. 2, which are also called gene space, or more generally, variable space. In these variable plots, the variables with highest weights (i.e., on the negative or positive ends of components) are the most influential variables and these define the co-inertia components. Variables and samples that are projected in the same direction from the origin have a strong association (i.e. the variables are increased or upregulated in those samples), whereas variables projected at the opposite direction to a sample are frequently have low values in those samples. Therefore, the variables with highest weights in each of components can be extract and these facilitate the biological interpretation of components. We will extract the highest weighted variables using the function `topVar`:

```
topVar(coin, axis = 1, end = "neg", topN = 10)

##    ax1_df1_negative ax1_df2_negative
## 1          TMEM45B  CALML4_5696_prot
## 2            OVOL1 UGT1A10_3875_prot
## 3           TSPAN8   AZGP1_3143_prot
## 4            POF1B    DLG1_2351_prot
## 5            MACC1   AKR7A3_4110_prot
## 6              DDC  SDCBP2_4458_prot
## 7          SLC27A2     FABP1_844_prot
## 8             LAD1    LGALS4_779_prot
## 9             FBP1   TMEM62_5329_prot
## 10         ANKRD22     MUC13_920_prot

topVar(coin, axis = 1, end = "pos", topN = 10)

##    ax1_df1_positive ax1_df2_positive
## 1           COL1A2  NCOA7_5477_prot
## 2              FAP COL6A2_3829_prot
## 3           CNRIP1  HSPB7_4887_prot
## 4            SPARC COL5A1_5880_prot
## 5         PPAPDC1A  MXRA8_5851_prot
## 6            LAMA4 PCOLCE_4369_prot
## 7           IGFBP7   COPZ2_724_prot
## 8           SPOCK1 LRRC15_7108_prot
## 9      A_24_P554882 COL3A1_1636_prot
## 10            PDPN    BMP1_704_prot
```

```
topVar(coin, axis = 2, end = "neg", topN = 10)

##   ax2_df1_negative  ax2_df2_negative
## 1            S100B  SH2D1A_2507_prot
## 2              TYR PTPRCAP_1858_prot
## 3         C14orf34    CD1C_5757_prot
## 4           TRIM63     CD5_2006_prot
## 5            TYRP1    CD3E_1036_prot
## 6            MLANA    GRAP_1115_prot
## 7            EDNRB    FLI1_4768_prot
## 8           BCL2A1    RHOH_1471_prot
## 9          ST8SIA1      NA_7341_prot
## 10           PTPRC    TRAT1_698_prot

topVar(coin, axis = 2, end = "pos", topN = 10)

##    ax2_df1_positive  ax2_df2_positive
## 1             ABCC3   CLCN1_7161_prot
## 2              MAL2   AKTIP_7273_prot
## 3           TACSTD2 TM4SF18_2667_prot
## 4              MMP7  PI4K2A_7076_prot
## 5              ELF3  SHISA2_3899_prot
## 6             ALPK2   PAPLN_6608_prot
## 7               NHS   HLA.A_5614_prot
## 8              CST6  DHDPSL_6037_prot
## 9          KRT8P20 DNTTIP2_3974_prot
## 10             MALL  HAVCR1_2844_prot
```

The results suggest that the positive end of the first component captures several collagens, including COL1A1, COL6A2, which are the major components of the extracellular matrix and connective tissues. In tumors, collagens are associated with cancer cell metastasis and poor prognosis in patients. Therefore, we can infer that the first dimension reflects the mechanism related to different potential of metastasis of the cell lines. The negative end of the second component is associated with leukemia and melanoma cell lines. Accordingly, genes (see "ax2_df1_negative") from transcriptomic data with high weights in this component include several melanogenesis genes, such as S100B and TRY, explaining why melanoma cell line LOXIMVI, a cell line that lacks melanin is projected closer to the origin in the proteome sample space. By contrast these proteins are absent in the proteomics data, which highlighted several immune cell markers, such as CD1C, CD5, and CD3E, which are highly expressed in the leukemia cell lines. Therefore, the projection of leukemia cell lines is more influenced by the proteomic data whereas the weights of melanoma cell lines are determined by the transcriptomic data. But both are separated from other cell lines.

2.3 Exploration of Three or More Datasets

2.3.1 Multiple Co-inertia Analysis

CIA can be used to explore the concordance and discrepancy between two datasets. MCIA is a generalization of CIA to analyze more than two datasets [4]. In MCIA, the multiple omics data are represented by K blocks of matrices $(X_1, X_2, \ldots, X_K)$. Similar with CIA, for the ith dimension, MCIA defines a set of block components using the linear combination of variables in each of the matrices. The goal of MCIA is to find a synthetic component, s^i, so as to maximize the sum of squared covariance between the block components and the synthetic components, that is

$$argmax_{u_k^i, s^i} \sum_{k=1}^{K} cov^2(X_k u_k^i, s^i) \tag{2}$$

where $X_k u_k^i$ are the set of block component and the s^i is the synthetic component; u_k^i are the loading for the variables in the kth matrix. In PCA, the principal components are the optimal lower rank approximation of a high dimensional dataset, whereas the block components in MCIA are sub-optimal in terms of representing the individual matrices, but they represent the best covariant structures across multiple datasets. Similar to CIA, the block components and synthetic components can be visualized in a two dimensional space to facilitate the interpretation of multiple high dimensional datasets.

2.3.2 Case Study 2: Cross Comparison of Gene Expression Data Obtained on Four Different Microarray Platforms

Cross-platform comparison is often used in EDA, such as meta-analysis or as part of cross-validation of findings. In this example, we explore the consensus in data from four transcriptomic studies of NCI60 cell line using different microarray platform (Affymetrix HG U95, U133, U133 plus2.0 and Agilent; downloaded from [7]). The goal of this analysis is to explore the concordances and discrepancies in mRNA expression measurements obtained using different platforms for each cell lines.

First, we load the data

```
load("../data/NCI60_4arrays.RDA")
```

and get an overview of dimensions of datasets using the following functions:

```
summary(NCI60)

##           Length Class  Mode
## agilent   640958 -none- numeric
## hgu95     510574 -none- numeric
## hgu133    524552 -none- numeric
## hgu133p2 602156 -none- numeric

sapply(NCI60, dim)

##      agilent hgu95 hgu133 hgu133p2
## [1,]   11051  8803   9044    10382
## [2,]      58    58     58       58

names(NCI60)

## [1] "agilent"  "hgu95"    "hgu133"   "hgu133p2"

tumorType <- sapply(strsplit(colnames(NCI60$agilent), "\\."), "[", 1)
tumorColor <- as.factor(tumorType)
levels(tumorColor) <- c("red", "green", "blue", "cyan", "pink",
                        "brown", "gray25", "orange", "gray75")
tumorColor <- as.character(tumorColor)
```

We draw a boxplot to explore the distribution of datasets.

```
layout(matrix(1:4, 2, 2))
boxplot(NCI60$agilent, main="Agilent", col=tumorColor)
boxplot(NCI60$hgu95, main="Affy HG U95", col=tumorColor)
boxplot(NCI60$hgu133, main="Affy HG U133", col=tumorColor)
boxplot(NCI60$hgu133p2, main="Affy HG U133 plus2.0", col=tumorColor)
```

MCIA is performed using the function `mcia` in *omicade4* package and is visualized using the function `plot`, for example

```
mcoin <- mcia(NCI60)

## 'svd' fail to convergence, 'eigen' used to perform singular value decomposition

plot(mcoin, df.color = 2:5, sample.color=tumorColor,
     sample.legend = FALSE,
     sample.lab=FALSE)
legend("bottomright", col = unique(tumorColor), pch=20, legend = unique(tumorType))
```

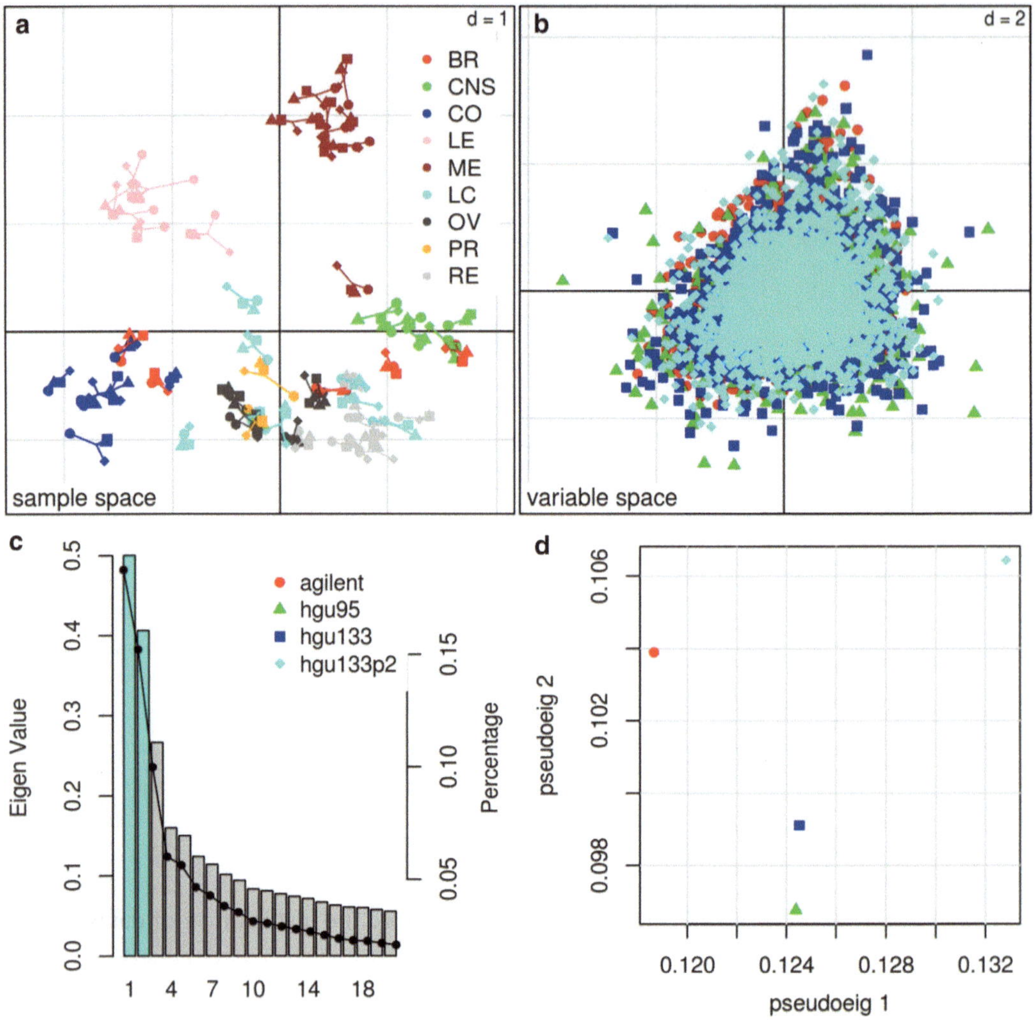

Fig. 4 A plot showing results of an MCIA which integrated and compared four different microarray gene expression datasets. (**a**): the sample space; (**b**): the variable space, variables from four different platforms are shown as different colors; (**c**): the scree plot shows the variance associated with each dimension. (**d**): Dataset space

The function `mcia` returns an object of class `mcia`. A typical visualization of the plot is shown in Fig. 2.4. Similar to CIA, the plot consists of the sample space (Fig. 4a) and variables spaces (Fig. 4b). In the sample space, samples from different datasets are shown as point with different shapes and also the same cell lines in each datasets are linked to the synthetic components. The short lines in this plot indicate a good correlation for the four datasets. The pairwise RV coefficient may be extracted by

```
round(mcoin$mcoa$RV, 3)

##          agilent hgu95 hgu133 hgu133p2
## agilent    1.000 0.953  0.955    0.955
## hgu95      0.953 1.000  0.988    0.965
## hgu133     0.955 0.988  1.000    0.969
## hgu133p2   0.955 0.965  0.969    1.000
```

All the RV coefficients are higher than 0.95, indicating a good correlation between datasets generated by different platforms.

Figure 4b shows the variable loadings, variables from different datasets are shown with different colors. The variables and samples projected on the same direction are highly associated with each other. Figure 4c shows the variance associated with each dimension. In MCIA, multiple datasets contribute to the variance of components. Therefore, the global variance can be decomposed into the contributions of each individual datasets. This information is shown in Fig. 4d, the pseudo-eigenvalue for each dimension indicates the decomposed variance of a single dataset, so this panel may also be called "dataset space." It shows that all the datasets contribute roughly equal to the first and second components as indicated by the small range of x and y axes in Fig. 4d. Strictly, data from HGU133plus2.0 contributes slightly higher than others to both first and second components, whereas Agilent data have a lower contribution to first component, but a higher contribution to the second dimension in comparison with HGU95 and HGU133 data.

Next, we will show how to use MCIA to detect outlier or "abnormal" samples using MCIA cross-platform comparison. To do so, we swap the names of two samples in the Agilent data to simulate a mis-labeling problem. We exchange the leukemia cell line SR and melanoma cell line LOXIMVI in Agilent data and run MCIA on the datasets

```
NCI60_rand <- NCI60
NCI60_rand$agilent[, c("ME.LOXIMVI", "LE.SR")] <-
  NCI60_rand$agilent[, c("LE.SR", "ME.LOXIMVI")]
```

```
mcoin_rand <- mcia(NCI60_rand)

## 'svd' fail to convergence, 'eigen' used to perform singular value decomposition

plot(mcoin_rand, df.color = 2:5, sample.color=tumorColor,
     sample.legend = FALSE,
     sample.lab=FALSE)
```

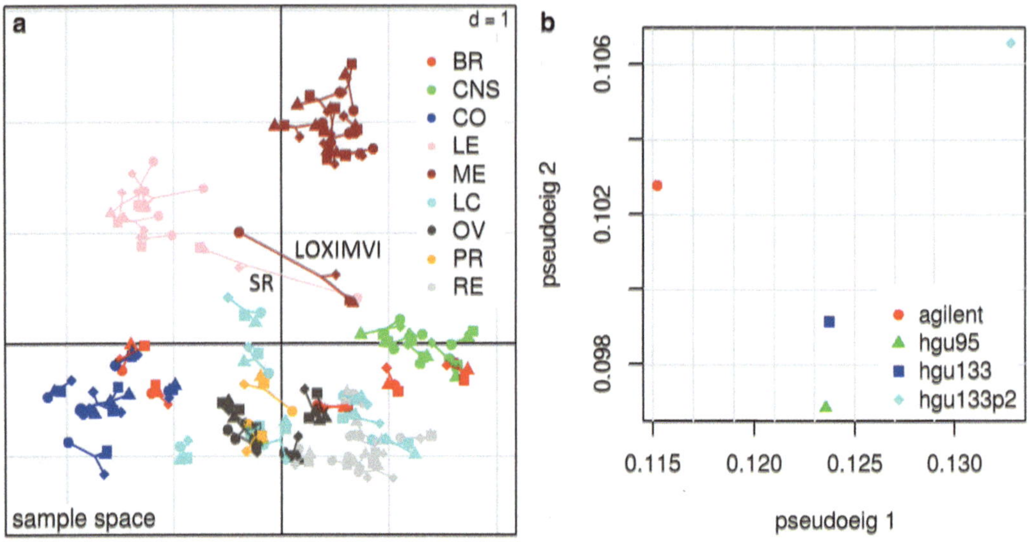

Fig. 5 Demonstration showing the application of MCIA to detecting outliers or "problem" samples. Here, the same analysis (MCIA) is performed as in Fig. 4. However, the names of the melanoma cell lines LOXIMVI and the leukemia cell lines SR are swapped. (**a**): the sample space; (**b**): Dataset space

Figure 5 shows the corresponding sample space and dataset space. In the sample space, it is clearly shown that the Agilent data for SR and LOXIMVI are projected away from their counterpart in other datasets. The relative long line in this plot suggests a mis-labeling problem. In addition, the exchange of the labels of cell lines in Agilent data results in that the covariate structure in both dimensions are less data, which can be seen from the decreased pseudo-eigenvalue for Agilent data in Fig. 5b (compare with Fig. 4d).

2.3.3 Example 2: Integrative Analysis of Transcriptome, Proteome, and Phosphoproteome of Stem Cell Lines

In this example, we use the data generated by Phanstiel et al. [9]. The goal of this research was to compare protein expression and phosphorylation between embryonic stem cell (ES) and induced pluripotent stem (iPS) cell lines. In the study, 4 ES and 4 iPS cell lines were selected and their transcriptome, proteome, and phosphoproteome data were measured in triplicates. Here, we only analyze one of the replicates as the aim to illustrate how to use MCIA to integrative analysis of multiple omics data of different levels.

First, load the data

```
load("../data/iPSES8ples.RDA")
```

and summarize the data

```
summary(iPSES)

##          Length Class  Mode
## mrna     124648 -none- numeric
## protein   30928 -none- numeric
## phospho   62272 -none- numeric

sapply(iPSES, dim)

##        mrna protein phospho
## [1,] 15581    3866    7784
## [2,]     8       8       8
```

perform the MCIA and plot the result

```
mcoin <- mcia(iPSES)
plot(mcoin, df.color = 2:4, sample.color=rep(c("cyan", "orange"), each=4),
     sample.legend = FALSE, sample.lab=FALSE)
```

Figure 6a–d shows the sample space, variable space, component variance, and dataset space, respectively. Panel c suggests that the first two dimensions have significantly higher variance than others. The corresponding sample space suggests that the iPS and ES cell lines could be distinguished by the first dimension. Particularly, iPS cell lines DF6.9 and DF4.7, DF19.7 and DF19.11 are highly correlated. The ES cell lines are more dispersely projected onto the space. H14 is on the negative end of the first component and H9 is highly weighted on the negative end of the second component. The relative short lines between samples indicate a good correlation between the datasets. However, the dataset space (panel d) suggests that the protein data have more variance on the first dimension, whereas the protein phosphorylation data contribute more variance on the second one. This finding is inconsistent with the variable space, where the proteomic data are more spread on the first component and the phosphoproteomic data have a wider range on the second component. The RV coefficients between datasets are

```
round(mcoin$mcoa$RV, 3)

##          mrna protein phospho
## mrna    1.000   0.705   0.845
## protein 0.705   1.000   0.457
## phospho 0.845   0.457   1.000
```

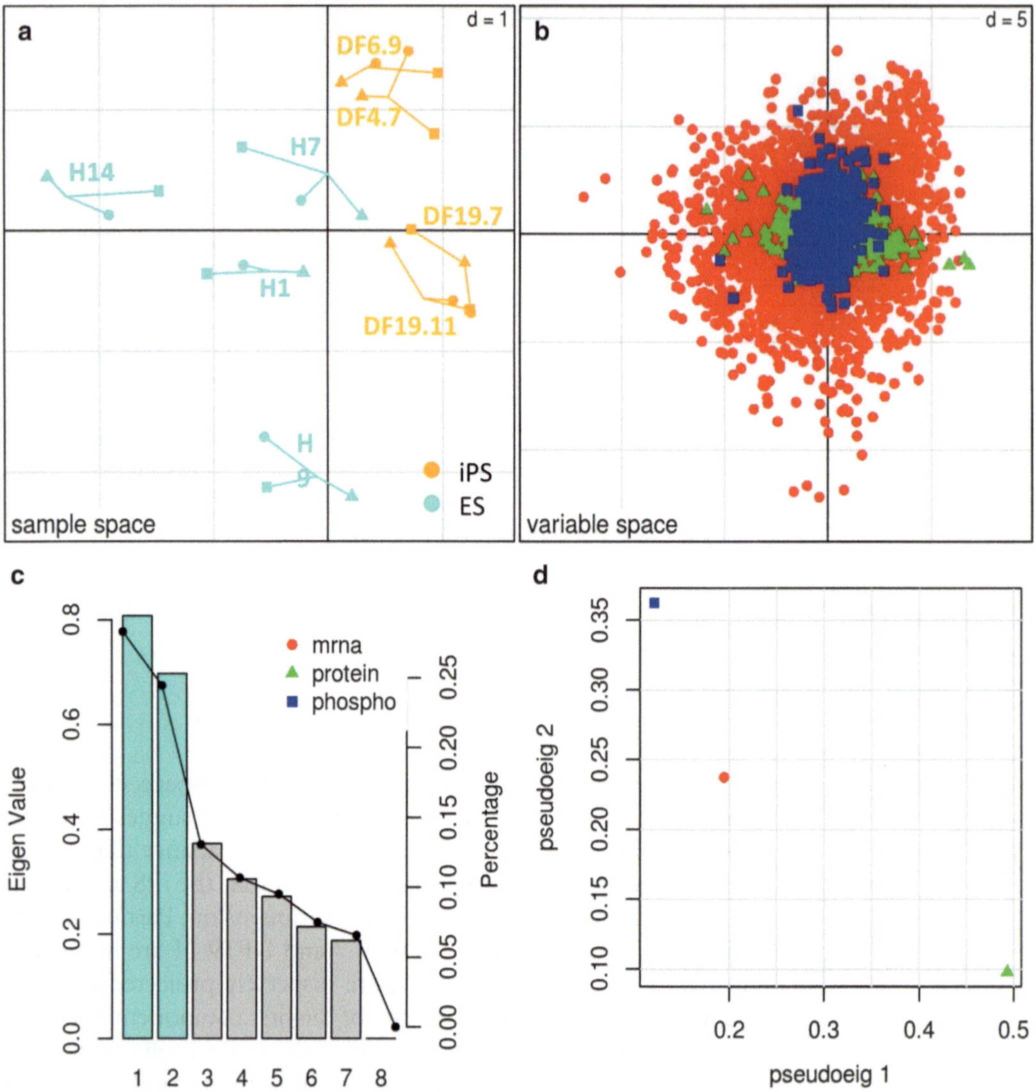

Fig. 6 Plot showing results of an MCIA of iPS and ES data. Three datasets were integrated in this analysis, including mRNA expression data (RNA sequencing), protein expression and phosphorylation data (Mass spectrometry based proteomic) (**a**): the sample space; (**b**): the variable space, variables from four different platforms are shown as different colors; (**c**): the scree plot shows the variance associated with each dimension. (**d**): Dataset space

Unexpectedly we find that the phosphorylation data have a better correlation with the mRNA data rather than the protein data. In addition, similarly to our previous analyses, we can select the variables with the greatest weights on each of the dimensions using the function `topVar`

```
topVar(mcoin, axis = 1, end = "neg", topN = 5)

##   ax1_mrna_negative ax1_protein_negative ax1_phospho_negative
## 1              ZIM2         IPI00946792.1  IPI00022628.5_s387
## 2              MMP1         IPI00555956.2  IPI00657687.1_s182
## 3           CYP4F11         IPI00012989.2   IPI00969114.1_s55
## 4              PEG3         IPI00219774.3 IPI00742682.2_s2155
## 5            LGALS4         IPI00873459.3  IPI00186139.8_s181

topVar(mcoin, axis = 1, end = "pos", topN = 5)

##   ax1_mrna_positive ax1_protein_positive       ax1_phospho_positive
## 1          C17orf50        IPI00026993.1      IPI00066543.2_s40.t44
## 2          C12orf39        IPI00306642.3    IPI00798034.2_s280.s281
## 3              IAPP        IPI00026219.4           IPI00304935.6_s6
## 4             IL4I1        IPI00915008.1    IPI00008422.5_s239.s242
## 5           GTF2H2D        IPI00607808.2 IPI00008422.5_s239.s242.s245

topVar(mcoin, axis = 2, end = "neg", topN = 5)

##   ax2_mrna_negative ax2_protein_negative ax2_phospho_negative
## 1             OLIG1        IPI00442171.4 IPI00292975.4_s1012
## 2          HLA.DQA1        IPI00444452.3   IPI00871890.1_s16
## 3             OR7D2        IPI00843802.2  IPI00217467.3_s104
## 4             OLIG2        IPI00218271.5  IPI00657687.1_s182
## 5           PSTPIP1        IPI00828098.2  IPI00011913.1_s188

topVar(mcoin, axis = 2, end = "pos", topN = 5)

##   ax2_mrna_positive ax2_protein_positive ax2_phospho_positive
## 1            ZNF560        IPI00306406.4  IPI00010800.2_s352
## 2            VWA5B1        IPI00472164.2  IPI00292059.2_s240
## 3              CSH1        IPI00549381.5  IPI00640088.1_s228
## 4            CYP2E1        IPI00386731.3  IPI00337315.1_s873
## 5               FGA        IPI00183002.6   IPI00005978.8_y23
```

To reveal the biological meaning of different components, additional methods, including gene set enrichment analysis (GSEA), may be applied to further analyze the selected genes on each component.

3 Session Info

```
toLatex(sessionInfo())
```

- R version 3.2.2 (2015-08-14), x86_64-pc-linux-gnu
- Locale: LC_CTYPE=en_US.UTF-8, LC_NUMERIC=C, LC_TIME=en_US.UTF-8, LC_COLLATE=en_US.UTF-8, LC_MONETARY=en_US.UTF-8, LC_MESSAGES=en_US.UTF-8, LC_PAPER=en_US.UTF-8, LC_NAME=C, LC_ADDRESS=C, LC_TELEPHONE=C, LC_MEASUREMENT=en_US.UTF-8, LC_IDENTIFICATION=C
- Base packages: base, datasets, graphics, grDevices, methods, stats, utils
- Other packages: ade4 1.7-2, gplots 2.17.0, knitr 1.11, made4 1.44.0, omicade4 1.10.0, RColorBrewer 1.1-2, scatterplot3d 0.3-36
- Loaded via a namespace (and not attached): BiocStyle 1.8.0, bitops 1.0-6, caTools 1.17.1, codetools 0.2-14, digest 0.6.8, evaluate 0.8, formatR 1.2.1, gdata 2.17.0, gtools 3.5.0, highr 0.5.1, KernSmooth 2.23-15, magrittr 1.5, stringi 1.0-1, stringr 1.0.0, tcltk 3.2.2, tools 3.2.2

References

1. Fellenberg K, Hauser NC, Brors B, Neutzner A, Hoheisel JD, Vingron M (2001) Correspondence analysis applied to microarray data. Proc Natl Acad Sci USA 98:10781–10786
2. Raychaudhuri S, Stuart JM, Altman RB (2000) Principal components analysis to summarize microarray experiments: application to sporulation time series. In: Pacific Symposium on Biocomputing, pp 455–466
3. Culhane AC, Perriere G, Higgins DG (2003) Cross-platform comparison and visualisation of gene expression data using co-inertia analysis. BMC Bioinformatics 21(4):59
4. Meng C, Kuster B, Culhane A, Gholami AM (2014) A multivariate approach to the integration of multi-omics datasets. BMC Bioinformatics 29(15):162
5. Dolédec S, Chessel D (1994) Co-inertia analysis: an alternative method for studying species-environment relationships. Freshw Biol 31:277–294
6. Culhane AC, Fagan A, Higgins DG (2007) A multivariate analysis approach to the integration of proteomic and gene expression data. Proteomics 7:2162–2171
7. Reinhold WC, Sunshine M, Liu H, Varma S, Kohn KW, Morris J, Doroshow J, Pommier Y (2012) Cellminer: a web-based suite of genomic and pharmacologic tools to explore transcript and drug patterns in the NCI-60 cell line set. Cancer Res 72(14):3499–511
8. Moghaddas Gholami A, Hahne H, Wu Z, Auer FJ, Meng C, Wilhelm M, Kuster B (2013) Global proteome analysis of the NCI-60 cell line panel. Cell Rep 4:609–620
9. Phanstiel DH, Brumbaugh J, Wenger CD, Tian S, Probasco MD, Bailey DJ, Swaney DL, Tervo MA, Bolin JM, Ruotti V, Stewart R, Thomson JA, Coon JJ (2011) Proteomic and phosphoproteomic comparison of human ES and iPS cells. Nat Methods 8:821–827

Chapter 3

Study Design for Sequencing Studies

Loren A. Honaas, Naomi S. Altman, and Martin Krzywinski

Abstract

Once a biochemical method has been devised to sample RNA or DNA of interest, sequencing can be used to identify the sampled molecules with high fidelity and low bias. High-throughput sequencing has therefore become the primary data acquisition method for many genomics studies and is being used more and more to address molecular biology questions. By applying principles of statistical experimental design, sequencing experiments can be made more sensitive to the effects under study as well as more biologically sound, hence more replicable.

Key words Reproducible research, Randomization, Block design, Replication, Sample size, Multiplexing, Pooling, Precision, Pilot study, Sequencing depth

1 Introduction

Second-generation sequencing (SGS, next-generation sequencing, NGS) is now used for a variety of high-throughput studies—from SNP and binding site detection to differential expression analysis to metagenomic profiling of entire ecosystems. Although individual samples, which may be as small as single cells, yield millions of measurements in the form of sequenced fragments of RNA or DNA, it is still critical to account for variability due to biological variation in the experimental material and as well measurement error. A good study design maximizes and quantifies the precision of the measurements while controlling costs.

Any research endeavor requires rational management of costs—material, money, time, the ethical use of subjects and resources, timeliness and opportunity—in addition to the intangible costs associated with making errors. Ideally, a statistically sound study design will aim to jointly optimize as many of these as possible—considering them all is generally impractical. In this chapter, we focus on ways to enhance precision and accuracy of our results for quantitative biological inference, discuss the impact

Ewy Mathé and Sean Davis (eds.), *Statistical Genomics: Methods and Protocols*, Methods in Molecular Biology, vol. 1418, DOI 10.1007/978-1-4939-3578-9_3, © Springer Science+Business Media New York 2016

of less-than-optimal study design decisions, and mention how some common problems can be mitigated.

Broadly, there are two kinds of sequencing studies—those that characterize the sequence and those that count the number of specific or related sequences present in a cell or sample preparation. Examples of the former in this volume include motif finding (Chapter 12), genotyping studies (Chapters 7, 11), structural variation discovery (Chapter 9), and variant calling (Chapter 11). The latter is exemplified by studies in which the objective is to quantify features such as gene expression (Chapters 7, 8), proportion of methylation (Chapter 10) and protein occupancy at specific sites (Chapter 10) over a population of similar samples. This chapter primarily addresses the latter—quantification studies—with a focus on comparative (differential) studies. We point out where our discussion is also applicable to identification studies.

Our discussion focuses on basic aspects of sequence-based experimental design and its application to typical sequencing studies. We begin in Subheading 2 with an introduction of the core statistical concepts required to understand experimental design and then apply these to the experimental steps prior to sequencing. In Subheading 3, we discuss some considerations for the sequencing steps. Subheading 4 wraps up with additional comments and conclusions.

2 Statistical Considerations

2.1 Statistical Background

The main objective of quantification studies is to determine a quantitative measure for selected characteristics (features) of the experimental system. Examples of common features of sequence-based quantification studies are transcripts (mRNA or small RNA), exon usage, protein occupancy at binding sites, or methylation at methylation sites. The basic unit of measurement is a read, which represents the nucleotide order of a target nucleic acid molecule (i.e. target library fragment). Reads are derived from RNA or DNA samples called libraries, which may yield thousands to millions of reads that may be assembled or mapped (aligned). Mapping can be to a reference genome or specific features such as exons or transcripts (Fig. 1). Quantification studies aim to count the reads mapped to a feature, which is a number that varies from library to library. For example, the number of reads derived from transcripts of the gene might quantify its expression. For comparative purposes, the proportion of reads in the library mapped to the feature can be considered the quantity of interest and is usually reported as counts per million reads (CPM) that is, 1,000,000 $\times$ counts/library size. For ease of discussion, we use the counts per total library size, which is the measured quantity for the remainder of the discussion. However, for statistical analysis we still require

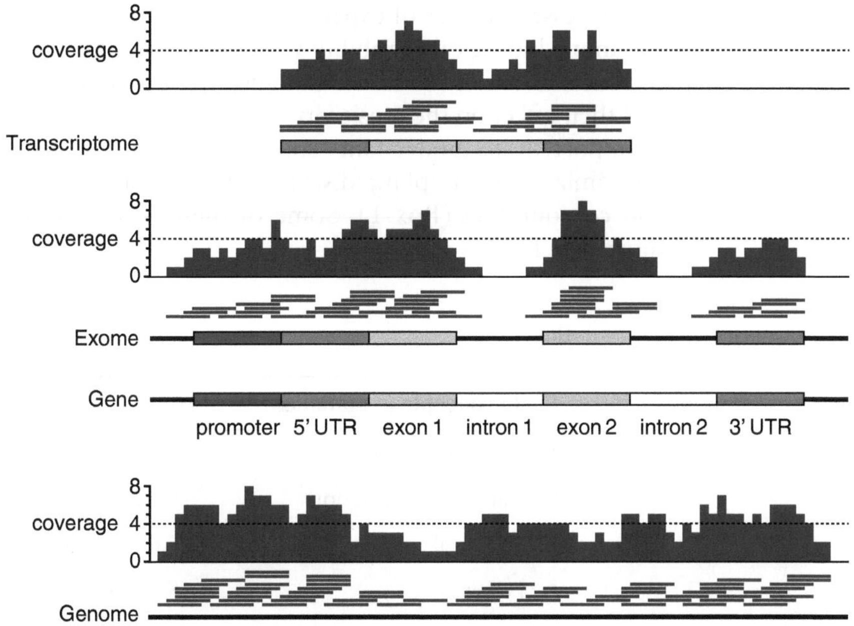

Fig. 1 Sequenced reads may be derived from the genome, exome, and transcriptome. In quantification studies these reads are mapped to features of interest and the number of reads falling within a feature are counted, depicted here as histograms. Exome sequencing reads span exon-intron boundaries. Shown is fourfold coverage of randomly distributed reads. Note that the depth profile is correlated across the length of a read, which makes the profile appear to have a periodic variation

both the actual counts per feature and the library size as these are critical for quantifying variability.

It is worth noting another quantity that is often discussed in the literature. Many analysis programs report fragments per kilobase per million (FPKM), which can be expressed as CPM/(feature length in kilobases), where the feature may be a gene transcript or another region of interest. Both CPM and FPKM are comparable across libraries of different sizes.

We use the term experimental unit (EU) to refer to a basic sample, such as a patient, mouse, plant, or cell line. In comparative studies, factors of interest may be attributes, such as genotype or gender, or experimental conditions, such as type of diet, temperature, or drug regimen. The different values of a factor are called its levels and any given combination of factor levels is called a treatment. To clarify both the issues and the technical jargon, we discuss a concrete example—a study of gene expression in the liver of two genotypes of mice exposed to a low or high fat diet. In this study, the features are the mRNA transcripts, the EUs are the mice and the factors are genotype with levels "wild type" and "mutant" and diet with levels 10 and 60 % fat content. In this case, "wild type mouse on a 10 % fat diet" would be one of the treatments.

An essential part of experimental design is balancing the need to sample biological variability with measuring and accounting for technical variation. To understand how to approach these elements and their effect on the sensitivity and specificity of an experiment, it is important to understand some basic concepts in statistics—randomization, sampling distributions, statistical testing, power and confounding (Box 1). Some of these issues are discussed in more detail in [1–3].

Box 1

Basic concepts in statistics	
• Randomization	Subheading 2.1.1
• Sampling distributions	Subheading 2.1.2
• Statistical testing	Subheading 2.1.3
• Power	Subheading 2.1.4
• Confounding	Subheading 2.1.5

2.1.1 Randomization

In our example study we apply randomization in one of two ways—to select EUs from the attribute populations and to assign controllable factors to EUs.

In an experiment with two controllable factors with two levels each (thus, four treatments), if our design called for three mice for each treatment, we would need a randomly selected pool of 12 mice (Fig. 2a). A simple way to assign mice to treatments would be to assign the treatments to unique toss outcomes of two coins (HH, HT, TH, or TT). Treatments would be assigned to each of the 12 mice in succession based on a toss, with the restriction that once any given treatment has been assigned to three mice, additional outcomes with the same toss are ignored.

If one of the factors is an attribute, such as genotype, the randomization approach must be different because attributes cannot be randomly assigned. Instead, we randomly select six mice from each of the two genotypes and toss a single coin to assign the second factor within each group (Fig. 2b). This is a restricted randomization, because the second controllable factor (e.g., diet) is assigned within the first attribute factor (e.g., genotype). The other panels in Fig. 2 are discussed in Subheading 2.2.3.

2.1.2 Sampling Distributions

In statistical terminology, the population is the set of all EUs we could possibly measure for a particular treatment and the sample is the set we actually measured. In our example study, we consider four different populations as defined by the treatments (e.g., "wild

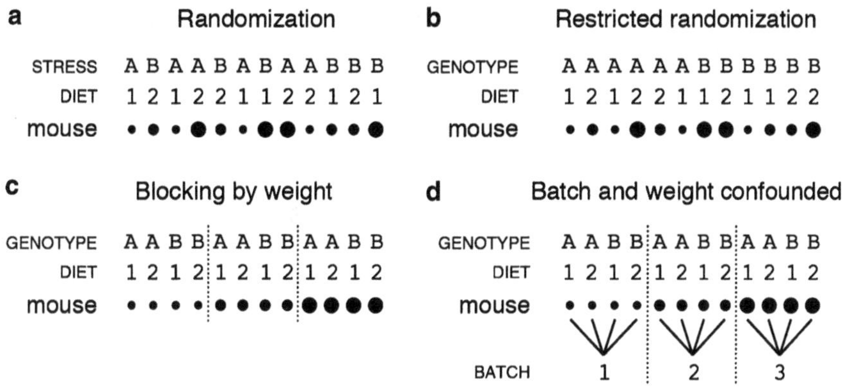

Fig. 2 Blocking increases sensitivity to treatment effects by accounting for variation in the data that is not due to treatment. (**a**) Two controllable factors (e.g., stress and diet) are assigned randomly without consideration of weight, indicated by the *circle size*. As a result, weight classes may be unrepresented for some treatments (e.g., low-weight mice and genotype B). (**b**) Controllable factors (e.g., diet) are assigned within attribute factors (e.g., genotype) in restricted randomization. (**c**) Blocking on weight generates a complete subexperiment for each weight class—all treatments are represented in a block. Within a block, randomization is carried out as in (**a**) or (**b**) depending on whether both factors are controllable. (**d**) Blocking can be performed on combinations of confounded factors (e.g., weight and batch)

type mouse on a 10 % fat diet," "mutant mouse on a 10 % fat diet"). Given that our features of interest are mRNA transcripts, we reduce information about the population and sample to a few statistical descriptive summaries for each feature.

The most commonly used summaries are the mean, variance, and standard deviation. The mean is the numerical average of all the values of the feature observed in the EUs. The population mean is usually denoted by the Greek letter mu (μ). If the sample value is denoted by x, then the sample mean is $\overline{x}$. Because we do not have access to the full population, we use the sample mean to estimate the population mean.

The variance is the average squared deviation of the measurements from their mean, where the deviation is $d = x - \mu$ for the population and $x - \overline{x}$ for the sample (Fig. 3a). The square root of the variance is the standard deviation and is usually denoted by σ for the population value and s for the sample value. The spread of the population (or sample) histogram is summarized by σ (or s) because it has the same units as the mean. However, the variance is generally easier to work with mathematically because independent sources of noise and variability can often be described as additive components of variance.

For a selected feature (e.g., number reads of a given mRNA transcript), lets assume that for one of the treatments (e.g., genotype A and low fat diet), the population mean proportion is π, expressed as a fraction of all mRNA transcripts. To measure the actual number of transcripts in an EU, we construct a sequencing

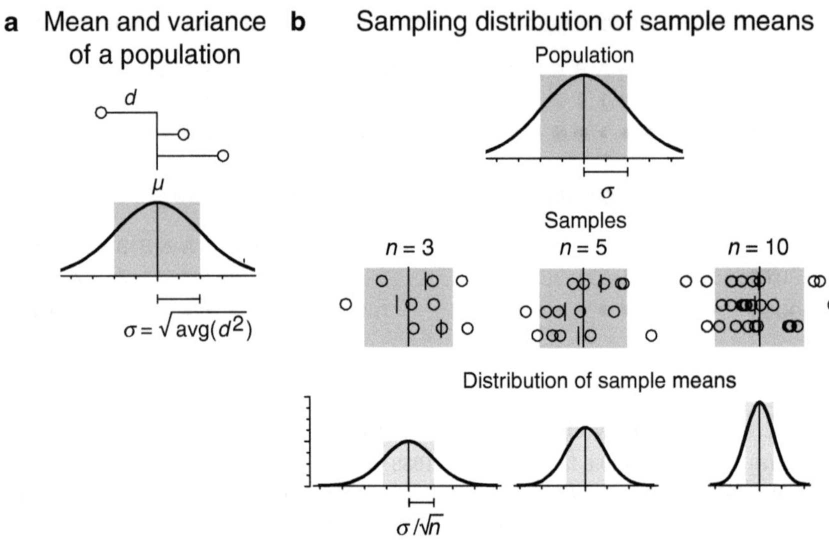

Fig. 3 Statistical inference is often applied to the mean and variance of a population. (**a**) The mean, μ, is a measure of centrality. The variance, σ^2, is a measure of the population spread and is given by the average square of the distance, *d*, also called the deviation, between a randomly selected value from the population and the population mean. The square root of the variance is called the standard deviation, σ. (**b**) Samples of size *n* drawn from a population with mean μ and variance σ^2 can be used to estimate the parameters of the population. The distribution of sample means has the same mean as the population and a variance that is smaller in proportion to the sample size, σ^2/n. Shown populations are normal but real populations often have other shapes

library, collect reads and enumerate those that correspond to the feature. If we collect N reads from a library then the population mean of the number of reads for this feature is $\mu = N\pi$. The actual observed read counts for this feature should be within a few standard deviations of $N\pi$. Statistical methods allow us to generalize to libraries of different sizes (and all of our data should be used in the analysis) but for simplicity of exposition in this chapter we assume that we obtain the same number N reads per library for all of our EUs.

Of course, we do not actually know μ or π—we can only estimate them by generalizing our observations from sequencing and counting the transcripts in each. For each treatment, we use the statistical summaries of our sample counts to estimate the corresponding properties of the population, keeping in mind that if we had taken a different sample, we would get a different summary. The set of all possible sample values of the summaries is called the sampling distribution [1]. Unless we replicate the experiment, we only get to observe one value from the sampling distribution—the value for the EUs we measured.

Statistical theory tells us two important facts about the sampling distribution of the sample mean: if the original population has mean μ and variance σ^2 then the sampling distribution of the sample mean has mean μ and variance σ^2/n where n is the number of EUs or, generally, the sample size (Fig. 3b). Consequently, the sample mean is always less variable than the original observations, and the larger the sample size the smaller this variance. For example, drawing from a population with variance σ^2, the means of samples of size $n = 4$ will have a variance of $\sigma^2/4$. If the sample size is increased to $n = 16$, this variance decreases to $\sigma^2/16$. The standard deviation of the sampling distribution (e.g., $\sigma/\sqrt{n}$) is called the standard error (SE). Most commonly, we speak about the sampling distribution of the sample mean and therefore the term "standard error" usually refers to the standard error of the sample mean.

After collecting the sample, we report the sample summaries with their corresponding estimated SE. When statistical theory does not provide a formula for estimating the SE of a summary, other methods are available, such as resampling with the bootstrap [4].

2.1.3 Statistical Testing

In high-throughput assays, observations for each feature are subjected to statistical testing to identify features that may be biologically relevant to the factors under study [2, 5]. For example, we might want to know whether gene expression in the liver is impacted by genotype and diet. To this end, we might measure and calculate the difference in the mean expression for mice with different genotypes on the same diet and in mice with the same genotype on different diets. We might also test whether the effects of genotype and diet interact—that is, if any effect attributable to the diet differs between the genotypes.

In this kind of study, we are assuming that the genotype and diet induce systematic effects, which we hope to detect. Since our goal is to determine the effect of diet and genotype on the biology of mice in general, we are not interested in the particular mice we measured. Instead, we use the sample mice as representative samples (hence the requirement for careful randomization) of the general population of mice and use our observations to make inferences about that population. One common type of inference is to pose the null hypothesis that there is no systematic effect of the treatments on the population summary. For example, in our study, for each gene, the null hypothesis would be there are no differences in mean expression in the four treatment groups.

To test a hypothesis about a population summary, we use the sample estimate of the summary. Any systematic effects will always be perturbed by biological differences among the EUs and noise induced by the measurement system. We refer to these perturbations as biological and technical variation, respectively. Because of this variation, the sampling distribution of the sample summaries needs to be taken into account.

Statistical testing proceeds by determining whether the differences between the observed sample summaries is larger than expected due to the biological and technical variation in the measurements alone if there was no systematic effect. If such differences are observed, they are ascribed to systematic differences between the treatments. Statistical tests suitable for various types of sequence data are described for example in Chapter 19.

In classical statistical testing, the concept of "larger than expected" in the absence of a systematic effect is quantified by the *P* value—the probability of observing a result at least as extreme as the sample result if the differences between treatment sample means are due only to biological and technical noise. A small *P* value indicates that the observed difference is unlikely due to noise alone [2]. If the *P* value is sufficiently small, we treat it as evidence towards a discovery—the smaller the *P* value cutoff, the stronger the evidence. The false discovery rate is the proportion of discoveries for which there is really no systematic effect due to the treatment.

P values are themselves sample summaries and are subject to sampling variation [6]. *P* values are more meaningful if accompanied by an estimate of the effect, preferably expressed as a confidence interval (CI) [7], which indicates the estimated size of the difference in means (or $\log_2$ of fold change) and thus adds biological interpretability to the result. This provides information about both statistical significance, captured by the *P* value, and biological significance, which may be gleaned from the confidence interval. For example, $P = 0.001$, would be traditionally considered statistically significant but a 95 % CI of 0.02–0.05 might indicate that the inferred change is below a biologically relevant threshold.

2.1.4 Statistical Power

Statistical power is the probability that a systematic effect is detected in our study, given that it is actually present. The proportion of features that are not statistically significant but actually have a systematic effect is called the false nondiscovery rate or FNR. Clearly there is a higher probability of detecting systematic effects if they are larger—thus, power is usually calculated for a given effect size (Fig. 4a). Power is also improved if the noise in the sampling distribution is lower—thus, power is increased with sample size and by use of a study design that controls some sources of variation. In sequencing experiments, power is also higher for features with higher sequencing depth (Fig. 4b). Using a less stringent *P* value threshold also increases power, but this undesirably increases the FDR. One of the objectives of good study design is to optimize power and thus have a low FNR, while still maintaining a low FDR [3, 8].

2.1.5 Confounding

We say that there is confounding in our experiment when two or more factors are applied in a way that makes it impossible to determine which factor could be linked to an observed effect. For example, suppose that we assign the diet factor so that the mice on the low fat diet all have the same genotype, and those on the high

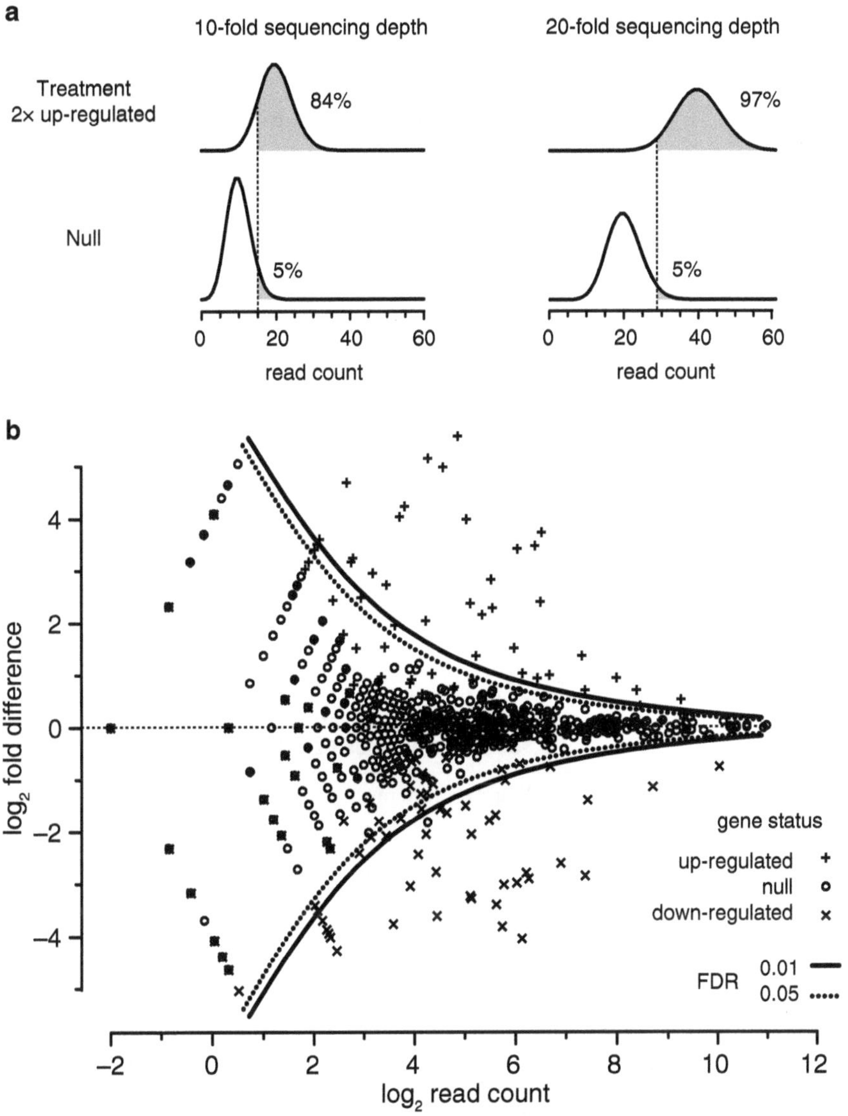

Fig. 4 The relationship between false discovery rate (FDR) and power is complementary. (**a**) Read counts are distributed with a mean proportional to the value of the feature in a quantification study (e.g., gene expression as measured by number of reads from transcripts). The power of detecting a given effect size (e.g., 2× upregulation) increases with sequencing depth. Read count distributions are Poisson. (**b**) Effect size ($\log_2$ of fold difference) as a function of coverage ($\log_2$ of read count). *Open circles* are estimated effects for genes with no differential expression (null). Estimated effects for upregulated genes are denoted with *plus sign* and for downregulated with *cross sign*. Overplotting indicates two or more genes with the same coverage and estimated effect size. Genes declared statistically significant are those outside the boundaries—the *dashed boundary* indicates significance at 5 % FDR and the *solid boundary* indicates significance at 1 % FDR. Up- and down-regulated genes inside the boundaries are false nondiscoveries. Null genes outside the boundaries are false discoveries. False nondiscoveries are more common for low coverage genes due to low power

fat diet all have the other genotype. Any differences in liver gene expression could be due to genotype, diet or both—we cannot know which. It is important to avoid this kind of confounding between factors of interest. It is also good practice to design studies in which factors of interest are not confounded with factors that are not of interest—for example, we should not have one lab handle one of the genotypes and another lab handle the other. Protocols for handling EUs and samples should take care to avoid systematic confounding for known nuisance factors, such as lab effects. Randomization helps avoid confounding with nuisance factors we are unaware of, such as accidentally assigning more active mice to the low fat diet.

2.2 Statistical Principles of Design

Statistical principles of study design include at least five components: pilot experiments, strong protocols for sample collection, randomization (discussed above), blocking and replication (Box 2). These components are applied at several stages of the study to avoid unwanted variation and ensure that any findings are valid and replicable.

Box 2

Statistical principles of design	
• Pilot experiments	Subheading 2.2.1
• Protocols	Subheading 2.2.2
• Randomization	See above (Subheading 2.1.1)
• Blocking	Subheading 2.2.3
• Replication	Subheading 2.2.4

2.2.1 Pilot Experiments

There are many decisions to be made in study design before sequencing can begin. How old should the mice be at the start of the diet and how long should they be on the diet? What amount and type of fat constitute a "high fat" diet? Should the mice be allowed to eat freely or should their intake be restricted to some preset amount? Will the results be the same if the mice are housed locally or by a collaborator? How should the liver samples be taken and stored? Should we take one sample per liver or multiple samples? During sample preparation and sequencing, additional decisions need to be made, which we discuss later.

The pilot experiment helps determine the answers to some of our design questions. It also provides training for personnel, tests the experimental protocol and provides an estimate of the size of the systematic effects of interest and the biological and technical variation. Finally, the pilot study can help determine if there are other nuisance factors that might influence the results. For example, mice on the high fat diet may have behavioral changes that affect gene expression. Depending on the experience of the lab, pilot studies might be required at several stages of the experiment to determine appropriate sample handling methods and sequencing protocols. Potentially costly steps in sample preparation, such as clean-up of impure samples, or amplification of low abundance samples can also be identified during the pilot experiment.

2.2.2 Protocols

The protocol describes everything from how the EUs are handled and how the treatments are applied to the details of the sequencing runs. While it is clearly difficult to think of and record every detail of a study, it is important to keep track of details that might affect the outcomes, such as the time the mice were last fed before sacrifice and how the livers were harvested. Well-written protocols enhance the reproducibility of the research in the case the study is replicated at another lab, samples fail and need to be redone at a later date, there is a change in lab personnel and so on.

Protocols for sample preparation and for sequencing are often maintained at the sequencing facility. The sequencing facility is an important resource that should be contacted early, possibly before the EUs are selected, to assist in protocol development for the sample preparation and sequencing steps. The reagent and sequencing technology manufacturers can also provide valuable assistance and are likely to have the most information about their products.

2.2.3 Blocking

Blocking is a restriction on randomization to ensure that the treatments are evenly distributed among factors that are known to be important contributors to the response variable [9]. For example, if we think that initial weight has a strong influence on gene expression, we would rank our 12 mice within genotype by weight and create three blocks of 4 mice each (Fig. 2c). Within each weight block, genotype would be assigned randomly (two mice per genotype) and subsequently within each genotype diet would be assigned randomly. Each weight block acts as a subexperiment—every combination of genotype and diet is represented. If we can only prepare four liver samples in a batch, we might block by batch and have one mouse of each genotype by diet in each block, so that the block is an entire replicate of the experiment. If we need to block by factors that are not of interest to us, it is convenient to confound them. For example, if neither weight nor batch is of interest, we might process all the mice in a weight block in the same batch so that these two factors are simultaneously controlled (Fig. 2d).

A common mistake is to confound batches with the factors of interest in the experiment. For example, if the experiment is run in collaborating labs, it might seem convenient to have each lab handle one genotype. However, if there prove to be genotype effects, it will be impossible to determine whether these are really due to genotype or due to differences between the labs. If confounding is absolutely necessary, for example if the mice require special care facilities, then a strong protocol can help in minimizing batch effects. Another common confounding factor is time. Whenever possible, it is best to run complete replicates of the experiment at the same time, using several time blocks if more replication is required, or if time effects appear to be important confounders.

When an entire replicate of the experiment can be run within a single block, the experiment is called a randomized complete block design (RCBD). RCBDs are an excellent means of reducing noise for comparative tests while maintaining a realistic amount of variability present in the population. For example, suppose that weight is an important source of variability in gene expression. If we do not group mice by weight, we will not be able to distinguish variation due to weight from variation due to genotype or diet, such as in the completely randomized design (Fig. 2a). This will make our analysis less sensitive to detecting treatment effects as the within-treatment variation being caused by weight may be considerable when compared to the between-treatment variation. If we account for weight, the weight blocks will serve to remove variability (noise) due to weight from comparisons (Fig. 2c). The total variation in the experiment will still be the same, but now we will be able to account for the portion due to weight. Thus, sensitivity to treatment effects will be improved because the effects of genotype and diet on gene expression are calculated within weight groups and subsequently averaged. If we restrict our experiment to only one weight group, we also reduce the variability due to weight from comparisons, but we cannot be sure that the genotype and diet effects can be generalized to the population of mice.

Randomized complete block designs require each block to have enough units to run the entire experiment (without replication). When there are not enough units to do this, more complicated incomplete block designs can be devised [10]. For example, two-channel microarrays can be used as blocks of two EUs (channels). The loop design [11, 12] is an example of an incomplete block design that protects against channel (dye) bias, while providing some variance reduction due to the blocking.

Appropriate statistical analysis requires information about how the randomization was done. When there are restrictions, such as blocking, this needs to be recorded in the data. So, for example, if there are multiple labs, reagent batches, processing dates or other blocking factors, these should be noted with sample information such as factor levels.

2.2.4 Replication

If there were no biological or technical variability, it would be sufficient to measure only one EU for each factor combination. Since this is not the case, we need to have replication, which provides both a more precise estimate of the population quantities and an estimate of variability. While discussion of sampling often focuses on the differences between biological and technical replication, the differences between them depend on the nature of the study. This is discussed in [13] in more detail. For our example, biological replicates are multiple mice with the same genotype on the same diet while technical replicates are multiple samples from the same mouse liver. However, if the liver lobe were a factor in our experiment, biological replicates would be one lobe from multiple mice with the same genotype on the same diet while technical replicates are multiple samples from the same lobe. Note that the mouse would be a blocking factor in the second design because samples within the same liver would have the same biological noise component due to mouse, while samples from different livers would have a different value of this component.

In a completely randomized design, the EUs with the same treatments are the replicates. Although it is common to use balanced designs in which the number of replicates is the same for every factor level combination, this is not required. Unequal replication might occur due to the loss of some EUs, but might also be part of the design—for example if some levels of an attribute factor are rare or some treatments are expensive to sample we might use fewer replicates of these.

In a block design, the blocks are the replicates. Although replication is sometimes done within block, this is inefficient, as averaging replicates within a block reduces only the within block variation.

Replication provides both a measure of variability and a decrease in the standard error of the sample mean or difference between sample means. The statistical analysis of the data requires knowledge of the blocking and replication scheme in order to estimate variability for effects and comparisons. In a completely randomized design, there is one only source of variability—the differences between EUs with the same treatment. However, in block designs, the within block variability and between block variability are expected to differ. For this reason it is important to record the blocking strategy and the block for each EU and to use this information in the statistical analysis.

Every measurement we take includes both biological variation and technical noise—these can be quantified only with replication. In the simplest case, we assume that the difference between the measurement and the population mean is the sum of the noise sources. The biological "noise" can be considered the difference between the population mean and the EU being measured, if the measurement had no technical noise. The biological variation is due to the fact that each EU is a distinct biological entity and will have slightly different proportions of reads from each feature (in our example, mRNA transcripts from genes of interest). The technical noise is the difference between the measurement and the true noise-free value for an EU—the value we would obtain if the measuring device were perfect. The variance of our observations is the *sum* of the all sources of variability and noise. We do not necessarily need to quantify the individual sources of variation, but to distinguish between systematic effects and random variation, we need to quantify the total variability. Replication strategies can be manipulated to both quantify and reduce variation.

In our sequencing study example, the technical variation has two components—variation due to sample preparation and count variation. The former is due to the fact that different sample preparations from the same EU will detect slightly different proportions of reads from each feature due to sample handling, instrumentation, etcetera. Count variation is due to the capture of a random set of reads from all the nucleic acids in the sample.

Count variation is a feature of any procedure that involves the capture of items at random—when the items are rare, the counts will have a Poisson distribution. The mean and variance of a Poisson distribution with N reads and mean proportion π are the same—$N\pi$. Features that compose a large proportion of the reads in a sample also have variation due to read capture, but the distribution of counts need not be Poisson. This count variation, which we'll call σ_c^2, is always present and represents the variability between aliquots taken from the same well-mixed sample.

Both the biological variation and the technical variation not due to sampling molecules are often called the extra-Poisson variation. The corresponding variance components are usually modeled as the $\sigma_b^2 = (N\pi)^2\phi_b$ and $\sigma_t^2 = (N\pi)^2\phi_t$ where ϕ_b and ϕ_t are called the biological and technical dispersion respectively. The total variance of observed counts is the sum of these components, $\sigma_c^2 + \sigma_b^2 + \sigma_t^2$ (Fig. 5a).

If the number of reads per sample N is increased, both the mean and variance increase. To understand the effects of increasing N, we should consider the squared "noise-to-signal" ratio—the variance divided by the square of the mean $(\sigma_c^2 + \sigma_b^2 + \sigma_t^2)/(N\pi)^2 = 1/(N\pi) + \phi_b + \phi_t$. We can see that as N increases, the Poisson count variation becomes negligible relative to the other sources of variation, but increasing the library size cannot reduce the two dispersion terms. To reduce the effects of dispersion, we need to average replicates as discussed below.

If we have m replicate liver samples per mouse, we can quantify the technical noise. For a single mouse, averaging m replicates reduces the technical variation by a factor of $1/m$ so that we obtain a more precise measure for this mouse. This is equivalent to reducing the technical dispersion to ϕ_t/m. However, each of these

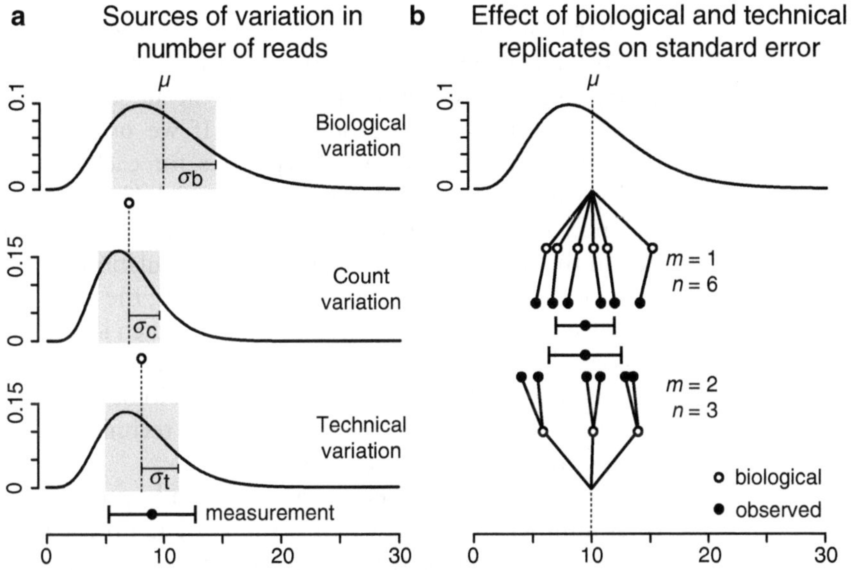

Fig. 5 Biological, count and technical variation contribute to the measurement of read counts and can be controlled using replication. (**a**) Three sources of variation in a sequencing experiment: biological, count and technical. The biological sample (e.g., 7) is derived from the biological population (e.g., $\mu = 10$, $\sigma^2 = 20$). The measurement of the sample is subject to count variation ($\mu = 7$, $\sigma^2 = 7$) and yields a value of 8, which is further affected by technical variation ($\mu = 8$, $\sigma^2 = 10$) for a final measurement of 9. The precision of the measurement has a variance of 37. Error bar is s.d. $= \sqrt{37}$. (**b**) Biological replication samples both the biological and technical variation and is therefore preferred over technical replication to reduce uncertainty. A sample with $n = 6$ biological replicates has a SE of $\sqrt{(37/6)} = 2.48$ and variance from all sources is reduced by a factor of 1/6. A sample with $n = 3$ biological replicates with $m = 2$ technical replicates each has an SE of $\sqrt{(20/3 + 17/6)} = 3.08$. Biological variance is reduced by 1/3

measurements has identical biological noise, which is not reduced by averaging. If instead we measure n mice and take the average, the total variance is reduced by a factor of $1/n$, so that all the sources of variation are reduced by a factor of $1/n$. If we take n biological replications and m technical replicates for each, the sample average will have an SE that reflects that we have reduced the biological variation by a factor of $1/n$ and the technical variation by a factor of $1/nm$.

To see how this applies to the context of sequencing, suppose we are able to sequence a total of nm libraries. If we use one EU and nm technical replicates and average, the variance of the mean is $\sigma_b^2 + (\sigma_c^2 + \sigma_t^2)/nm$. If we use n biological replicates with m technical replicates for each, the variance of the mean is $\sigma_b^2/n + (\sigma_c^2 + \sigma_t^2)/nm$. If we use nm biological replicates with no technical replicates, the variance of the mean is $(\sigma_c^2 + \sigma_b^2 + \sigma_t^2)/nm$. It is clear that case with the smallest variance is the last one—when all of our sequencing resources are committed to biological replication (Fig. 5b).

For example, suppose that we can afford to take six measurements (all costing the same). If we use six biological replicates, the average has SE reflecting a reduction in all sources of variance, $\sigma_b^2 + \sigma_c^2 + \sigma_t^2$, by a factor of 1/6. If we use three biological replicates with two technical replicates for each, we reduce the biological variation, σ_b^2, by a factor of 1/3 and the technical variance, $\sigma_c^2 + \sigma_t^2$, by a factor of 1/6 (Fig. 5b). Notice that we obtain the most precise estimate of the population mean by taking only biological replicates, no matter how big the technical variance is. However, technical replication is often much less expensive than biological replication due to the cost of obtaining EUs. Reference 13 discusses balancing the number of biological and technical replicates to optimize precision when technical replication is cheaper than biological replication.

The total cost of sequencing is usually proportional to the total number of reads, which is the average library size N, times the number of replicates, nm. Expressing the noise to signal ratio as a function of library size together with effects of biological and technical replication on total variance, we can see that for quantification studies, there is a trade-off in the number of reads per sample and the number of replicates. In particular, once N is large enough to identify most features present in the samples, there is little to be gained by greater read depth compared to increasing the replication. Increasing the number of biological replicates n is preferable to increasing the number of technical replicates m.

2.3 Design of Sequencing Studies

Sequencing studies have three main design stages (Box 3)

Box 3

Pre-sequencing design stages	
• Treatment design	Subheading 2.3.1
• Sample design	Subheading 2.3.2
• Sample prep. design	Subheading 2.3.3
Sequencing design stage	
• Sequencing design	Subheading 3

Treatment design—the selection of the experimental conditions under study.

Sample design—the selection of the samples for each experimental condition.

Sample preparation design—the assignment of samples to labs and/or personnel, batches of reagents, preparation times, etcetera.

While the pilot study should include all three stages of the design, the statistical considerations discussed above (protocols for sample collection, randomization, blocking and replication) are applied separately to sample design and sample preparation design and the sequencing steps, which can lead to very complicated experimental designs. Blocking at some stage and then using the same blocks for the subsequent stages can help contain the complexity of the design. Note that all biological replication happens at the sample design stage.

Sequencing design—the allocation of samples to the sequencer and selection of sequencing parameters such as read length and target library size—has many considerations and is discussed in Subheading 3.

2.3.1 Treatment Design

The treatment design defines the systematic effects of interest. The remaining steps all use statistical principles to improve our ability to detect these systematic effects in the presence of biological and technical variability. Appropriate design in sample and sample preparation allow the investigator to quantify variation that cannot be explained by treatment effects—such as that caused by technical and biological variability. This makes it possible to identify variation that is larger than expected by chance in the absence of a treatment effect and thus infer that a treatment effect is present. In experiments that are descriptive, rather than comparative, quantifying the random variation allows the investigator to obtain some idea of what to expect from other similar samples (using frequentist inferential methods) or to update prior estimates using the current data (using Bayesian inferential methods) [14].

The simplest treatment design for comparative studies is a one-way design with a single factor at several levels. With multiple factors, the simplest design is the factorial design, in which all combinations of factor levels are used. However, factorial designs are not always feasible—for example, for a design with three time points, three genotypes and three conditions, there are 27 combinations which may be too many to practically handle. Efficient designs for purposes such as finding the optimal response (response surface designs) or screening large numbers of factors (e.g., fractional factorial designs) are available [10] although they have seldom been used in high-throughput studies. As well, some of the treatment combinations might be extraordinarily expensive or may be lethal to the EUs—for example later time points under high stress conditions for a susceptible genotype. Designs with missing treatment combinations or special treatments such as factorial designs with a control are available, although care must be used to ensure that all of the comparisons of interest can actually be performed.

Factors and levels are generally selected based on prior information and the hypotheses of interest in the experiment. When the prior information is too sparse, a pilot experiment can be implemented to ensure that the features of most interest are being captured. For continuous factors such as time and dose, it is important to note that the levels need not be equally spaced—instead the levels should be selected to capture the most important changes in response.

2.3.2 Sample Design

Sample design determines what is sampled, how the sampling is performed, and how many samples are used. A well-designed study balances internal validity (the ability to replicate the results with similar starting material) with external validity (the ability to replicate design at another time or in a different lab) while controlling sampling cost. Blocking is an important component of sampling that assists in balancing internal and external validity by sampling all the expected variability in the material in different blocks while restricting comparisons to blocks of less variable EUs.

When the study objective is to characterize or optimize technical processes such as sample preparation methods and/or sequencing technologies, only the ability of the investigator to replicate technical aspects of the study are of interest. For these studies, differences between EUs are not of interest and only technical replication is required.

For biological studies, we have to start with collecting the biological units, whether they are aliquots from a tissue culture, mice or samples from cadavers. To induce internal validity, the samples should be as similar as possible before the treatments are applied so that any observed differences between treated samples can be attributed to the treatments. To maintain external validity, the biological variability among the EUs must be representative of the population—this can be achieved with randomization. Finally, for a given sample size and replication scheme, blocking can be used optimize both internal and external validities.

A special type of block design for attribute factors is the matched design. Typically in this type of study, there is an attribute factor of primary interest, and other attributes that are known confounders. EUs with different values of the principal factor are matched on the other attributes to control the variation in response due to the confounding factors. For example, in studying the genetic component of drug addiction in human subjects, addicts may be matched to one or more controls of the same age, gender and education level. Or, soil samples from rehabilitated waste sites might be matched to samples from uncontaminated sites with similar soil and climate. This type of design is usually used when factor levels cannot be assigned at random (e.g., long-term exposure to drugs) or when practical, ethical or time constraints make it impossible to subject the same EU to different treatments.

Time course studies are a common study design that require some care in analysis. There are two types of time course studies: randomized and repeated measures. In a randomized time course study, time is treated like a controllable factor. EUs are assigned at random to each time point and measured only once. In a repeated measures design, the same EU is measured at each time point, effectively making the EU a blocking factor for time and reducing the biological variance of the system. Suppose in our genotype by diet study, we were also interested in the length of time on the diet—say 1 week and 1 month. If the mice need to be sacrificed to obtain the liver sample, an EU cannot be measured at both time points, so we need to use a randomized time course design and assign three mice of each genotype to each diet and time point. This doubles the number of mice required. However, if instead we can take the samples by needle biopsy, and if the time between biopsies is sufficient for return of the mice to normal gene expression, then we can measure each EU at each time point, thus maintaining the original sample size while reducing biological variation due to mouse for comparisons of the same treatment at different times.

An important part of the sample design is determining the sample size to achieve a target precision for estimates or power for detecting differences. The principles of sample size estimation for a single response measure are discussed in [3]. However the problem is difficult for any high throughput "omics" study, as the mean, biological variance and technical variance differ among the features. We actually need to determine the number of biological replicates, the number of technical replicates per biological replicate and the amount of sequencing for each sample. As well, we need to account for the count variation, which is a function of sequencing depth that may differ among features. Sample size estimation is discussed in Chapter 18 for RNA-seq studies and is similar for other comparative studies.

2.3.3 Sample Preparation Design

Sample preparation includes preparing the biological material for RNA or DNA extraction, PCR or rt-PRC, manipulations such as cross-linkage, sample cleaning, amplification, and target sequence enrichment. Replication at this stage is technical replication and is done primarily to determine the variability of technical steps in the sample processing.

Cost is an important consideration—the ideal experimental design is often too expensive. For example, although blocking for date of processing will help minimize confounding of effects of interest with processing date, it is often cost-effective to process samples as they arrive. Additionally, labor-intensive protocols or sample instability may limit the number of samples that can be processed in a batch, so that processing may need to be spread over several days. Other batch effects including those imposed by, for example, different brands of reagents, different lots of the same reagents, and different lots or sources of water used to reconstitute concentrated or dried reagents are easily minimized by blocking.

The ability to accurately and precisely reproduce an experiment can depend on well-written and appropriately detailed protocols. Protocols should be developed with built-in contingencies to adjust to programs such as sample loss, insufficient sample quantity, failure of standard procedures, changes of personnel and technology changes. For example, when biological sampling is inexpensive, the protocol might call for generating excess material or additional biological replicates in case of failures at later stages.

It is generally accepted that a single expert technician has lower variability than a group of technicians with similar skill level. Therefore, a good protocol may include divisions of labor that reflect this assumption. For example, in the mouse study, if a single person cannot reliably execute all steps, one expert should handle the task of sacrificing the mice, another the surgery to remove liver samples, and another to process the livers for analysis. While this type of protocol can greatly reduce the variability for the study at hand, it can be problematic for reproducing the results with different personnel. Reproducibility can be enhanced if each person involved produces a detailed protocol detailing the steps taken. In some cases information that might be difficult to describe with text, may be easily and efficiently conveyed with a video record or illustration of procedures.

Sample handling steps should be uniform across all samples. For example, if amplification or additional cleanup of impure nucleic acid is required for some samples, then even "clean" samples must be subjected to the same procedures. This avoids confounding of the treatments with effects that might be introduced by sample handling.

When changes to the protocol are unavoidable due to, for example, changes in lab personnel or new batches of reagents, it is best to isolate the effects by blocking. For example, if there are insufficient remaining reagents to complete an entire replicate of the experiment, it is preferable to start a new reagent kit rather than switching in mid-block, thus making the reagent batch part of the block effect. If a technician that usually performs a sample preparation step is unavailable, then the replacement technician should perform this step for the entire replicate, even if the original technician returns to work.

Since sample preparation is generally the most expensive step in sequencing, [15] suggest pooling replicate samples (before barcoding) and sequencing the pools. In Fig. 6, a series of replication and pooling schemes for mouse RNA samples is illustrated. The highest cost experiment involves two technical replicates of each biological replicate for a total of eight samples. The lowest cost experiment is all units combined in a single tissue (t_{1234}) or preparation (p_{1234}) that represents the mean signal, with no possibility for estimate of variance. We can adjust the strategy to save cost while minimizing information loss.

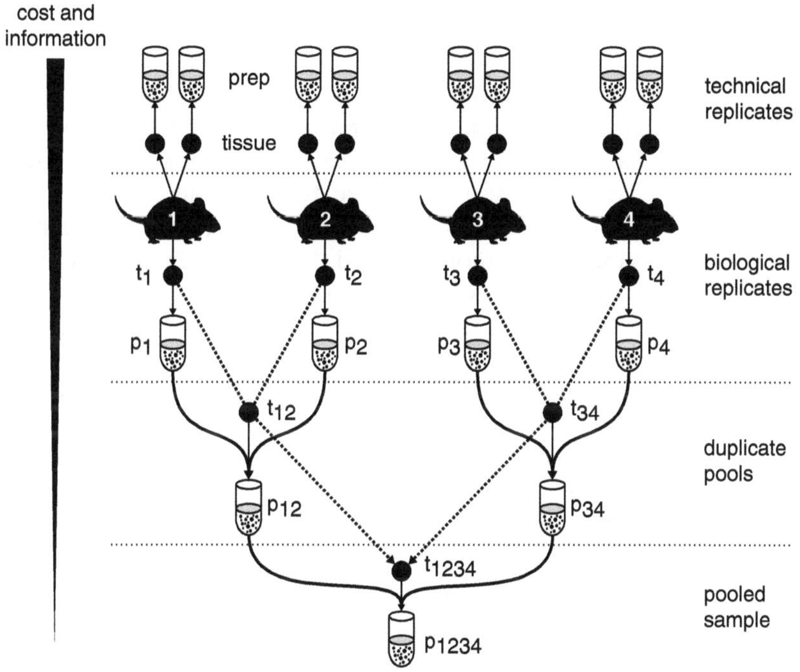

Fig. 6 Relationship between technical replicates, biological replicates, and sample pooling. When technical variability is much lower than biological variability, as often is the case, technical replication only slightly increases the amount of information in the samples but doubles the cost of sample preparation and sequencing. Pooling slightly decreases information and halves the cost of sample prep and sequencing at each pooling step. The lowest cost is realized when all samples are pooled, but then statistical analysis cannot be done, as no information about biological variability can be obtained. Pooling may be done at the tissue sample stage (*dotted line*) or sample preparation stage (*solid line*)

For example, pooling mouse tissues t_1 and t_2 from replicates 1 and 2 into tissue duplicate pool t_{12} and, similarly t_3 and t_4 from replicates 3 and 4 into t_{34}, prior to sample preparation will halve the preparation cost compared to a strategy where each biological replicate is prepared separately (e.g., p_1 and p_2) and then pooled (e.g., p_{12}). Pooling equal masses of tissue is less accurate than pooling equal masses of RNA or DNA, though reduction in cost may justify noisier pooling, especially if the cost is diverted to increasing biological replication.

In theory, pooling equal masses of RNA or DNA from biological replicates averages the biological noise and the technical noise from sample preparation and hence has a similar effect on noise reduction to the more expensive option of sequencing each biological replicate [15, 16]. However, for statistical inference, it is necessary to have replicate pools from independent biological replicates [17] in order to quantify the biological variability.

3 Sequencing Design

The sequencing design determines the type of sequencing and preparation of the samples for sequencing as well as how the samples are assigned to sequencing units (i.e., lanes, cells, plates). To understand this more fully, we need a quick sketch of the measurement process for sequencing studies, starting from tissue sampling. A more detailed description can be found in [18].

Early steps in sample preparation, namely steps taken to isolate nucleic acids, do not vary greatly between technologies. However, the extent of modifications to nucleic acids to prepare them for sequencing varies between sequencing technologies. For massively parallel sequencing-by-synthesis technologies (MPSS, e.g., Illumina™), library preparation is a many-step process that typically begins with intact RNA or DNA and results in a double stranded DNA or cDNA (complementary DNA derived from mRNAs) flanked by adapter sequences. The adapters contain sequences that facilitate clonal enrichment of fragments, the sequencing-by-synthesis reaction, and barcode based cluster identification during multiplexed runs. For third generation technologies (e.g., PacBio™) that generate a signal from a single molecule, rather than a clonal array of molecules (as in MPSS), the extent of nucleic acid preparation is reduced and is generally amplification free.

Sequencing typically proceeds sequentially from one end of a fragment; each base is determined by a signal that is specific to each of the four bases, modified bases, or combinations thereof. A signal sufficient to indicate a base is recorded, along with a quality score that roughly quantifies the confidence in the base call. For MPSS, a predetermined number of bases less than 1 kbp are sequenced, with costs increasing proportionally with read length. MPSS sequencing may proceed from one end only (single end) or both ends (paired end). For third generation technologies, read lengths are much longer, up to 30 kbp [19] and are not predetermined, but a function of fragment size.

In quantification studies, the sequenced reads are mapped to features of interest and the numbers of reads falling within a feature are counted (Fig. 1). In identification studies, the reads are mapped to a reference to identify features such as genetic variants or used to create a reference by de novo assembly.

Sequencing design begins with a careful definition of the questions at hand. If the goal is to discover which genes are differentially regulated in the experimental groups of mice in the example diet study, mRNA quantification in liver samples from biological replicates is a good approach to discovering those genes via a global scale mRNA survey (i.e., RNA-Seq). If genotypes (e.g., single nucleotide polymorphisms, SNPs) that are correlated with a phenotype are desired, the experimental design would be changed to sample many more mice and either DNA or RNA could be assessed to search for SNPs in each EU. If gene or variant discovery is desired, target sequence enrichment (i.e., exome capture [20]) or transcriptome sequencing may be done, with an emphasis on data structure (such as sufficiently long reads) that allows resolution of the features. In experiments in which both identification *and* quantification are goals, then sequencing strategies with complementary data structures may be desirable [21].

3.1 Library Construction

Starting a library preparation with nucleic acids that are of the desired length (full length mRNAs or high molecular weight DNA), of high purity (free of interfering molecules) and of accurate quantity increases the likelihood of successful library preparation and low sequencing error. Success at this stage results in sequencing data that represents the original nucleic acid content and sequences with high fidelity.

For DNA, typically high molecular weight fragments are desired, either as direct input for single molecule technologies or as a starting point for flexibility of fragmentation to a desired length. For RNA, molecule integrity is essential; degradation is often caused by environmentally ubiquitous and robust RNase enzymes [22]. The RNA Integrity Number (RIN) is calculated by an algorithm that assesses rRNA peak ratios, and is a reliable RNA quality indicator [22]. The RIN number is reported on a scale of 1 (worst) to 10 (best) and is often calculated during RNA analysis by commercially available machines.

Purity of the RNA or DNA must be assessed via spectrophotometric methods that provide A_{260}/A_{280} and A_{260}/A_{230} ratios. For a discussion of scanning spectrophotometric methods of nucleic acid analysis see ref. 23. Finally, accurate measurements of sample quantity are desirable, and the use of methods with complementary error (e.g., spectrophotometric vs. fluorometric) can increase confidence in estimates of sample quantity.

3.2 Library Validation

Methods for MPSS library validation have improved in recent years, beginning with an assessment of library fragment length distribution and cDNA concentration, to quantitative real-time PCR (qPCR) based analysis to quantify "sequenceable" library fragments, to small-scale pilot sequencing runs. Standard validation steps of MPSS libraries typically include a qPCR analysis of the library. Just as a pilot experiment can improve the outcome of the full experiment, a pilot sequencing analysis can improve the outcome of the full-scale sequencing effort. A discussion with core facility personnel about sequencing library validation will help users more accurately predict sequencing experiment outcomes, and may also present options for cost management.

3.3 Read Length

Optimal read length depends on the objective of the study. For quantification, the read needs to be long enough to unambiguously assign it to a feature while also being in sufficient quantity to detect rare features. Thus, MPSS is preferred owing to low cost per base and low error for reads between 100 and 350 bp. MPSS sequencing currently allows either single end or paired end sequencing (i.e., sequencing fragments from both ends). Paired-end sequencing produces redundant sequence data if the total sequencing length is greater than the size of the average the part of the library fragment that can be sequenced. For example, if the average insert length is 180 bases, a 150 + 150 paired-end sequencing protocol will result in an average ~120 bp of redundant sequence data for each read pair. A 100 + 100 paired-end protocol, or 200 bp single-end reads will efficiently sequence most library inserts while limiting redundant bases.

In our work, we have found that 100 bp reads (or 50 bp paired end reads) are sufficient for gene expression quantification. For feature discovery, longer reads are better, especially if the read captures an entire feature. This can alleviate challenges of, or abate the need for, de novo assembly. The ENCODE project maintains guidelines for appropriate read lengths for various types of study [http://genome.ucsc.edu/ENCODE/protocols/dataStandards].

Updates in instrument chemistry occur regularly that drive longer and more accurate reads, and changes in sample preparation protocols change the structure of the library. For example, a recent update in a popular instrument's chemistry now allows 250 bp reads, making paired-end sequencing protocols obsolete for typical transcriptome libraries with average insert lengths of about 180 bp. As instrumentation allows longer and longer reads, paired-end sequencing may become obsolete unless insert lengths also increase. Generally the sequencing center, the sequencer manufacturer, and data analysis software developers are knowledgeable about the supported technologies and are excellent sources of advice.

3.4 Multiplexing

Since current sequencers can produce up to hundreds of millions of reads per lane (and this is increasing rapidly), it is cost-effective to multiplex. In a multiplex experiment, the nucleic acid samples are prepared and labeled individually with a unique barcode identifier, and then pooled for sequencing in a single run. Multiplexed sequencing experiments also present a convenient approach to avoid "lane effects." Multiplex runs require an additional validation step, often a pilot run, to adjust the library ratios in the final pool and to assess the quality (i.e., complexity) of individual libraries. The pooled library can be rebalanced as needed and low quality libraries replaced to improve the evenness of sample coverage.

Reference 24 suggests that if possible, *all* samples should be sequenced in a multiplexed run with (technical) replicate lanes if greater read depth is required. By contrast [25], notes that this makes it impossible to determine if there are labeling effects, and recommends using lanes as blocks.

We have found that the strategy of multiplexing all the samples is wise for several reasons. First, as mentioned previously, this controls for "lane effects." Second, a tuning step can be built in to correct variations in the library pool ratios to ensure greater evenness of data volume for each sample, thereby enhancing statistical power. Third, intermediate data analysis can identify low quality libraries allowing the researcher to substitute libraries before the full sequencing effort. Fourth, if a single lane is lost due to error, other lanes will still contain *all* libraries, and while the experiment may suffer loss of statistical power, it will not suffer loss of critical comparisons. Finally, if the data volume for an experiment is abnormally high or low, researchers can add or remove sequencing runs to optimize the experimental outcome. Depending on sequencing platforms, state of the art multiplexing allows pools of tens to hundreds of samples. The trend is towards increased density and more barcodes, making it possible to sequence entire experiments in a single multiplex run.

3.5 Sequencing Depth

Assuming relatively uniform distribution of reads among the multiplexed samples, the average number of reads per sample is determined by the number of reads produced by a lane, the number of samples that are multiplexed, and the number of lanes run. The number of reads required per sample will depend on the type of study and the proposed analysis methods. The average number of reads per feature is summarized by "sequence depth" or "coverage." This is, on the surface, a simple concept but in reality coverage for individual features is deeply confounded by technical and biological variability [26].

For whole genome sequencing, the statistics that describe total coverage and coverage at a given (or higher) depth can be determined analytically, if simplifying assumptions are made. Various publications describe this [27–31] and we summarize the results here.

Suppose that we have a genome of size G which is sampled by N reads of length L, where $L << G$. The average depth of sequencing is given by $\rho = NL/G$. Conversely, the number of reads required to attain depth ρ is $N = \rho G/L$. However, due to count variation, coverage varies by locus—in particular, the average fraction of the genome with no coverage is $e^{-\rho}$. For example, if we assume a haploid human genome, $G = 3 \times 10^9$, then if we sequence at a depth $\rho > 22$, the idealized model tells us that the average number of unrepresented bases will be less than 1. However, because of bias and the fact that many reads do not have unique mappability, the actual value will be larger.

A useful quantity is the average fraction of the genome that is covered by at least k reads, which can also be estimated from the average depth. Significant coverage penalty is incurred in diploid

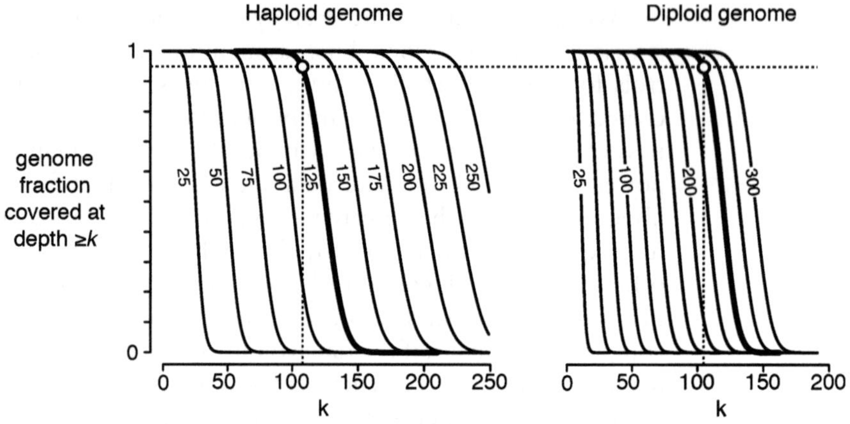

Fig. 7 Cumulative coverage for sequencing of haploid and diploid genomes at depths 25–300×. For a given depth, each trace shows the fraction of the genome covered at a depth of $\geq k$. For example, sequencing at 125× provides at least 107× coverage for 95 % of a haploid genome and 250× provides at least 104× coverage for 95 % of a diploid genome (shown as *hollow circles*)

sequencing. For example, if we require that 99.9 % of the genome is covered by at least 1 read, the average depth needed for a haploid organism is 7 and for a diploid is 16. Diploid sequencing at $\rho = 50$, 100, 150, and 200 coverage achieves 99.9 % coverage by at least $k = 10$, 28 and 48 and 69 reads (Fig. 7).

For a given sequencing depth, coverage across the genome will vary randomly about the average depth. However, across well-mapping regions that are similar in size to a read, coverage will appear to smoothly vary due to the fact that a given read contributes coverage to an interval—the total depth between positions x and $x + L - 1$ will be correlated (Fig. 1).

3.6 Targeted Sequencing

When the target region is not the entire genome, it is useful to reduce G by enriching for the target region. For example, for exome capture of genomic DNA [20], the target genomic DNA is manipulated to increase the fraction of exons. Similarly, transcriptome sequencing enriches for transcribed genomic loci. For ChIP-seq and methylation (Chapter 10) studies, the target regions are the features of interest. Fragment abundance varies because of differential occupancy of the binding or methylation sites. Targeted sequencing induces coverage variability leading to difficulty in predicting coverage.

For transcriptome sequencing transcript abundance spans orders of magnitude [32, 33] and MPSS library preparation introduces cDNA bias by artifacts such as non-random fragmentation [34] causing feature (i.e., transcript) coverage to be heterogeneous. For these reasons, estimates of coverage must be considered on a feature-by-feature basis. Uneven coverage can also be an artifact of low confidence in mapping. For example, chance similarity in low-complexity genome repeats, or duplicated genome regions, can cause reads to map to multiple genome locations. Repetitive sequences in the target genome and sequencing errors in the reads can also lead to mapping errors.

For comparative studies, the total reads for a feature for a given combination of factor levels is the most critical issue in determining the power to detect differences. Both statistical theory and experimental data [35] support the use of more biological replication rather than more sequencing depth. However, the depth should be great enough to ensure that even features producing only a small number of reads can be detected in most samples. There seems to be general agreement that 25 million reads per library is sufficient for most purposes and some studies suggest as few as 10 million are sufficient if there are enough biological replicates [35].

For sequence identification studies, more sequencing depth is required, since genomic variants must be distinguished from sequencing errors. For example, at least 50× coverage at each target base is recommended for genotyping, although lower coverage may be equally effective for more accurate sequencing methods. From our discussion above, we can achieve at least 50× coverage for 99.9 % of the diploid human genome if we sequence at a depth of about 150× (4.5×10^9 200 bp reads). References 36, 37 estimated similar coverage requirements for exome capture sequencing.

4 Conclusion

High-throughput sequencing has become the technology of choice for studying biological processes that can be identified by or characterized by nucleic acid identification. New ideas for processes that can be studied by enriching nucleic acid samples are being proposed at a rapid pace. For these reasons, it is not possible to give a complete set of guidelines for designing sequencing studies.

On the other hand, the same guidelines for study and experimental design that have proven so effective in the past continue to be effective in the high-throughput era. Use of high quality samples and reagents, well-defined protocols, randomization, blocking, replication and avoiding confounding have long been the hallmarks of high quality biological research.

References

1. Krzywinski M, Altman N (2013) Points of significance: importance of being uncertain. Nat Methods 10(9):809–810
2. Krzywinski M, Altman N (2013) Points of significance: significance, P values and t-tests. Nat Methods 10(11):1041–1042
3. Krzywinski M, Altman N (2013) Points of significance: power and sample size. Nat Methods 10(12):1139–1140
4. Kulesa A et al (2015) Points of significance: sampling distributions and the bootstrap. Nat Methods 12(6):477–478

5. Krzywinski M, Altman N (2014) Points of significance: designing comparative experiments. Nat Methods 11(6):597–598
6. Halsey LG et al (2015) The fickle P value generates irreproducible results. Nat Methods 12 (3):179–185
7. Krzywinski M, Altman N (2013) Points of significance: error bars. Nat Methods 10 (10):921–922
8. Anders S, Huber W (2010) Differential expression analysis for sequence count data. Genome Biol 11(10):R106
9. Krzywinski M, Altman N (2014) Points of significance: analysis of variance and blocking. Nat Methods 11(7):699–700
10. Montgomery DC (2012) Design and analysis of experiments, 8th edn. Wiley, Hoboken, NJ
11. Kerr MK, Churchill GA (2001) Experimental design for gene expression microarrays. Biostatistics 2(2):183–201
12. Kerr MK, Churchill GA (2001) Statistical design and the analysis of gene expression microarray data. Genet Res 77(2):123–128
13. Blainey P, Krzywinski M, Altman N (2014) Points of significance: replication. Nat Methods 11(9):879–880
14. Puga JL, Krzywinski M, Altman N (2015) Points of significance: Bayesian statistics. Nat Methods 12(12):277–278
15. Biswas S et al (2013) Biological averaging in RNA-Seq. arXiv:1309.0670v2
16. Williams AG et al (2014) RNA-seq data: challenges in and recommendations for experimental design and analysis. Curr Protoc Hum Genet 83:11 13 1–11 13 20
17. Altman N (2005) Replication, variation and normalisation in microarray experiments. Appl Bioinformatics 4(1):33–44
18. Normand R, Yanai I (2013) An introduction to high-throughput sequencing experiments: design and bioinformatics analysis. Methods Mol Biol 1038:1–26
19. Kremkow BG, Lee KH (2015) Sequencing technologies for animal cell culture research. Biotechnol Lett 37(1):55–65
20. Chilamakuri CS et al (2014) Performance comparison of four exome capture systems for deep sequencing. BMC Genomics 15: 449
21. Wall PK et al (2009) Comparison of next generation sequencing technologies for transcriptome characterization. BMC Genomics 10:347
22. Schroeder A et al (2006) The RIN: an RNA integrity number for assigning integrity values to RNA measurements. BMC Mol Biol 7:3
23. Wilfinger WW, Mackey K, Chomczynski P (1997) Effect of pH and ionic strength on the spectrophotometric assessment of nucleic acid purity. Biotechniques 22(3):474–476, 478–481
24. Auer PL, Doerge RW (2010) Statistical design and analysis of RNA sequencing data. Genetics 185(2):405–416
25. Nettleton D (2014) Design of RNA sequencing experiments. In: Nettleton D, Datta S (eds) Statistical analysis of next generation sequencing data. Springer, New York
26. Sims D et al (2014) Sequencing depth and coverage: key considerations in genomic analyses. Nat Rev Genet 15(2):121–132
27. Wendl MC et al (2001) Theories and applications for sequencing randomly selected clones. Genome Res 11(2):274–280
28. Wendl MC, Waterston RH (2002) Generalized gap model for bacterial artificial chromosome clone fingerprint mapping and shotgun sequencing. Genome Res 12(12):1943–1949
29. Wendl MC, Wilson RK (2008) Aspects of coverage in medical DNA sequencing. BMC Bioinformatics 9:239
30. Lander ES, Waterman MS (1988) Genomic mapping by fingerprinting random clones: a mathematical analysis. Genomics 2(3):231–239
31. Balding DJ, Torney DC (1991) Statistical analysis of DNA fingerprint data for ordered clone physical mapping of human chromosomes. Bull Math Biol 53(6):853–879
32. Marioni JC et al (2008) RNA-seq: an assessment of technical reproducibility and comparison with gene expression arrays. Genome Res 18(9):1509–1517
33. Honaas LA, Wafula EK, Wickett NJ, Der JP, Zhang Y, et al. (2016) Selecting Superior De Novo Transcriptome Assemblies: Lessons Learned by Leveraging the Best Plant Genome. PLoS ONE 11(1): e0146062. doi: 10.1371/journal.pone.0146062
34. Hansen KD, Brenner SE, Dudoit S (2010) Biases in Illumina transcriptome sequencing caused by random hexamer priming. Nucleic Acids Res 38(12):e131
35. Liu Y, Zhou J, White KP (2014) RNA-seq differential expression studies: more sequence or more replication? Bioinformatics 30 (3):301–304
36. Ajay SS et al (2011) Accurate and comprehensive sequencing of personal genomes. Genome Res 21(9):1498–1505
37. Meynert AM et al (2014) Variant detection sensitivity and biases in whole genome and exome sequencing. BMC Bioinformatics 15:247

Chapter 4

Genomic Annotation Resources in R/Bioconductor

Marc R.J. Carlson, Hervé Pagès, Sonali Arora, Valerie Obenchain, and Martin Morgan

Abstract

Annotation resources make up a significant proportion of the Bioconductor project (Huber et al., Nat Methods 12:115–121, 2015). And there are also a diverse set of online resources available which are accessed using specific packages. Here we describe the most popular of these resources and give some high level examples on how to use them.

Key words Annotation, Next-generation sequencing, R, Bioconductor, Genomics

1 Introduction

Annotations in Bioconductor have traditionally been used near the end of an analysis. After the bulk of the data analysis, annotations would be used interpretatively to learn about the most significant results. But increasingly, they are also used as a starting point or even as an intermediate step to help guide a study that is still in progress. In addition to this, what it means for something to be an annotation is also becoming less clear than it once was. It used to be clear that annotations were only those things that had been established after multiple different studies had been performed (such as the primary role of a gene product). But today many large data sets are treated by communities in much the same way that classic annotations once were: as a reference for additional comparisons.

Another change that is underway with annotations in Bioconductor is in the way that they are obtained. In the past annotations existed almost exclusively as separate annotation packages [2–4]. Today packages are still an enormous source of annotations. The current repository contains over 800 annotation packages (http://bioconductor.org/packages/release/BiocViews.html#___AnnotationData). Here is a table that summarizes some of the more

Ewy Mathé and Sean Davis (eds.), *Statistical Genomics: Methods and Protocols*, Methods in Molecular Biology, vol. 1418, DOI 10.1007/978-1-4939-3578-9_4, © Springer Science+Business Media New York 2016

Table 1
Summary of popular Annotation Objects

Object type	Example package name	Contents
OrgDb	org.Hs.eg.db	Gene based information for *Homo sapiens*
TxDb	TxDb.Hsapiens.UCSC.hg19.knownGene	Transcriptome ranges for *Homo sapiens*
OrganismDb	Homo.sapiens	Composite information for *Homo sapiens*
BSgenome	BSgenome.Hsapiens.UCSC.hg19	Genome sequence for *Homo sapiens*

important classes of annotation objects that are often accessed using packages (Table 1).

But in spite of the popularity of annotation packages, annotations are increasingly also being pulled down from Web services like biomaRt [5–7] or from the AnnotationHub [8]. And both of these represent enormous resources for annotation data.

In part because of the rapidly evolving landscape, it is currently impossible in a single chapter to cover every possible annotation or even every kind of annotation present in Bioconductor. So this chapter instead goes over the most popular annotation resources and describe them in a way intended to expose common patterns used for accessing them. The hope is that a user with this information will be able to make educated guesses about how to find and use additional resources that will inevitably be added later. Topics that are covered include the following:

- Using the AnnotationHub
- OrgDb Objects
- TxDb Objects
- OrganismDb Objects
- BSgenome Objects
- Using the biomaRt Web service
- Creating annotation objects

2 Materials

In this chapter we make use of several Bioconductor packages. You can install them with by using biocLite() like so:

```
source("http://bioconductor.org/biocLite.R")
biocLite("AnnotationHub")
```

```
biocLite("Rattus.norvegicus")
biocLite("TxDb.Dmelanogaster.UCSC.dm3.ensGene")
biocLite("BSgenome.Dmelanogaster.UCSC.dm3")
biocLite("biomaRt")
```

The usage of the installed packages is described in detail within Subheading 3.

3 Methods

3.1 Using the AnnotationHub

The top of the list for learning about annotation resources is the relatively new AnnotationHub package [8]. The AnnotationHub was created to provide a convenient access point for end users to find a large range of different annotation objects for use with Bioconductor. Resources found in the AnnotationHub are easy to discover and are presented to the user as familiar Bioconductor data objects. Because it is a recent addition, the AnnotationHub allows access to a broad range of annotation like objects, some of which may not have been considered annotations even a few years ago. To get started with the AnnotationHub users only need to load the package and then create a local AnnotationHub object like this:

```
library("AnnotationHub")
ah <- AnnotationHub()
```

The very first time that you call the AnnotationHub, it will create a cache directory on your system and download the latest metadata for the hubs current contents. From that time forward, whenever you download one of the hubs data objects, it will also cache those files in the local directory so that if you request the information again, you will be able to access it quickly.

The show method of an AnnotationHub object will tell you how many resources are currently accessible using that object as well as give a high level overview of the most common kinds of data present.

```
ah
## AnnotationHub with 19268 records
## # snapshotDate(): 2015-03-26
## # $dataprovider: UCSC, Ensembl, BroadInstitute, NCBI, Haemcode, dbSNP, ...
## # $species: Homo sapiens, Mus musculus, Bos taurus, Pan troglodytes, Da...
## # $rdataclass: GRanges, FaFile, OrgDb, ChainFile, CollapsedVCF, Inparan...
## # additional mcols(): taxonomyid, genome, description, tags,
## #   sourceurl, sourcetype
## # retrieve records with, e.g., 'object[["AH169"]]'
##
##          title
##  AH169   | Meleagris_gallopavo.UMD2.69.cdna.all.fa
##  AH170   | Meleagris_gallopavo.UMD2.69.dna.toplevel.fa
```

```
## AH171    | Meleagris_gallopavo.UMD2.69.dna_rm.toplevel.fa
## AH172    | Meleagris_gallopavo.UMD2.69.dna_sm.toplevel.fa
## AH173    | Meleagris_gallopavo.UMD2.69.ncrna.fa
## ...        ...
## AH28851 | Tursiops_truncatus.turTru1.77.gtf
## AH28852 | Vicugna_pacos.vicPac1.77.gtf
## AH28853 | Xenopus_tropicalis.JGI_4.2.77.gtf
## AH28854 | Xiphophorus_maculatus.Xipmac4.4.2.77.gtf
## AH28855 | RNA-Sequencing and clinical data for 7706 tumor samples fro...
```

As you can see from the object above, there are a LOT of different resources available. So normally when you get an AnnotationHub object the first thing you want to do is to filter it to remove unwanted resources.

Fortunately, the AnnotationHub has several different kinds of metadata that you can use for searching and subsetting. To see the different categories all you need to do is to type the name of your AnnotationHub object and then tab complete from the "$" operator. And to see all possible contents of one of these categories you can pass that value in to unique like this:

```
unique(ah$dataprovider)
## [1] "Ensembl"                          "EncodeDCC"
## [3] "UCSC"                             "dbSNP"
## [5] "Inparanoid8"                      "NCBI"
## [7] "BroadInstitute"                   "NHLBI"
## [9] "ChEA"                             "Pazar"
## [11] "NIH Pathway Interaction Database" "RefNet"
## [13] "Haemcode"                         "GEO"
```

One of the most valuable ways in which the data is labeled is according to the kind of R object that will be returned to you.

```
unique(ah$rdataclass)
## [1] "FaFile"        "GRanges"            "CollapsedVCF"
## [4] "Inparanoid8Db" "OrgDb"              "TwoBitFile"
## [7] "ChainFile"     "SQLiteConnection"   "data.frame"
## [10] "biopax"       "VcfFile"            "ExpressionSet"
```

Once you have identified which sorts of metadata you would like to use to find your data of interest, you can then use the subset or query methods to reduce the size of the hub object to something more manageable. For example you could select only those records where the string "GRanges" was in the metadata. As you can see GRanges are one of the more popular formats for data that comes from the AnnotationHub.

```
grs <- query(ah, "GRanges")
grs
## AnnotationHub with 12390 records
## # snapshotDate(): 2015-03-26
```

```
## # $dataprovider: UCSC, BroadInstitute, Haemcode, Ensembl, Pazar, EncodeDCC
## # $species: Homo sapiens, Mus musculus, Bos taurus, Pan troglodytes, Ca...
## # $rdataclass: GRanges
## # additional mcols(): taxonomyid, genome, description, tags,
## #  sourceurl, sourcetype
## # retrieve records with, e.g., 'object[["AH3166"]]'
##
##        title
##  AH3166  | wgEncodeRikenCageSknshraCellPapTssHmm
##  AH3912  | wgEncodeUwDgfTregwb78495824Hotspots
##  AH3913  | wgEncodeUwDgfTregwb78495824Pk
##  AH4368  | wgEncodeUwDnaseWi38PkRep1
##  AH4369  | wgEncodeUwDnaseWi38PkRep2
##  ...      ...
##  AH28850 | Tupaia_belangeri.TREESHREW.77.gtf
##  AH28851 | Tursiops_truncatus.turTru1.77.gtf
##  AH28852 | Vicugna_pacos.vicPac1.77.gtf
##  AH28853 | Xenopus_tropicalis.JGI_4.2.77.gtf
##  AH28854 | Xiphophorus_maculatus.Xipmac4.4.2.77.gtf
```

Or you can use subsetting to only select for matches on a specific field

```
grs <- ah[ah$rdataclass == "GRanges",]
```

The subset function is also provided.

```
orgs <- subset(ah, ah$rdataclass == "OrgDb")
orgs
```

```
## AnnotationHub with 1145 records
## # snapshotDate(): 2015-03-26
## # $dataprovider: NCBI
## # $species: 'Nostoc azollae'_0708, Acaryochloris marina_MBIC11017, Acet...
## # $rdataclass: OrgDb
## # additional mcols(): taxonomyid, genome, description, tags,
## #  sourceurl, sourcetype
## # retrieve records with, e.g., 'object[["AH12818"]]'
##
##        title
##  AH12818 | org.Pseudomonas_mendocina_NK-01.eg.sqlite
##  AH12819 | org.Streptomyces_coelicolor_A3(2).eg.sqlite
##  AH12820 | org.Cricetulus_griseus.eg.sqlite
##  AH12821 | org.Streptomyces_cattleya_NRRL_8057_=_DSM_46488.eg.sqlite
##  AH12822 | org.Cavia_porcellus.eg.sqlite
##  ...      ...
##  AH13958 | org.Ochotona_princeps.eg.sqlite
##  AH13959 | org.Aeromonas_veronii_B565.eg.sqlite
##  AH13960 | org.Oryctolagus_cuniculus.eg.sqlite
##  AH13961 | org.Tetraodon_nigroviridis.eg.sqlite
##  AH13962 | org.Burkholderia_gladioli_BSR3.eg.sqlite
```

And if you really need access to all the metadata you can extract it as a DataFrame using mcols() like so:

```
meta <- mcols(ah)
```

Also if you are a fan of GUI's you can use the display method to look at your data in a browser and return selected rows back as a smaller AnnotationHub object like this:

```
sah <- display(ah)
```

Calling this method will produce a Web based interface like the one pictured here (Fig. 1).

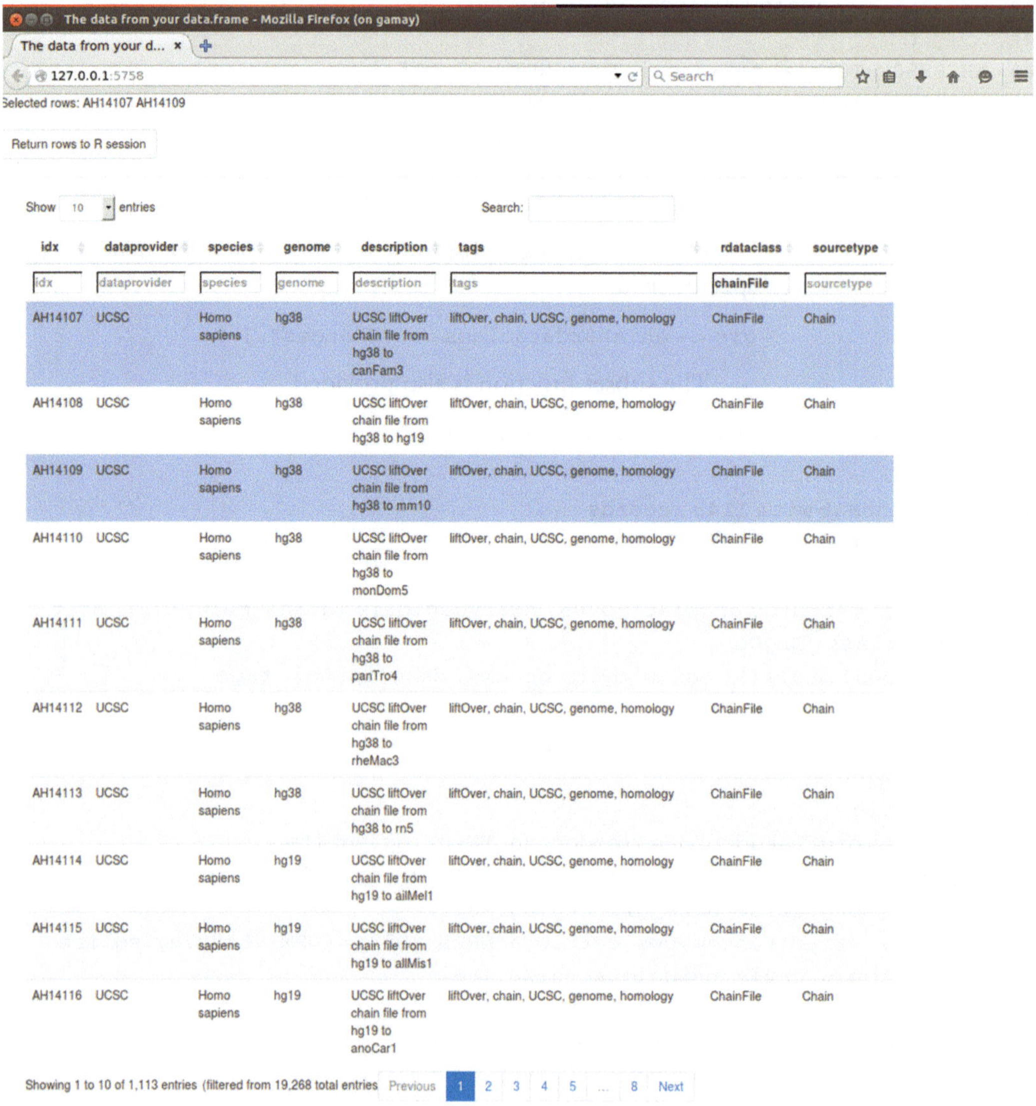

Fig. 1 The display() method will open a user friendly GUI in a local Web browser (if available)

Once you have the AnnotationHub object pared down to a reasonable size, and are sure about which records you want to retrieve, then you only need to use the '[[' operator to extract them. Using the '[[' operator, you can extract by numeric index (1,2,3) or by AnnotationHub ID. If you choose to use the former, you simply extract the element that you are interested in. So for our chain example, you might just want to first one like this:

```
res <- grs[[1]]
## require("GenomicRanges")
head(res, n=3)
## GRanges object with 3 ranges and 5 metadata columns:
##     seqnames            ranges strand |
##        <Rle>         <IRanges>  <Rle> |
##  [1]   chr11 [65266509, 65266572]     + |
##  [2]    chr1 [156675257, 156675415]    - |
##  [3]   chr10 [33247091, 33247233]     - |
##                          name     score     level
##                   <character> <integer> <numeric>
##  [1] chr11:65266509:65266572:+:0.04:1.000000     0 20993.670
##  [2] chr1:156675257:156675415:-:0.35:1.000000    0 11580.471
##  [3] chr10:33247091:33247233:-:0.32:1.000000     0  8098.093
##       signif    score2
##    <numeric> <integer>
##  [1] 3.36e-10         0
##  [2] 3.54e-10         0
##  [3] 3.82e-10         0
##  -------
##  seqinfo: 24 sequences from hg19 genome
```

Or you might have decided that you want to see the data for the green spotted pufferfish by that you spotted in the orgs subset under the name "AH13961". That data could also be extracted like this:

```
tetra <- orgs[["AH13961"]]
## Loading required package: AnnotationDbi
## Loading required package: Biobase
## Welcome to Bioconductor
##
##     Vignettes contain introductory material; view with
##     'browseVignettes()'. To cite Bioconductor, see
##     'citation("Biobase")', and for packages 'citation("pkgname")'.
##
##
## Attaching package: 'Biobase'
##
## The following object is masked from 'package:AnnotationHub':
```

```
##
##   cache
```

```
tetra
```

```
## OrgDb object:
## | DBSCHEMAVERSION: 2.1
## | DBSCHEMA: NOSCHEMA_DB
## | ORGANISM: Tetraodon nigroviridis
## | SPECIES: Tetraodon nigroviridis
## | CENTRALID: GID
## | TAXID: 99883
## | Db type: OrgDb
## | Supporting package: AnnotationDbi
##
## Please see: help('select') for usage information
```

3.2 OrgDb Objects

At this point you may be wondering: What is this OrgDb object about? OrgDb objects are one member of a family of annotation objects that all represent hidden data through a shared set of methods. So if you look closely at the tetra object in the example above you can see that it contains data for Tetraodon nigroviridis (taxonomy ID = 99883). You can learn a little more about it by learning about the columns method.

```
columns(tetra)
## [1] "ACCNUM"     "ALIAS"       "CHR"         "ENTREZID"   "GENENAME"
## [6] "SYMBOL"     "GID"         "GO"          "EVIDENCE"   "ONTOLOGY"
## [11] "GOALL"     "EVIDENCEALL" "ONTOLOGYALL" "PMID"       "REFSEQ"
```

The columns method gives you a vector of data types that can be retrieved from the object that you call it on. So the above call indicates that there are several different data types that can be retrieved from the tetra object.

A very similar method is the keytypes method, which will list all the data types that can also be used as keys.

```
keytypes(tetra)
## [1] "ACCNUM"     "ALIAS"       "ENTREZID"   "GENENAME"   "SYMBOL"
## [6] "GID"       "GO"          "EVIDENCE"   "ONTOLOGY"   "GOALL"
## [11] "EVIDENCEALL" "ONTOLOGYALL" "PMID"       "REFSEQ"
```

In many cases most of the things that are listed as columns will also come back from a keytypes call, but since these two things are not guaranteed to be identical, we must maintain two separate methods.

Now that you can see what kinds of things can be used as keys, you can call the keys method to extract out all the keys of a given key type.

```
keys(tetra, keytype="ENTREZID")
```

```
## [1] "3453248" "3453249" "3453250" "3453251" "3453252" "3453253" "3453254"
## [8] "3453255" "3453256" "3453257" "3453258" "3453259" "3474988"
```

This is useful if you need to get all the IDs of a particular kind but the keys method has a few extra arguments that can make it even more flexible. For example, using the keys method you could also extract the gene SYMBOLS that start with "COX" like this:

```
keys(tetra, keytype="SYMBOL", pattern="COX")
## [1] "COX1" "COX2" "COX3"
```

Or if you really needed an other keytype, you can use the column argument to extract the ENTREZ GENE IDs for those gene SYMBOLS that start with "COX":

```
keys(tetra, keytype="ENTREZID", pattern="COX", column="SYMBOL")
## [1] "3453250" "3453251" "3453252"
```

But often, you will really want to extract other data that matches a particular key or set of keys. For that there are two methods which you can use. The more powerful of these is probably select. Here is how you would look up the gene SYMBOL, and REFSEQ id for the entrez gene ID "3453250".

```
select(tetra, keys="3453250", columns=c("SYMBOL","REFSEQ"),
   keytype="ENTREZID")
##  ENTREZID SYMBOL   REFSEQ
## 1 3453250  COX1 YP_254663.1
```

When you call it, select will return a data.frame that attempts to fill in matching values for all the columns you requested. However, if you ask select for things that have a many to one relationship to your keys it can result in an expansion of the data object that is returned. For example, watch what happens when we ask for the GO terms for the same entrez gene ID:

```
select(tetra, keys="3453250", columns="GO", keytype=
"ENTREZID")
##  ENTREZID    GO
## 1 3453250 GO:0046686
## 2 3453250 GO:0051597
## 3 3453250 GO:0004129
## 4 3453250 GO:0009055
## 5 3453250 GO:0020037
## 6 3453250 GO:0009060
## 7 3453250 GO:0016021
```

Because there are seven GO terms associated with the gene "3453250", you end up with seven rows in the data.frame... This can become problematic if you then ask for several columns that have a many to one relationship to the original key. If you were to do that, not only would the result multiply in size, it would also become really hard to use. So the recommended strategy is to be selective when using select.

Sometimes you might want to look up matching results in a way that is simpler than the data.frame object that select returns. This is especially true when you only want to look up one kind of value per key. For these cases, we recommend that you look at the mapIds method. Let us look at what happens if request the same basic information as the last select call, but instead using the mapIds method:

```
mapIds(tetra, keys="3453250", column="GO", keytype=
"ENTREZID")
```

```
##   3453250
## "GO:0046686"
```

As you can see, the mapIds method allows you to simplify the result that is returned. And by default, mapIds only returns the first matching element for each key. But what if you really need all those GO terms returned when you call mapIds? Well then you can make use of the mapIds multiVals argument. There are several options for this argument, we have already seen how by default you can return only the "first" element. But you can also return a "list" or "CharacterList" object, or you can "filter" out or return "asNA" any keys that have multiple matches. You can even define your own rule (as a function) and pass that in as an argument to multiVals. Let us look at what happens when you return a list:

```
mapIds(tetra, keys="3453250", column="GO", keytype=
"ENTREZID",
   multiVals="list")
```

```
## $'3453250'
##    [1]    "GO:0046686"    "GO:0051597"    "GO:0004129"
"GO:0009055" "GO:0020037"
## [6] "GO:0009060" "GO:0016021"
```

Now you know how to extract information from an OrgDb object, you might find it helpful to know that there is a whole family of other AnnotationDb derived objects that you can also use with these same five methods (keytypes(), columns(), keys(), select(), and mapIds()). For example there are ChipDb objects, InparanoidDb objects and TxDb objects which contain data about microarray probes, inparanoid homology partners or transcript range information respectively. And there are also more specialized objects like GODb or ReactomeDb objects which offer access to data from GO and reactome. In the next section, we look at one of the more popular classes of these objects: the TxDb object.

3.3 TxDb Objects

As mentioned before, TxDb objects can be accessed using the standard set of methods: keytypes(), columns(), keys(), select(), and mapIds(). But because these objects contain information about a transcriptome, they are often used to compare range based information to these important features of the genome [3, 4]. As a result they also have specialized accessors for extracting out ranges that correspond to important transcriptome characteristics.

Let us start by loading a TxDb object from an annotation package based on the UCSC ensembl genes track for *Drosophila*. A common practice when loading these is to shorten the long name to “txdb” (just as a convenience).

```
library("TxDb.Dmelanogaster.UCSC.dm3.ensGene")
## Loading required package: GenomicFeatures
txdb <- TxDb.Dmelanogaster.UCSC.dm3.ensGene
txdb
## TxDb object:
## # Db type: TxDb
## # Supporting package: GenomicFeatures
## # Data source: UCSC
## # Genome: dm3
## # Organism: Drosophila melanogaster
## # UCSC Table: ensGene
## # Resource URL: http://genome.ucsc.edu/
## # Type of Gene ID: Ensembl gene ID
## # Full dataset: yes
## # miRBase build ID: NA
## # transcript_nrow: 29173
## # exon_nrow: 76920
## # cds_nrow: 62135
## # Db created by: GenomicFeatures package from
Bioconductor
## # Creation time: 2015-03-19 14:00:50 -0700 (Thu, 19 Mar
2015)
## # GenomicFeatures version at creation time: 1.19.32
## # RSQLite version at creation time: 1.0.0
## # DBSCHEMAVERSION: 1.1
```

Just by looking at the TxDb object, we can learn a lot about what data it contains including where the data came from, which build of the fly genome it was based on and the last time that the object was updated. One of the most common uses for a TxDb object is to extract various kinds of transcript data out of it. So for example you can extract all the transcripts out of the TxDb as a GRanges object like this:

```
txs <- transcripts(txdb)
txs
## GRanges object with 29173 ranges and 2 metadata columns:
##      seqnames       ranges strand  |     tx_id     tx_name
##         <Rle>    <IRanges>  <Rle>  | <integer> <character>
```

```
##     [1]  chr2L  [7529, 9484]    +  |     1 FBtr0300689
##     [2]  chr2L  [7529, 9484]    +  |     2 FBtr0300690
##     [3]  chr2L  [7529, 9484]    +  |     3 FBtr0330654
##     [4]  chr2L  [21952, 24237]    +  |     4 FBtr0309810
##     [5]  chr2L  [66584, 71390]    +  |     5 FBtr0306539
##     ...   ...         ...  ......    ...    ...
##   [29169]  chrYHet [205196, 205372]  -   |     29166
FBtr0302914
##   [29170]  chrYHet [307129, 307365]  -   |     29167
FBtr0114289
##   [29171]  chrYHet [312456, 313714]  -   |     29168
FBtr0114243
##   [29172]  chrYHet [319739, 320997]  -   |     29169
FBtr0114244
##   [29173]  chrYHet [327052, 328489]  -   |     29170
FBtr0114245
## -------
## seqinfo: 15 sequences (1 circular) from dm3 genome
```

Similarly, there are also extractors for exons(), cds(), genes() and promoters(). Which kind of feature you choose to extract just depends on what information you are after. These basic extractors are fine if you only want a flat representation of these data, but many of these features are inherently nested. So instead of extracting a flat GRanges object, you might choose instead to extract a GRanges-List object that groups the transcripts by the genes that they are associated with like this:

```
txby <- transcriptsBy(txdb, by="gene")
txby
## GRangesList object of length 15682:
## $FBgn0000003
## GRanges object with 1 range and 2 metadata columns:
##      seqnames          ranges strand |     tx_id     tx_name
##         <Rle>       <IRanges>  <Rle> | <integer> <character>
##  [1]    chr3R [2648220, 2648518]     + |     17345 FBtr0081624
##
## $FBgn0000008
## GRanges object with 3 ranges and 2 metadata columns:
##      seqnames            ranges strand | tx_id    tx_name
##  [1]    chr2R [18024494, 18060339]   + |  7681 FBtr0100521
##  [2]    chr2R [18024496, 18060346]   + |  7682 FBtr0071763
##  [3]    chr2R [18024938, 18060346]   + |  7683 FBtr0071764
##
## $FBgn0000014
## GRanges object with 4 ranges and 2 metadata columns:
##      seqnames            ranges strand | tx_id    tx_name
##  [1]    chr3R [12632936, 12655767]   - | 21863 FBtr0306337
```

```
## [2] chr3R [12633349, 12653845] - | 21864 FBtr0083388
## [3] chr3R [12633349, 12655300] - | 21865 FBtr0083387
## [4] chr3R [12633349, 12655474] - | 21866 FBtr0300485
##
##...
## <15679 more elements>
## -------
## seqinfo: 15 sequences (1 circular) from dm3 genome
```

Just as with the flat extractors, there is a whole family of extractors available depending on what you want to extract and how you want it grouped. They include transcriptsBy(), exonsBy(), cdsBy(), intronsByTranscript(), fiveUTRsByTranscript(), and threeUTRsByTranscript().

When dealing with genomic data it is almost inevitable that you will run into problems with the way that different groups have adopted alternate ways of naming chromosomes. This is because almost every major repository has cooked up their own slightly different way of labeling these important features.

To cope with this, the Seqinfo object was invented and is attached to TxDb objects as well as the GenomicRanges extracted from these objects. You can extract it using the seqinfo() method like this:

```
si <- seqinfo(txdb)
si
## Seqinfo object with 15 sequences (1 circular) from dm3
genome:
## seqnames seqlengths isCircular genome
## chr2L     23011544   FALSE  dm3
## chr2R     21146708   FALSE  dm3
## chr3L     24543557   FALSE  dm3
## chr3R     27905053   FALSE  dm3
## chr4      1351857   FALSE  dm3
## ...        ...       ...    ...
## chr3LHet   2555491   FALSE  dm3
## chr3RHet   2517507   FALSE  dm3
## chrXHet    204112   FALSE  dm3
## chrYHet    347038   FALSE  dm3
## chrUextra  29004656   FALSE  dm3
```

And since the seqinfo information is also attached to the GRanges objects produced by the TxDb extractors, you can also call seqinfo on the results of those methods like this:

```
txby <- transcriptsBy(txdb, by="gene")
si <- seqinfo(txby)
```

The Seqinfo object contains a lot of valuable data about which chromosome features are present, whether they are circular or linear, and how long each one is. It is also something that will be checked against if you try to do an operation like "findOverlaps" to compute overlapping ranges etc. So it is a valuable way to make sure that the chromosomes and genome are the same for your annotations as the range that you are comparing them to. But sometimes you may have a situation where your annotation object contains data that is comparable to your data object, but where it is simply named with a different naming style. For those cases, there are helpers that you can use to discover what the current name style is for an object. And there is also a setter method to allow you to change the value to something more appropriate. So in the following example, we are going to change the seqlevelStyle from "UCSC" to "ensembl" based naming convention (and then back again).

```
seqlevels(txdb)

## [1] "chr2L"    "chr2R"    "chr3L"    "chr3R"    "chr4"
## [6] "chrX"     "chrU"     "chrM"     "chr2LHet" "chr2RHet"
## [11] "chr3LHet"  "chr3RHet"  "chrXHet"    "chrYHet"
"chrUextra"

seqlevelsStyle(txdb)

## [1] "UCSC"

seqlevelsStyle(txdb) <- "ensembl"
seqlevels(txdb)

## [1] "2L"                 "2R"
## [3] "3L"                 "3R"
## [5] "4"                  "X"
## [7] "U"                  "dmel_mitochondrion_genome"
## [9] "2LHet"              "2RHet"
## [11] "3LHet"             "3RHet"
## [13] "XHet"              "YHet"
## [15] "Uextra"

## then change it back

seqlevelsStyle(txdb) <- "UCSC"
seqlevels(txdb)

## [1] "chr2L"    "chr2R"    "chr3L"    "chr3R"    "chr4"
## [6] "chrX"     "chrU"     "chrM"     "chr2LHet" "chr2RHet"
## [11] "chr3LHet"  "chr3RHet"  "chrXHet"    "chrYHet"
"chrUextra"
```

In addition to being able to change the naming style used for an object with seqinfo data, you can also toggle which of the chromosomes are "active" so that the software will ignore certain chromosomes. By default, all of the chromosomes are set to be "active".

```
isActiveSeq(txdb)

##  chr2L  chr2R  chr3L  chr3R   chr4   chrX   chrU
##   TRUE   TRUE   TRUE   TRUE   TRUE   TRUE   TRUE
```

```
##    chrM chr2LHet chr2RHet chr3LHet chr3RHet  chrXHet
chrYHet
##    TRUE    TRUE    TRUE    TRUE    TRUE    TRUE    TRUE
## chrUextra
##    TRUE
```

But sometimes you might wish to ignore some of them. For example, let us suppose that you wanted to ignore the Uextra chromosome from our fly txdb. You could do that like so:

```
isActiveSeq(txdb)["chrUextra"] <- FALSE
isActiveSeq(txdb)
```

```
##   chr2L   chr2R   chr3L   chr3R    chr4    chrX    chrU
##    TRUE    TRUE    TRUE    TRUE    TRUE    TRUE    TRUE
##    chrM chr2LHet chr2RHet chr3LHet chr3RHet  chrXHet
chrYHet
##    TRUE    TRUE    TRUE    TRUE    TRUE    TRUE    TRUE
## chrUextra
##   FALSE
```

3.4 OrganismDb Objects

So what happens if you have data from multiple different Annotation objects. For example, what if you had gene SYMBOLS (found in an OrgDb object) and you wanted to easily match those up with known gene transcript names from a UCSC based TxDb object? There is an ideal tool that can help with this kind of problem and it is called an OrganismDb object [9]. On the one hand OrganismDb objects can seem more complex than OrgDb, GODb or TxDb objects because they can allow you to access all three object sources at once. But in reality they are not that complicated. What they do is just query each of these resources for you and then merge the results back together in way that lets you pretend that you only have one source for all your annotations.

To try one out let us load one that was made as an annotation package:

```
library("Rattus.norvegicus")
```

```
## Loading required package: OrganismDbi
## Loading required package: GO.db
## Loading required package: DBI
##
## Loading required package: org.Rn.eg.db
##
## Loading required package: TxDb.Rnorvegicus.UCSC.rn5.
refGene
```

```
Rattus.norvegicus
```

```
## class: OrganismDb
## Annotation resources:
## [1] "GO.db"                "org.Rn.eg.db"
```

```
## [3] "TxDb.Rnorvegicus.UCSC.rn5.refGene"
## Annotation relationships:
##    xDbs       yDbs                  xKeys
## [1,] "GO.db"     "org.Rn.eg.db"            "GOID"
##   [2,]    "org.Rn.eg.db"   "TxDb.Rnorvegicus.UCSC.rn5.
refGene" "ENTREZID"
##    yKeys
## [1,] "GO"
## [2,] "GENEID"
## For more details, please see the show methods for the
component objects listed above.
```

And that is it. You can now use these objects in the same way that you use OrgDb or TxDb objects. It works the same as the base objects that it contains:

```
select(Rattus.norvegicus, keys="24152", columns=c("SYM-
BOL","TXNAME"), keytype= "ENTREZID")
```

```
##  ENTREZID SYMBOL  TXNAME
## 1   24152  Asip NM_052979
```

In fact the five methods that worked for all of the other Db objects that we have discussed (keytypes(), columns(), keys(), select (), and mapIds()) should all work for OrganismDb objects. And if the OrganismDb object was composed to include a TxDb, then the range based accessors should also work:

```
txs    <-    transcripts(Rattus.norvegicus,    columns=c
("TXNAME","SYMBOL"))
txs
```

```
## GRanges object with 18762 ranges and 2 metadata columns:
##                 seqnames           ranges strand  |
##                    <Rle>        <IRanges>  <Rle>  |
##     [1]             chr1   [388173, 401149]     +  |
##     [2]             chr1   [391729, 401149]     +  |
##     [3]             chr1   [688298, 696950]     +  |
##     [4]             chr1   [3400376, 3428890]     +  |
##     [5]             chr1   [3468471, 3478304]     +  |
##     ...              ...              ... ......
##   [18758]              chrX [153807459, 153839833]     -  |
##   [18759]              chrX [154351133, 154467649]     -  |
##   [18760]              chrX [154511679, 154518145]     -  |
##   [18761] chrX_AABR06110762_random [119,    711]     +  |
##  [18762] chrX_AABR06110835_random [5214,   10526]     +  |
##               TXNAME       SYMBOL
##         <CharacterList> <CharacterList>
##     [1]  NM_001099460       Vom2r3
##     [2]  NM_001099457       Vom2r2
##     [3]  NM_001099462       Vom2r5
```

```
##     [4]  NM_001106217      Lrp11
##     [5]  NM_001128191      Nup43
##     ...          ...         ...
##  [18758]  NM_001271327      Map7d3
##  [18759]  NM_001005565     Arhgef6
##  [18760]  NM_001025663        Rbmx
##  [18761]  NM_001037554        Bex4
##  [18762]  NM_001025288     Dmrtc1a
##  -------
##  seqinfo: 2739 sequences (1 circular) from rn5 genome
```

3.5 BSgenome Objects

Another important annotation resource type is a BSgenome package [10]. There are many BSgenome packages in the repository for you to choose from. And you can learn which organisms are already supported by using the available.genomes() function.

```
library("BSgenome")
## Loading required package: Biostrings
## Loading required package: XVector
## Loading required package: rtracklayer
head(available.genomes())
## [1] "BSgenome.Alyrata.JGI.v1"
## [2] "BSgenome.Amellifera.BeeBase.assembly4"
## [3] "BSgenome.Amellifera.UCSC.apiMel2"
## [4] "BSgenome.Amellifera.UCSC.apiMel2.masked"
## [5] "BSgenome.Athaliana.TAIR.04232008"
## [6] "BSgenome.Athaliana.TAIR.TAIR9"
```

Unlike the other resources that we have discussed here, these packages are meant to contain sequence data for a specific genome build of an organism. You load one of these packages in the usual way, and each of them normally has an alias for the primary object that is shorter than the full package name (as a convenience):

```
library("BSgenome.Dmelanogaster.UCSC.dm3")
ls(2)
## [1] "BSgenome.Dmelanogaster.UCSC.dm3" "Dmelanogaster"
Dmelanogaster
## Fly genome:
## # organism: Drosophila melanogaster (Fly)
## # provider: UCSC
## # provider version: dm3
## # release date: Apr. 2006
## # release name: BDGP Release 5
## # 15 sequences:
## #  chr2L   chr2R   chr3L   chr3R   chr4    chrX    chrU
## #  chrM    chr2LHet chr2RHet chr3LHet chr3RHet chrXHet
chrYHet
```

```
### chrUextra
### (use 'seqnames()' to see all the sequence names, use the
'$' or '[['
### operator to access a given sequence)
```

The getSeq method is a useful way of extracting data from these packages. This method takes several arguments but the important ones are the first two. The first argument specifies the BSgenome object to use and the second argument (names) specifies what data you want back out. So for example, if you call it and give a character vector that names the seqnames for the object then you will get the sequences from those chromosomes as a DNAStringSet object.

```
seqNms <- seqnames(Dmelanogaster)
head(seqNms)

## [1] "chr2L" "chr2R" "chr3L" "chr3R" "chr4" "chrX"

getSeq(Dmelanogaster, seqNms[1:2])

## A DNAStringSet instance of length 2
##     width seq
## [1] 23011544
CGACAATGCACGACAGAGGAAGCAGAACAG...TGCAAATTTTGAT-
GAACCCCCCT TTCAAA
## [2] 21146708
GACCCGCTAGGAGATGTTGAGATTGTGAGT...CAACGTGACTGTT TGCATTCTA
GGAATTC
```

Whereas if you give the a GRanges object for the second argument, you can instead get a DNAStringSet that corresponds to those ranges. This can be a powerful way to learn what sequence was present from a particular range. For example, here we can extract the range of a specific gene of interest from *Drosophila*.

```
txby <- transcriptsBy(txdb, by="gene")
geneOfInterest <- txby[["FBgn0000008"]]
res <- getSeq(Dmelanogaster, geneOfInterest)
res

## A DNAStringSet instance of length 3
##   width seq
## [1] 35846
CGCGGCGGTCGCATCGGAGTCGAGAACTCGA...CATACAACCTTAATATATTTA
CCTGAAAAGC
## [2] 35851
CGGCGGTCGCATCGGAGTCGAGAACTCGAAG...CCTTAATATATTTACCTGAAA
AGCAATATAC
## [3] 35409
CGAATACACAAATCAAAGCAAGTGTCTGTGT...CCTTAATATATTTACCT GAAA
AGCAATATAC
```

Additionally, the Biostrings [11] package has many useful functions for finding a pattern in a string set etc. You may not have noticed when it happened, but the Biostrings package was loaded when you loaded the BSgenome object, so these functions will already be available for you to explore.

3.6 biomaRt

Another great annotation resource is the biomaRt package [5–7]. The biomaRt package exposes a huge family of different online annotation resources called marts. Each mart is another of a set of online Web resources that are following a convention that allows them to work with this package. So the first step in using biomaRt is always to load the package and then decide which "mart" you want to use. Once you have made your decision, you will then use the useMart() method to create a mart object in your R session. Here we are looking at the marts available and then choosing to use one of the most popular marts: the "ensembl" mart.

```
library("biomaRt")
head(listMarts())
```

```
##         biomart                    version
## 1        ensembl     ENSEMBL GENES 79 (SANGER UK)
## 2            snp  ENSEMBL VARIATION 79 (SANGER UK)
## 3     regulation  ENSEMBL REGULATION 79 (SANGER UK)
## 4           vega          VEGA 59  (SANGER UK)
## 5    fungi_mart_26        ENSEMBL FUNGI 26 (EBI UK)
## 6 fungi_variations_26 ENSEMBL FUNGI VARIATION 26 (EBI UK)
```

```
ensembl <- useMart("ensembl")
ensembl
```

```
## Object of class 'Mart':
## Using the ensembl BioMart database
## Using the dataset
```

Each "mart" can contain datasets for multiple different things. So the next step is that you need to decide on a dataset. Once you have chosen one, you will need to specify that dataset using the dataset argument when you call the useMart() constructor method. Here we will point to the dataset for chicken.

```
head(listDatasets(ensembl))
```

```
##                  dataset
## 1     oanatinus_gene_ensembl
## 2    cporcellus_gene_ensembl
## 3    gaculeatus_gene_ensembl
## 4     lafricana_gene_ensembl
## 5 itridecemlineatus_gene_ensembl
## 6    choffmanni_gene_ensembl
##                         description version
## 1   Ornithorhynchus anatinus genes (OANA5)  OANA5
```

```
## 2        Cavia porcellus genes (cavPor3) cavPor3
## 3   Gasterosteus aculeatus genes (BROADS1) BROADS1
## 4      Loxodonta africana genes (loxAfr3) loxAfr3
## 5 Ictidomys tridecemlineatus genes (spetri2) spetri2
## 6     Choloepus hoffmanni genes (choHof1) choHof1
```

```
ensembl <- useMart("ensembl",dataset="ggallus_gene_
ensembl")
ensembl
```

```
## Object of class 'Mart':
## Using the ensembl BioMart database
## Using the ggallus_gene_ensembl dataset
```

Next we need to think about attributes, values, and filters. Let us start with attributes. You can get a listing of the different kinds of attributes from biomaRt buy using the listAttributes method:

```
head(listAttributes(ensembl))
```

```
##                  name        description
## 1     ensembl_gene_id     Ensembl Gene ID
## 2 ensembl_transcript_id Ensembl Transcript ID
## 3   ensembl_peptide_id  Ensembl Protein ID
## 4     ensembl_exon_id     Ensembl Exon ID
## 5       description        Description
## 6     chromosome_name     Chromosome Name
```

And you can see what the values for a particular attribute are by using the getBM method:

```
head(getBM(attributes="chromosome_name", mart=ensembl))
```

```
##  chromosome_name
## 1          1
## 2         10
## 3         11
## 4         12
## 5         13
## 6         14
```

Attributes are the things that you can have returned from biomaRt. They are analogous to what you get when you use the columns method with other objects.

In the biomaRt package, filters are things that can be used with values to restrict or choose what comes back. The "values" here are treated as keys that you are passing in and which you would like to know more information about. In contrast, the filter represents the kind of key that you are searching for. So for example, you might choose a filter name of "chromosome_name" to go with specific value of "1". Together these two argument values would request whatever attributes matched things on the first chromosome. And just as there here is an accessor for attributes, there is also an accessor for filters:

```
head(listFilters(ensembl))
##            name    description
## 1 chromosome_name Chromosome name
## 2           start Gene Start (bp)
## 3             end   Gene End (bp)
## 4    marker_start   Marker Start
## 5      marker_end     Marker End
## 6          name_2      Name 2033
```

So now you know about attributes, values, and filters, you can call the getBM method to put it all together and request specific data from the mart. So for example, the following requests gene symbols and entrez gene IDs that are found on chromosome 1 of flies:

```
res <- getBM(attributes=c("hgnc_symbol", "entrezgene"),
          filters = "chromosome_name",
          values = "1", mart = ensembl)
head(res)
##   hgnc_symbol entrezgene
## 1                425783
## 2                430443
## 3      GOLGB1        NA
## 4                426867
## 5        NME6    426866
## 6                426865
```

Of course you may have noticed that a lot of the arguments for getBM are very similar to what you do when you call select. So if it is your preference you can now also use the standard select methods with mart objects.

```
head(columns(ensembl))
## [1] "ensembl_gene_id"    "ensembl_transcript_id" "ensembl_peptide_id"
## [4] "ensembl_exon_id"    "description"           "chromosome_name"
```

3.7 Creating Annotation Objects

By now you are aware that Bioconductor has a lot of annotation resources. But it is still completely impossible to have every annotation resource prepackaged for every conceivable use. Because of this, almost all annotation objects have special functions that can be called to create those objects (or the packages that load them) from generalized data resources or specific file types. Below is a table with a few of the more popular options (Table 2).

In most cases the output for resource creation functions will be an annotation package that you can install.

And there is unfortunately not enough space to demonstrate how to call each of these functions here. But to do so is actually pretty straightforward and most such functions will be well documented with their associated manual pages and vignettes [3, 4, 10, 12]. As usual, you can see the help page for any function right inside of R.

Table 2
How to create popular Annotation Objects

If you want this	And you have this	Then you could call this to help
TxDb	Tracks from UCSC	GenomicFeatures:: makeTxDbPackageFromUCSC
TxDb	Data from biomaRt	GenomicFeatures:: makeTxDbPackageFromBiomaRt
TxDb	gff or gtf file	GenomicFeatures:: makeTxDbFromGFF
OrgDb	Custom data.frames	AnnotationForge::makeOrgPackage
OrgDb	Valid Taxonomy ID	AnnotationForge:: makeOrgPackageFromNCBI
ChipDb	org package and data. frame	AnnotationForge:: makeChipPackage
BSgenome	fasta or twobit sequence files	BSgenome:: forgeBSgenomeDataPkg

```
help("makeTxDbPackageFromUCSC")
```

If you plan to make use of these kinds of functions then you should expect to consult the associated documentation first. These kinds of functions tend to have a lot of arguments and most of them also require that their input data meet some fairly specific criteria. Finally, you should know that even after you have succeeded at creating an annotation package, you will also have to make use of the install.packages() function (with the repos argument=NULL) to install whatever package source directory has just been created.

4 Notes

The bioconductor project represents a very large and active codebase from an active and engaged community. Because of this, you should expect that the software described in this chapter will change over time and often in dramatic ways. As an example, the getSeq function that is described in this chapter is expected to a big overhaul in the coming months. When this happens the older function will be deprecated for a full release cycle (6 months) and then labeled as defunct for another release cycle before it is removed. This cycle is in place so that active users can be warned about what is happening and where they should look for the appropriate replacement functionality. But obviously, this system cannot warn end users if they have not been vigilant about updating their software to the latest version. So please take the time to always update your software to the latest version.

In order to stay abreast of new developments users are also encouraged to explore the bioconductor website which contains many current walkthroughs and vignettes (http://bioconductor.org/), and also to ask questions on the forums (https://support.bioconductor.org/).

Acknowledgments

Research reported in this chapter was supported by the National Human Genome Research Institute of the National Institutes of Health under Award Number U41HG004059 and by the National Cancer Institute of the National Institutes of Health under Award Number U24CA180996. We also want to thank the numerous institutions who produced and maintained the data that are used for generating and updating the annotation resources described here.

References

1. Huber W, Carey VJ, Gentleman R, Anders S, Carlson M, Carvalho BS, Bravo HC, Davis S, Gatto L, Girke T, Gottardo R, Hahne F, Hansen KD, Irizarry RA, Lawrence M, Love MI, MacDonald J, Obenchain V, Oleś AK, Pagès H, Reyes A, Shannon P, Smyth GK, Tenenbaum D, Waldron L, Morgan M (2015) Orchestrating high-throughput genomic analysis with Bioconductor. Nat Methods 12:115–121
2. Pages H, Carlson M, Falcon S, Li N. AnnotationDbi: annotation database interface. R package version 1.30.0. http://bioconductor.org/packages/AnnotationDbi/. Accessed May 2015
3. Carlson M, Pages H, Aboyoun P, Falcon S, Morgan M, Sarkar D, Lawrence M. GenomicFeatures: tools for making and manipulating transcript centric annotations R package version 1.19.38. http://bioconductor.org/packages/GenomicFeatures/. Accessed May 2015
4. Lawrence M, Huber W, Pagès H, Aboyoun P, Carlson M, Gentleman R, Morgan M, Carey V (2013) Software for computing and annotating genomic ranges. PLoS Comput Biol 9, http://dx.doi.org/10.1371/journal.pcbi.1003118http://www.ploscompbiol.org/article/info%3Adoi%2F10.1371%2Fjournal.pcbi.1003118
5. Durinck S, Huber W. biomaRt: interface to BioMart databases (e.g. Ensembl, COSMIC, Wormbase and Gramene) R package version 2.23.5. http://bioconductor.org/packages/biomaRt/. Accessed May 2015
6. Durinck S, Spellman P, Birney E, Huber W (2009) Mapping identifiers for the integration of genomic datasets with the R/Bioconductor package biomaRt. Nat Protoc 4:1184–1191
7. Durinck S, Moreau Y, Kasprzyk A, Davis S, De Moor B, Brazma A, Huber W (2005) BioMart and Bioconductor: a powerful link between biological databases and microarray data analysis. Bioinformatics 21:3439–3440
8. Morgan M, Carlson M, Tenenbaum D, Arora S. AnnotationHub: client to access AnnotationHub resources. R package version 2.0.1. http://bioconductor.org/packages/AnnotationHub/. Accessed May 2015
9. Carlson M, Pages H, Morgan M, Obenchain V. OrganismDbi: software to enable the smooth interfacing of different database packages. R package version 1.10.0. http://bioconductor.org/packages/OrganismDbi/. Accessed May 2015
10. Pages H. BSgenome: infrastructure for Biostrings-based genome data packages. R package version 1.36.0. http://bioconductor.org/packages/BSgenome/. Accessed May 2015

11. Pages H, Aboyoun P, Gentleman R, DebRoy S. Biostrings: string objects representing biological sequences, and matching algorithms. R package version 2.36.0. http://bioconductor.org/packages/Biostrings/. Accessed May 2015

12. Carlson M, Pages H. AnnotationForge: code for building annotation database packages. R package version 1.10.0. http://bioconductor.org/packages/AnnotationForge/. Accessed May 2015

Part II

Public Genomic Data

Chapter 5

The Gene Expression Omnibus Database

Emily Clough and Tanya Barrett

Abstract

The Gene Expression Omnibus (GEO) database is an international public repository that archives and freely distributes high-throughput gene expression and other functional genomics data sets. Created in 2000 as a worldwide resource for gene expression studies, GEO has evolved with rapidly changing technologies and now accepts high-throughput data for many other data applications, including those that examine genome methylation, chromatin structure, and genome–protein interactions. GEO supports community-derived reporting standards that specify provision of several critical study elements including raw data, processed data, and descriptive metadata. The database not only provides access to data for tens of thousands of studies, but also offers various Web-based tools and strategies that enable users to locate data relevant to their specific interests, as well as to visualize and analyze the data. This chapter includes detailed descriptions of methods to query and download GEO data and use the analysis and visualization tools. The GEO homepage is at http://www.ncbi.nlm.nih.gov/geo/.

Key words Microarray, High-throughput sequencing, Gene expression, Functional genomics, Database, Data mining

1 Introduction

Gene Expression Omnibus (GEO) is a database supported by the National Center for Biotechnology Information (NCBI) at the National Library of Medicine (NLM) that accepts raw and processed data with written descriptions of experimental design, sample attributes, and methodology for studies of high-throughput gene expression and genomics. The introduction of DNA microarrays and the Serial Analysis of Gene Expression (SAGE) protocol as methods of simultaneously assaying gene expression of multiple genes in 1995 enabled scientists to study gene expression of hundreds to thousands of genes, thereby vastly increasing the experimental scale and providing a far more complete understanding of biological processes compared to earlier single-gene studies [1, 2]. Microarray technology quickly dominated the field of high-throughput gene expression studies and with the genome sequencing of humans [3] and many model organisms [4–7], genome-wide

Ewy Mathé and Sean Davis (eds.), *Statistical Genomics: Methods and Protocols*, Methods in Molecular Biology, vol. 1418, DOI 10.1007/978-1-4939-3578-9_5, © Springer Science+Business Media New York 2016

gene expression and other functional genomic studies became commonplace by the early 2000s. The accelerating pace of genomic-level data production and the bulky raw and processed data files they generated created a challenge for individual labs or journals to make the data available to the research community. In 2000, NCBI launched the GEO database as a repository for high-throughput gene expression data [8]. In 2002, major journals started to require deposit of microarray data into public repositories [9], and consequently, the content of GEO grew quickly. Furthermore, the nature of high-throughput genomic experiments expanded rapidly since the first microarrays used to analyze gene expression, and thus the GEO database similarly evolved to keep pace with the changing technologies and applications. Today, GEO accepts data from a wide variety of technologies, including DNA microarrays, protein or tissue arrays, high-throughput nucleic acid sequencing, SAGE, and RT-PCR. And while the majority, approximately 90 %, of the data in GEO are indeed gene expression data, the applications have also expanded to include studies on genome methylation, genome binding/occupancy, protein profiling, chromosome conformation studies, and genome variation/copy number [10].

It is serendipitous that the word "geo" is a prefix meaning "earth" because not only does GEO primarily host global gene expression data, GEO itself is indeed a global resource; at the time of this writing GEO contains submissions from 72 nations. There are no fees to submit data to GEO, download data, or use GEO tools. Scientists submit to GEO in order to share their data with the research community and/or as a requirement of publication or grant directives. GEO supports the Minimum Information About a Microarray Experiment (MIAME) [11] and Minimum Information about a high-throughput SEQuencing Experiment (MINSEQE) guidelines set forth by the Functional Genomics Data Society (http://www.fged.org/) for standardization of information about microarray and sequencing experiments that enable the data to be interpreted and replicated by the research community. The GEO database handles the majority of direct submissions from the research community and at the time of this writing holds 54,640 public studies, comprising over 1.3 million samples, derived from 2889 different organisms. An up-to-date summary of GEO data types and content is provided at http://www.ncbi.nlm.nih.gov/geo/summary/.

While the chief role of GEO is to serve as a public data archive, the database is not simply an online warehouse of data. GEO strives to make the data it contains accessible to the research community. Due to the complex nature of the data generated by genomic experiments most studies are analyzed by bioinformaticians and statisticians, or researchers with specialized analysis software. Researchers who lack these skills or software face a substantial challenge if they wish to analyze genomics experiments themselves. In order to make such data analysis accessible to all researchers, GEO has developed several tools for data query, visualization, and analysis that can be performed directly on the GEO website and do not require the download or manipulation of the data files.

2 Methods

2.1 Retrieve a Specific GEO Record

If the GEO accession number is already known, for example, it has been cited in a manuscript describing the data, the user can type the accession number into the "GEO accession" query box. It recognizes Series (GSExxx), Sample (GSMxxx), Platform (GPLxxx), and DataSet (GDSxxx) accession numbers (*see* **Note 1**), and returns the queried record. The "GEO accession" query box is located on the upper right-hand corner of the GEO homepage at http://www.ncbi.nlm.nih.gov/geo/, and is also present at the top of all GEO records for easy movement between accessions. Alternatively, searching the Web directly with a major engine like Google will usually retrieve the correct record.

2.2 Quick Search Using Keywords

NCBI uses a search and retrieval system that can be used to search the content of its entire network of integrated databases including PubMed, GenBank, Genomes, Taxonomy, and many others [12] (see http://www.ncbi.nlm.nih.gov/gquery/gquery.fcgi). GEO data are available in two separate NCBI databases:

1. *GEO DataSets*: Users should use this database to search for studies of interest. The database stores all original submitter-supplied Platform, Sample and Series records, as well as curated gene expression DataSet records. Retrievals include the title, summary, organism, and accession for each record, as well as links to related data (Fig. 1). The *GEO DataSets* search interface is available at http://www.ncbi.nlm.nih.gov/gds/ or can be selected from the dropdown databases menu from the main search box on the NCBI home page at http://www.ncbi.nlm.nih.gov/.
2. *GEO Profiles*: Users should use this database to search for expression profiles of genes. The database stores gene expression profiles derived from curated DataSet records. Retrievals include the gene name, DataSet title, and a thumbnail image that depicts the expression values of that gene across each Sample in that DataSet. Experimental context is provided in the blocks at the foot of the charts making it possible to see immediately whether that gene is differentially expressed across experimental conditions (Fig. 2). Clicking on the thumbnail image enlarges the chart to reveal the full profile details, expression values, and the DataSet subsets that reflect experimental

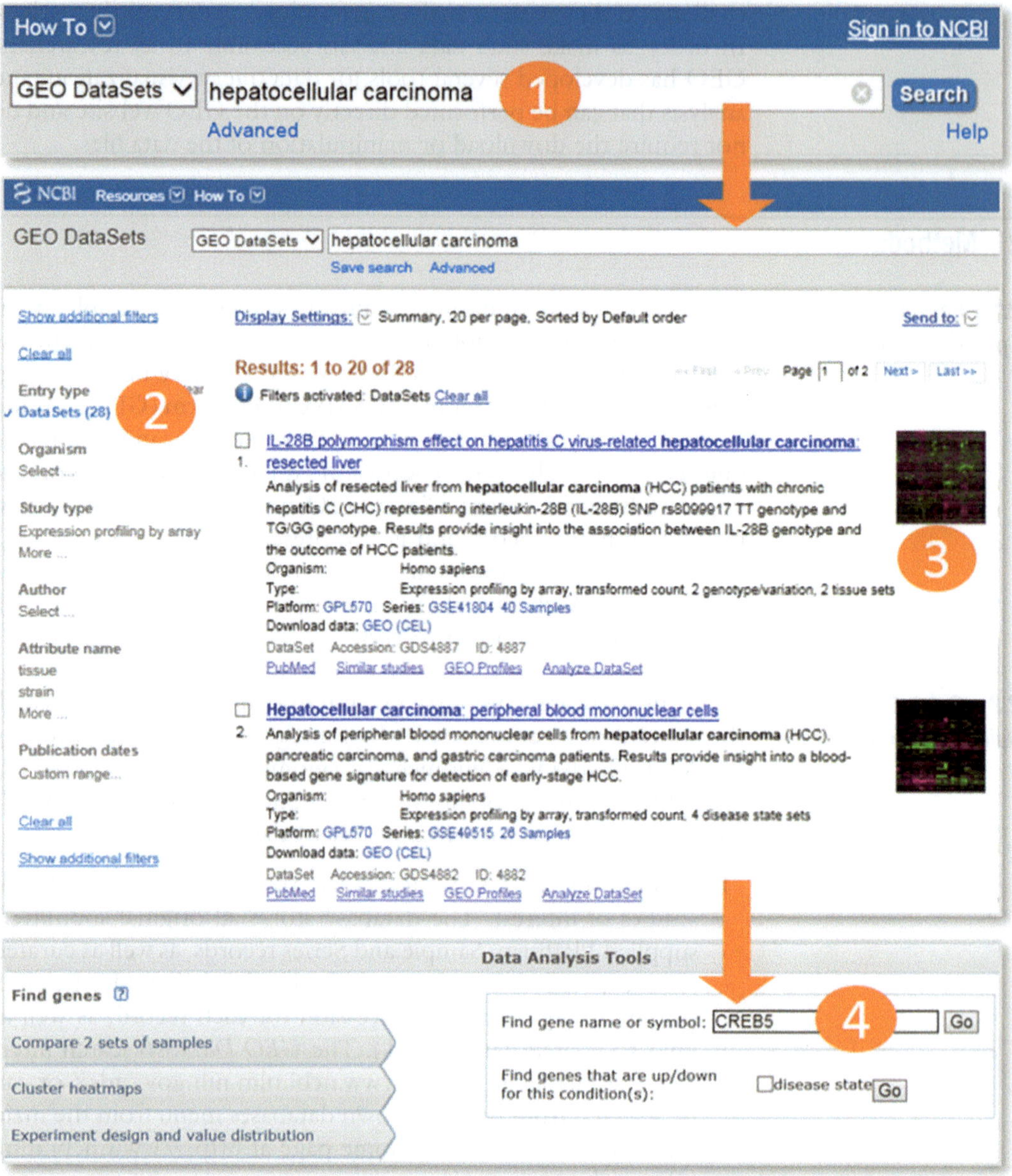

Fig. 1 Workflow screenshots. After typing a search term into the *GEO DataSets* search box (*1*), and using the filter feature to restrict to DataSet entries (*2*), the user retrieves 28 relevant records (*3*). The user selects the second DataSet, GDS4882, and uses the "Find genes" feature in the DataSet Analysis Tools to search for gene CREB5 in that DataSet (*4*). Workflow continues in Fig. 2

design (*see* Subheading 2.7 for more details). The *GEO Profiles* search interface is available at http://www.ncbi.nlm.nih.gov/geoprofiles/ or can be selected from the dropdown databases menu from the main search box on the NCBI home page at http://www.ncbi.nlm.nih.gov/.

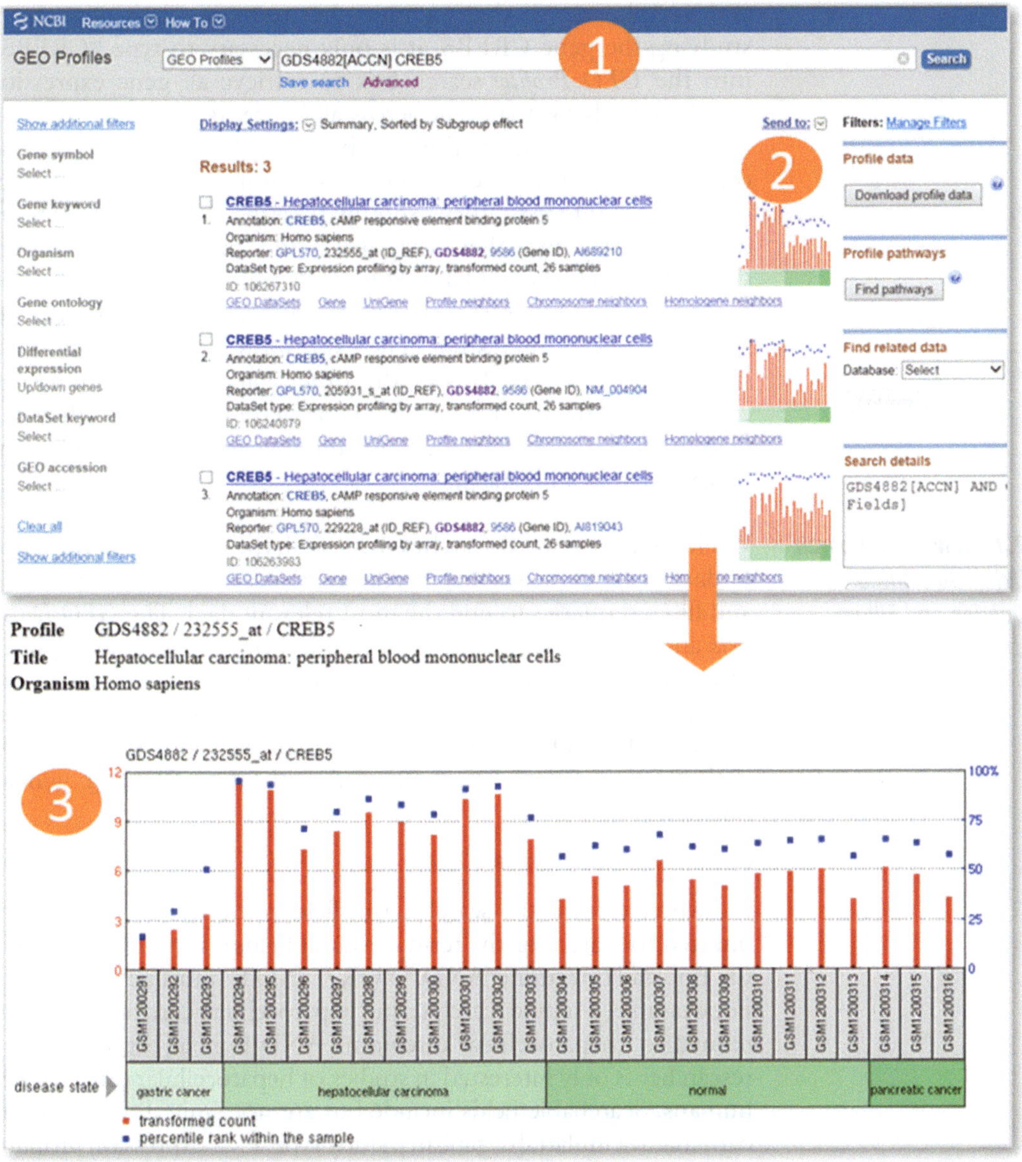

Fig. 2 Workflow screenshots (continued). The "Find genes" feature in the DataSet Analysis Tools (in Fig. 1) creates a search for gene CREB5 in DataSet GDS4882 (*1*). The user is presented with three results in *GEO Profiles* (*2*), meaning that the CREB5 gene is represented by three separate probesets on the Platform in GDS4882. Looking at the chart images, the user can immediately see that all three CREB5 probesets exhibit a similar expression pattern. Clicking on the *top chart* reveals a detailed graphic (*3*), where the user can see that CREB5 is more highly expressed in the hepatocellular Samples, compared to the other Samples examined in that DataSet

Simple keyword searches work very well in these databases. For example, if a user wants to find studies that examine hepatocellular carcinoma, it is only necessary to type "hepatocellular carcinoma" into the *GEO DataSets* search box to retrieve all the DataSet, Series, and Sample records that mention that term. Similarly, if a user is studying the gene CREB5, it is only necessary to type "CREB5" into the *GEO Profiles* search box to retrieve all gene expression profile records for that gene across all DataSets.

Typical workflows within and between these databases depend on the aims of the user. Generally, if users want to identify particular studies of interest, they should search the *GEO DataSets* database first, and then they have the option to use either GEO2R (Subheading 2.8), *GEO DataSets* analysis tools (Subheading 2.6) or *GEO Profiles* analysis tools (Subheading 2.7) to identify specific genes or interesting gene expression patterns within those studies (Figs. 1 and 2). On the other hand, users want to see expression patterns of a favorite gene across any study, they can search the *GEO Profiles* database directly to see how that gene behaves across all DataSets.

2.3 Advanced Search Using Structured Queries and Filters

While simple keyword searches work well, the ever-growing volume of data in GEO means it is increasingly necessary to use structured and filtered queries to find the most relevant data. The *GEO DataSets* and *GEO Profiles* databases enable both simple and sophisticated queries to identify data of interest. Basic keyword searches can be performed alone or in combination with Boolean operators (AND, OR, NOT) to refine the search. Keyword searches with multiple parameters are structured with the following general format:

```
term[field] OPERATOR term[field]
```

where term is the search term, field is the search field (can be omitted to search for the term across all fields), and OPERATOR is the Boolean operator ("AND," "OR," "NOT" must be capitalized). For example, a simple search term "hepatocellular carcinoma" can be refined to "hepatocellular carcinoma AND human[organism]" if a researcher is only interested in studies of hepatocellular carcinoma in humans. Searchable fields include but are not limited to: organism, e.g., macaca mulatta[organism]; study type, e.g., expression profiling by array[DataSet type]; number of samples, e.g., 100:300[number of samples]; author, e.g., smith, a[author]; supplementary file type, e.g., cel[Supplementary Files]. The "How to construct a query" link from the GEO home page provides detailed directions and examples for building queries (http://www.ncbi.nlm.nih.gov/geo/info/qqtutorial.html) (*see* **Note 2**).

The results presented in either *GEO DataSets* or *GEO Profiles* can be further filtered or refined in several ways. Clicking on the word "Advanced" under the search box displaying the original query takes the user to the "Advanced Search Builder" page where searches can be built from drop-down menus. Here queries can be expanded to include multiple fields and operators. The clickable text "show index list" provides the options for available terms to use. For example, the search "hepatocellular carcinoma AND human[organism]" can be further refined to identify only those studies assaying gene expression by array by adding AND DataSet Type and choosing "expression profiling by array" in the Builder.

Search results can also be refined on the same page as the returned results by using the filters available on the left sidebar (Fig. 1). Filters available on *GEO DataSets* and *GEO Profiles* search results are specific to the study- or gene-level type of data contained within each database. For the results in *GEO DataSets*, filters exist for entry type (DataSets, Series, Samples, Platforms), organism, study type, publication date, and author, among others. Filters can be applied by clicking on the text beneath each header (entry type, organism, etc.). For some filters, choosing "select" will open a dialog box to enter the text to be used in the filtering. Once a filter is applied, only the results matching the filter requirements will be available in the results page. Each filter that is applied can be removed by clicking the word "clear" available in small, gray text adjacent to the filter header. At the bottom of the left sidebar is the text "Clear all" which will remove all applied filters. The filters on *GEO Profiles* function in a similar manner but are specific to gene-level queries. Filters on *GEO Profiles* include gene symbol, gene keyword, gene ontology and a filter called "Differential expression" to identify genes that have been flagged as being differentially expressed based on the effect of treatment or condition in a DataSet. The filters provide a flexible way to restrict searches to drill down to relevant data.

2.4 Search Programmatically

Data in both *GEO DataSets* and *GEO Profiles* can be searched programmatically. A suite of NCBI programs called Entrez Programming Utilities (E-utils) are used to conduct queries. E-utils enable sophisticated queries to be performed similar to the nature of the keyword searches or filtering as described above. The utilities are designed to be called from within a computer program that can process their output, which is in XML format.

A typical E-utils workflow might have the following steps:

1. Use the qualifier fields in the *GEO DataSets* database to locate data of interest and construct the appropriate eSearch query in your script or program
2. Run the query, retrieve the results in the form of unique identifiers or history parameters as needed
3. Run eSummary or eFetch and/or eLink depending on specific needs to retrieve the final metadata or accessions.
4. If full data tables or supplementary files need to be downloaded, use the accession information to construct an FTP URL and download the data.

More information for constructing programmatic queries with E-utils can be found at http://www.ncbi.nlm.nih.gov/geo/info/geo_paccess.html.

2.5 Query with a Nucleotide Sequence

The GEO BLAST database contains all GenBank sequences represented on microarray Platforms that participate in DataSets. The GEO BLAST search function provides the opportunity to perform a sequence-based search against GEO using either a nucleotide sequence or a GenBank accession number. The output of GEO BLAST is standard BLAST format where each alignment contains a list of "Related Information" at the right side of the page. Therein is a link to *GEO Profiles* where gene expression on the BLASTed-sequence can be explored. A link to the GEO BLAST utility is available on the GEO home page, http://blast.ncbi.nlm.nih.gov/Blast.cgi?PROGRAM=blastn&BLAST_SPEC=GeoBlast&PAGE_TYPE=BlastSearch

2.6 DataSet Analysis Tools

The first analysis features developed at GEO were based on curated DataSet records which are created at periodic intervals by GEO staff from selected Series; over 3800 DataSets exist at the time of this writing. DataSet records are designed to provide both visualization and data analysis tools for normalized, array-based gene expression studies stored in GEO [13].

The top section of a DataSet record provides information about the study including title, study summary, organism, citation, and Platform and Series accession numbers upon which the DataSet is based. The lower portion of the DataSet record has four tabs encompassing customizable data analysis tools to assist with identification of genes of interest within that DataSet (Fig. 1) (*see* **Note 3**):

1. *Find genes*: Provides a search box for looking up specific gene names or symbols in this DataSet, as well as an option to identify genes that have been flagged as being differentially expressed according to the specific experimental variables in this study. Both of these types of searches take the user to the resulting gene(s) in *GEO Profiles*.
2. *Compare two sets of samples*: Enables a user to perform a customized Student's *t*-test of self-selected Samples in order to identify differentially expressed genes in this DataSet. In order to start the analysis, the user firsts select the test to perform and *P*-value significance level from the drop-down menus. Second, the Samples to be included in the analysis are selected. For example if a user is interested in genes with differential expression only in hepatocellular carcinoma compared to healthy controls from DataSet GDS4882, the user would select the hepatocellular carcinoma Samples for group A by clicking on the corresponding Sample GSM accession numbers under "Group A" and the "normal" Samples under group B. The analysis is

initiated by clicking "Query Group A vs. B". The calculation is performed and takes the user to the resulting genes in *GEO Profiles* (*see* **Note 4**).

3. *Cluster heatmaps*: Presents precalculated and interactive cluster heatmap images that help detect natural groups of coordinately regulated genes. Genes with high levels of expression are represented in pink while genes with low levels of expression are represented in green. This tool allows for choice of hierarchical and partitional (K-means/medians) clustering or clustering genes by chromosome position. The hierarchical and K-means/median clustering contain options so that users can specify the method for linkage and distance, number of clusters, and color display options using drop-down menus. After clicking on the heatmap, the user can zoom in to areas of the cluster, select and export underlying expression values, or view the genes in *GEO Profiles*.
4. *Experiment design and value distribution*: draws boxplots for the expression values for all Samples in a study with corresponding Sample identifiers and Sample subset labels (e.g., drug-treated or control). The boxplot provides a visual overview of the data distribution and Sample categories in this DataSet. Boxplot images that display the distribution of expression values together with experimental design are useful for quality control checks.

2.7 GEO Profiles Analysis Tools

Regardless of how a user arrives at *GEO Profiles* results, either through direct searches (Subheading 2.2) or as a result of performing DataSet analyses (Subheading 2.6), various features exist on *GEO Profiles* records to assist with further analysis and exploration.

Each entry in *GEO Profiles* displays the name of the gene and the title of the DataSet that the data are from, and additional annotation and information about the organism, Platform and probe identifier. Several links that enable further analysis are also presented on the page:

1. *Profile Neighbors*: Retrieves Profiles with similar patterns of expression within the same DataSet, as calculated by Pearson correlation coefficients between pairs of Profiles. The top 200 results are arbitrarily considered to be Profile Neighbors, which may help identify genes with coordinated regulation.
2. *Chromosome Neighbors*: Retrieves Profiles for up to 20 of the closest-found chromosome neighbors within the same DataSet, helping identify expression data for genes within the same chromosomal region.
3. *Sequence Neighbors*: Retrieves Profiles based on BLAST nucleotide sequence similarity across all DataSets, assisting in the identification of genes representing sequence homologs and orthologs.

4. *Homologene neighbors*: Retrieves Profiles that belong to the same HomoloGene group across all DataSets. HomoloGene is an NCBI resource for automated detection of homologs among the annotated genes of several completely sequenced eukaryotic genomes.
5. *Download Profile Data button*: Downloads the values, experimental factors, and gene annotations for each Profile on the page. Download files are tab-delimited and suitable for opening in a spreadsheet application such as Excel (*see* **Note 5**)
6. *Find Pathways button*: Maps the Profiles to a frequency weighted list of pathways in NCBI's BioSystems database. Knowing the pathways the set of Profiles participate in can help characterize that list of genes.

Each record also has a thumbnail image of a chart of gene expression across all Samples in that DataSet, and are useful for rapidly scanning and comparing multiple Profile retrievals (Fig. 2). Clicking this thumbnail image opens a new window with a large view of the same chart. The red bars represent expression values while the blue squares represent the percentile rank of that expression value within the Sample (*see* **Note 6**). The blocks at the bottom of the chart represent experimental variable subsets within the DataSet. Each subset has a type, e.g., "disease state," and a description, e.g., "hepatocellular carcinoma" so users can see at a glance how a gene is behaving across experimental variables (Fig. 2) (*see* **Note 7**).

2.8 Analyze with GEO2R

While the curated DataSets and Profiles records described in Subheadings 2.6 and 2.7 provide analysis and visualization tools for many GEO Series records, the pace of GEO submissions now greatly exceeds the rate at which DataSets can be produced. In order to provide immediate, Web-based, and user-driven analysis for GEO data, GEO2R was developed. GEO2R is an interactive tool that enables the analysis of approximately 90 % of GEO Series as soon as they are released. It uses a Web-based program that employs the Bioconductor [14] packages GEOQuery [15] and limma [16] in R, with the Benjamini–Hochberg false-discovery rate method [17] for multiple-testing correction as its default method. A typical GEO2R workflow would be:

1. Search *GEO DataSets* (*see* Subheading 2.3) to identify a suitable study of interest (*see* **Note 8**).
2. Access the GEO2R tool by clicking the text "Analyze with GEO2R" on the Series record. This brings the user to the GEO2R page which is organized with a table of all Samples in the Series complete with Sample accession numbers, titles, and attributes such as cell type and tissue. Below the Sample table is a set of tabs ("GEO2R," "Value Distribution," "Options,"

"Profile Graph," and "R Script") where analysis by GEO2R is initiated and data visualization and analysis options are available.

3. Check the distribution of the Sample values using the "Value Distribution" tab. Before performing an analysis with GEO2R it is important to check the distribution of the values that are to be analyzed. The quantitative data available for use with GEO2R comes directly from the user-provided Sample data tables with no GEO curation. These data may not be median-centered indicating that they have not been normalized and thus may not be suitable for cross-comparison. The value distributions can be viewed as graphically as a boxplot, or exported and saved as a number summary table.
4. Choose the groups of Samples to be analyzed. For example, GSE18388 is a study of gene expression in the thymuses of mice flown in space compared to those of earth-bound control mice. In order to identify upregulated and downregulated genes in the mouse thymus after space flight, the Samples are placed into two groups. The groups are created and named by clicking "Define groups" at the top of the list of Samples and entering a group name, such as "space flown" and "control". To assign Samples to groups, either click or drag the cursor on the Samples and once highlighted, click the group to which they belong ("space flown" or "control"). The Samples will be highlighted in the group color. More than two groups can be defined for testing across multiple factors.
5. Perform the analysis with default parameters by clicking "Top 250" on the "GEO2R" tab. The top 250 differentially expressed genes are presented in a new window, ordered by *P*-value. The expression pattern of each gene in the table can be visualized by clicking the row to depict expression profile graphs. The complete set of ordered results can be downloaded as a table by clicking "Save all results" (*see* **Note 9**).
6. If desired, default parameters GEO2R can be customized by choosing an alternative method for multiple-testing correction of *P*-values in the "Options" tab. Options are also available to skip or force a log transformation of the input data. If options are changed, GEO2R must be run again, performed by returning to the "GEO2R" tab and clicking "Recalculate".
7. If desired, the "R Script" tab provides all of the R commands used in that analysis, and this information can be saved and used as a reference for how a set of results were calculated.

Alternatively, if a user is not interested in performing a comparison to identify differentially expressed genes, but rather only wants to view the expression profile of a specific gene in the Series, the "Profile Graph" tab can be used to draw the gene expression profile graph for that gene. A YouTube video demonstrating GEO2R features is available at http://www.youtube.com/watch?v=EUPmGWS8ik0.

2.9 Visualize Data as a Genome Track

Data visualization is an important aspect of the analysis of many high-throughput sequencing experiments such as chromatin immunoprecipitation (ChIP-seq) and DNA methylation profiling (bisulfite-seq). Files with raw or normalized genome-wide signal are produced which can be loaded into online genome browsers. Once the data are entered into the genome browser, it is very easy to move to a gene of interest or zoom out to see signal across an entire chromosome. Since GEO receives files types appropriate for genome browser visualization (.WIG, .bedGraph, .BED, etc.) as processed data, GEO is in the process of making these files viewable as tracks in NCBI's Genome Data Viewer. GEO records with tracks include a button with the text "See the Data on the Genome Data Viewer". Clicking this button takes the user directly to the Genome Data Viewer where the track has been preloaded (Fig. 3). On the left side of the Genome Data Viewer page are several boxes with search options to view the desired genomic region: (1) An ideogram view of chromosomes to view signal over an entire chromosome (2) A search box for entering chromosome coordinates or gene name or accession, (3) A box called "Your Data" where

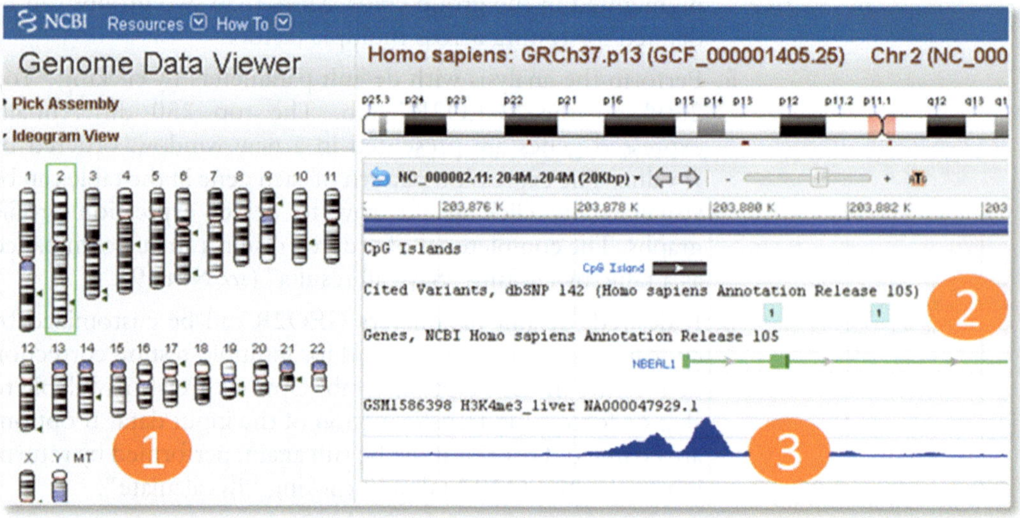

Fig. 3 Screenshot of NCBI Genome Data Viewer. The *left side* of the viewer has tools for locating specific regions of the genome (*1*). The tracks area depicts RefSeq gene, CpG island, and SNP tracks which are set as default for context (*2*), and a track for GEO Sample GSM1586398 which is a H4K3me3 histone ChIP-seq experiment performed on liver tissue (*3*). This track shows a typical H3K4me3 double peak with depletion at the transcriptional start site of gene NBEAL1

additional files can be uploaded for side-by-side viewing with the preloaded tracks from GEO. At the time of this writing there are almost 10,000 GEO Samples with links to tracks preloaded in the Genome Data Viewer. Track creation is an ongoing project and the number of GEO records with tracks is expected to increase in the near future.

2.10 Download GEO Data

All GEO data can be downloaded in various formats using a variety of mechanisms. A popular method for downloading data for specific studies is to download directly from Series pages. At the bottom of each Series page, there is a banner with the text "*Download family*" under which there are links for downloading the data for that Series in three different formats:

1. *SOFT formatted family file(s)* is a link for downloading all of the Series, Sample and Platform data in a single SOFT formatted file. SOFT is an acronym that stands for "Simple Omnibus Format in Text" and formats the data as line-based, plain text.
2. *MINiML formatted family file(s)* is a link for downloading all of the Series, Sample, and Platform data in MiNIML formatted files. MiNIML is an acronym that stands for MIAME Notation in Markup Language, and formats the data as XML with separate data tables. MINiML is essentially an XML rendering of SOFT format.
3. *Series Matrix File(s)* is a link for downloading a tab-delimited value-matrix table generated from the "VALUE" column of each Sample record, headed by Sample and Series metadata. This format is convenient for uploading into data programs such as Microsoft Excel or R.

The Series page also contains links to any supplementary files associated with the Series and a link to a tar archive of all supplementary files provided with the Samples, typically raw data files (*see* **Note 10**). If only a subset of the supplementary files are required there is an option to customize the set of files in the tar archive by clicking the word "custom" on same line as "GSExxx_-RAW.tar". Clicking the "custom" button expands the page to include a list of all Sample supplementary files in the Series with check boxes to select the desired files. Once the boxes next to the needed files have been selected, pressing "Download" initiates the download of a tar archive containing only the selected files.

Additional options for downloading data, including downloading specific portions of records, or programmatic approaches are described at http://www.ncbi.nlm.nih.gov/geo/info/download.html.

3 Methods/Conclusion

The GEO database is now 15 years old, and continues to serve as the leading public repository for direct deposits of high-throughput gene expression and other functional genomics data sets. GEO offers fast and efficient accessioning of raw and processed data with experimental descriptions for the worldwide research community at no cost to the submitter or user. GEO archives these data and makes the data available through flexible querying and download capabilities, and offers several Web-based tools and graphical renderings that facilitate data interpretation and exploration. These tools enable researchers to analyze GEO data with no prerequisite computational skills or software, and without time-consuming download or processing, thereby greatly increasing the utility of the data. GEO continues to evolve to accommodate new data types and increase and improve access to data.

The availability of the high-throughput data in GEO is driving new research. Several thousand publications exist where GEO data have been reused and reanalyzed to develop and test new hypotheses (http://www.ncbi.nlm.nih.gov/geo/info/citations.html). It is evident that the community is using GEO data to address matters far beyond those the initial studies were intended to tackle. Examples include using GEO data to test new or improved algorithms [18], create new subject-specific databases [19], identify disease biomarkers [20], and further characterize gene function [21]. These types of reuse of large data increase the pace and efficiency of scientific discovery and demonstrate the power of the GEO database as a resource for all scientists.

4 Notes

1. Record types, accession codes, and their relationships to each other are described in detail at http://www.ncbi.nlm.nih.gov/geo/info/overview.html. Three primary record types, referred to as Platform (GPLxxx), Sample (GSMxxx), and Series (GSExxx), are supplied by submitters. A Platform record is composed of a summary description of the array or sequencer and, for array-based Platforms, a data table defining the array template. A Sample record describes the conditions under which an individual Sample was handled, the manipulations it underwent, and the measurements derived from it. A Series record links together a group of related Samples and provides a focal point and description of the whole study. A fourth record type, referred to as DataSets (GDSxxx), are assembled by the GEO curation staff from the three primary records. A DataSet represents a curated collection of biologically and statistically

comparable GEO Samples and forms the basis of the *GEO DataSets* and *GEO Profiles* analysis tools. Only array-based expression data are currently considered for DataSet creation, and not all expression data qualify (for instance, due to having experimental designs incompatible with GEO tools). Furthermore, many expression studies have not yet been reviewed by the curation staff for DataSet creation.

2. It is advisable to be aware of full search capabilities including: using asterisks as a wild cards; the fact that some fields accept ranges of data; putting quotes around text to retrieve specific phrases; and proper placement of parentheses.
3. Analyses with *GEO DataSet* tools and GEO2R cannot be performed across multiple Series. Each data normalization is performed only for the Samples within a Series thus the normalized signals may be quite different across Samples from different Series, making direct cross-Series analysis without re-normalization invalid. However, by using *GEO DataSets* or GEO2R to identify differentially expressed genes in two independent Series in the same subject area and with similar experimental designs can be a powerful way to identify genes that are consistently identified in specific diseases, cell types, treatments, etc.
4. Not all the Samples within the DataSet need be included in the comparison.
5. The download file only includes Profiles shown on the current page; to get the maximum number of Profiles, go to the "Display Settings" link and set the "Items per page" to 500.
6. It is important to note that the values (red columns) and ranks (blue squares) are charted on different scales—the blue ranks are always on a scale of 1–100 % (right *Y* axis of the chart) while the red value scale slides to fit the values of a particular profile (left *Y* axis of the chart). This sliding value scale allows subtle differences in values to be more clearly visualized.
7. Clicking the subset type names resorts the chart according to a particular experimental variable—this can assist in clearer visualization of an expression trend in DataSets with multiple variables.
8. Data for use in GEO2R are provided directly by submitters. Data for all Samples in a Series may not meant to be comparable due to a loop design, non-standard or no normalization. Users should thoroughly read through the Series and Sample descriptions to make sure the planned analysis with GEO2R is appropriate for the Samples.
9. The GEO2R results table contains various additional categories of gene annotation that are not immediately visible in default

settings, including Gene Ontology (GO) terms and chromosome locations. Use the "Select columns" link to amend your table with this information.

10. The Series tar archive of supplementary files is a convenient way to download all supplementary files at one time, instead of individually downloading the files from each Sample record.

Acknowledgments

The authors acknowledge the expertise and efforts of the whole GEO curation and development team—Steve Wilhite, Pierre Ledoux, Carlos Evangelista, Irene Kim, Kimberly Marshall, Katherine Phillippy, Patti Sherman, Cynthia Robertson, Hyeseung Lee, Maxim Tomashevsky, Andrey Yefanov, Nadezhda Serova, Naigong Zhang, and Alexandra Soboleva.

Funding

This research was supported by the Intramural Research Program of the National Institutes of Health, National Library of Medicine.

References

1. Schena M, Shalon D, Davis RW, Brown PO (1995) Quantitative monitoring of gene expression patterns with a complementary DNA microarray. Science 270(5235):467–470
2. Velculescu VE, Zhang L, Vogelstein B, Kinzler KW (1995) Serial analysis of gene expression. Science 270(5235):484–487
3. Lander ES, Linton LM, Birren B, Nusbaum C, Zody MC, Baldwin J, Devon K, Dewar K, Doyle M, FitzHugh W, Funke R, Gage D, Harris K, Heaford A, Howland J, Kann L, Lehoczky J, LeVine R, McEwan P, McKernan K, Meldrim J, Mesirov JP, Miranda C, Morris W, Naylor J, Raymond C, Rosetti M, Santos R, Sheridan A, Sougnez C, Stange-Thomann N, Stojanovic N, Subramanian A, Wyman D, Rogers J, Sulston J, Ainscough R, Beck S, Bentley D, Burton J, Clee C, Carter N, Coulson A, Deadman R, Deloukas P, Dunham A, Dunham I, Durbin R, French L, Grafham D, Gregory S, Hubbard T, Humphray S, Hunt A, Jones M, Lloyd C, McMurray A, Matthews L, Mercer S, Milne S, Mullikin JC, Mungall A, Plumb R, Ross M, Shownkeen R, Sims S, Waterston RH, Wilson RK, Hillier LW, McPherson JD, Marra MA, Mardis ER, Fulton LA, Chinwalla AT, Pepin KH, Gish WR, Chissoe SL, Wendl MC, Delehaunty KD, Miner TL, Delehaunty A, Kramer JB, Cook LL, Fulton RS, Johnson DL, Minx PJ, Clifton SW, Hawkins T, Branscomb E, Predki P, Richardson P, Wenning S, Slezak T, Doggett N, Cheng JF, Olsen A, Lucas S, Elkin C, Uberbacher E, Frazier M, Gibbs RA, Muzny DM, Scherer SE, Bouck JB, Sodergren EJ, Worley KC, Rives CM, Gorrell JH, Metzker ML, Naylor SL, Kucherlapati RS, Nelson DL, Weinstock GM, Sakaki Y, Fujiyama A, Hattori M, Yada T, Toyoda A, Itoh T, Kawagoe C, Watanabe H, Totoki Y, Taylor T, Weissenbach J, Heilig R, Saurin W, Artiguenave F, Brottier P, Bruls T, Pelletier E, Robert C, Wincker P, Smith DR, Doucette-Stamm L, Rubenfield M, Weinstock K, Lee HM, Dubois J, Rosenthal A, Platzer M, Nyakatura G, Taudien S, Rump A, Yang H, Yu J, Wang J, Huang G, Gu J, Hood L, Rowen L, Madan A, Qin S, Davis RW, Federspiel NA, Abola AP, Proctor MJ, Myers RM, Schmutz J, Dickson M, Grimwood J, Cox DR, Olson MV, Kaul R, Raymond C, Shimizu N, Kawasaki K, Minoshima S, Evans GA, Athanasiou M, Schultz R, Roe BA, Chen F, Pan H, Ramser J, Lehrach H, Reinhardt R, McCombie WR, de la Bastide M, Dedhia N, Blocker H, Hornischer K, Nordsiek G, Agarwala R, Aravind L, Bailey JA, Bateman A, Batzoglou S, Birney E, Bork P, Brown DG, Burge CB, Cerutti L, Chen HC, Church D, Clamp

M, Copley RR, Doerks T, Eddy SR, Eichler EE, Furey TS, Galagan J, Gilbert JG, Harmon C, Hayashizaki Y, Haussler D, Hermjakob H, Hokamp K, Jang W, Johnson LS, Jones TA, Kasif S, Kaspryzk A, Kennedy S, Kent WJ, Kitts P, Koonin EV, Korf I, Kulp D, Lancet D, Lowe TM, McLysaght A, Mikkelsen T, Moran JV, Mulder N, Pollara VJ, Ponting CP, Schuler G, Schultz J, Slater G, Smit AF, Stupka E, Szustakowski J, Thierry-Mieg D, Thierry-Mieg J, Wagner L, Wallis J, Wheeler R, Williams A, Wolf YI, Wolfe KH, Yang SP, Yeh RF, Collins F, Guyer MS, Peterson J, Felsenfeld A, Wetterstrand KA, Patrinos A, Morgan MJ, de Jong P, Catanese JJ, Osoegawa K, Shizuya H, Choi S, Chen YJ, International Human Genome Sequencing Consortium (2001) Initial sequencing and analysis of the human genome. Nature 409(6822):860–921. doi:10.1038/35057062

4. Goffeau A, Barrell BG, Bussey H, Davis RW, Dujon B, Feldmann H, Galibert F, Hoheisel JD, Jacq C, Johnston M, Louis EJ, Mewes HW, Murakami Y, Philippsen P, Tettelin H, Oliver SG (1996) Life with 6000 genes. Science 274 (5287):546, 563–547
5. Mouse Genome Sequencing Consortium, Waterston RH, Lindblad-Toh K, Birney E, Rogers J, Abril JF, Agarwal P, Agarwala R, Ainscough R, Alexandersson M, An P, Antonarakis SE, Attwood J, Baertsch R, Bailey J, Barlow K, Beck S, Berry E, Birren B, Bloom T, Bork P, Botcherby M, Bray N, Brent MR, Brown DG, Brown SD, Bult C, Burton J, Butler J, Campbell RD, Carninci P, Cawley S, Chiaromonte F, Chinwalla AT, Church DM, Clamp M, Clee C, Collins FS, Cook LL, Copley RR, Coulson A, Couronne O, Cuff J, Curwen V, Cutts T, Daly M, David R, Davies J, Delehaunty KD, Deri J, Dermitzakis ET, Dewey C, Dickens NJ, Diekhans M, Dodge S, Dubchak I, Dunn DM, Eddy SR, Elnitski L, Emes RD, Eswara P, Eyras E, Felsenfeld A, Fewell GA, Flicek P, Foley K, Frankel WN, Fulton LA, Fulton RS, Furey TS, Gage D, Gibbs RA, Glusman G, Gnerre S, Goldman N, Goodstadt L, Grafham D, Graves TA, Green ED, Gregory S, Guigo R, Guyer M, Hardison RC, Haussler D, Hayashizaki Y, Hillier LW, Hinrichs A, Hlavina W, Holzer T, Hsu F, Hua A, Hubbard T, Hunt A, Jackson I, Jaffe DB, Johnson LS, Jones M, Jones TA, Joy A, Kamal M, Karlsson EK, Karolchik D, Kasprzyk A, Kawai J, Keibler E, Kells C, Kent WJ, Kirby A, Kolbe DL, Korf I, Kucherlapati RS, Kulbokas EJ, Kulp D, Landers T, Leger JP, Leonard S, Letunic I, Levine R, Li J, Li M, Lloyd C, Lucas S, Ma B, Maglott DR, Mardis ER, Matthews L, Mauceli E, Mayer JH, McCarthy M, McCombie WR, McLaren S, McLay K, McPherson JD, Meldrim J, Meredith B, Mesirov JP, Miller W, Miner TL, Mongin E, Montgomery KT, Morgan M, Mott R, Mullikin JC, Muzny DM, Nash WE, Nelson JO, Nhan MN, Nicol R, Ning Z, Nusbaum C, O'Connor MJ, Okazaki Y, Oliver K, Overton-Larty E, Pachter L, Parra G, Pepin KH, Peterson J, Pevzner P, Plumb R, Pohl CS, Poliakov A, Ponce TC, Ponting CP, Potter S, Quail M, Reymond A, Roe BA, Roskin KM, Rubin EM, Rust AG, Santos R, Sapojnikov V, Schultz B, Schultz J, Schwartz MS, Schwartz S, Scott C, Seaman S, Searle S, Sharpe T, Sheridan A, Shownkeen R, Sims S, Singer JB, Slater G, Smit A, Smith DR, Spencer B, Stabenau A, Stange-Thomann N, Sugnet C, Suyama M, Tesler G, Thompson J, Torrents D, Trevaskis E, Tromp J, Ucla C, Ureta-Vidal A, Vinson JP, Von Niederhausern AC, Wade CM, Wall M, Weber RJ, Weiss RB, Wendl MC, West AP, Wetterstrand K, Wheeler R, Whelan S, Wierzbowski J, Willey D, Williams S, Wilson RK, Winter E, Worley KC, Wyman D, Yang S, Yang SP, Zdobnov EM, Zody MC, Lander ES (2002) Initial sequencing and comparative analysis of the mouse genome. Nature 420 (6915):520–562. doi:10.1038/nature01262
6. Adams MD, Celniker SE, Holt RA, Evans CA, Gocayne JD, Amanatides PG, Scherer SE, Li PW, Hoskins RA, Galle RF, George RA, Lewis SE, Richards S, Ashburner M, Henderson SN, Sutton GG, Wortman JR, Yandell MD, Zhang Q, Chen LX, Brandon RC, Rogers YH, Blazej RG, Champe M, Pfeiffer BD, Wan KH, Doyle C, Baxter EG, Helt G, Nelson CR, Gabor GL, Abril JF, Agbayani A, An HJ, Andrews-Pfannkoch C, Baldwin D, Ballew RM, Basu A, Baxendale J, Bayraktaroglu L, Beasley EM, Beeson KY, Benos PV, Berman BP, Bhandari D, Bolshakov S, Borkova D, Botchan MR, Bouck J, Brokstein P, Brottier P, Burtis KC, Busam DA, Butler H, Cadieu E, Center A, Chandra I, Cherry JM, Cawley S, Dahlke C, Davenport LB, Davies P, de Pablos B, Delcher A, Deng Z, Mays AD, Dew I, Dietz SM, Dodson K, Doup LE, Downes M, Dugan-Rocha S, Dunkov BC, Dunn P, Durbin KJ, Evangelista CC, Ferraz C, Ferriera S, Fleischmann W, Fosler C, Gabrielian AE, Garg NS, Gelbart WM, Glasser K, Glodek A, Gong F, Gorrell JH, Gu Z, Guan P, Harris M, Harris NL, Harvey D, Heiman TJ, Hernandez JR, Houck J, Hostin D, Houston KA, Howland TJ, Wei MH, Ibegwam C, Jalali M, Kalush F, Karpen GH, Ke Z, Kennison JA, Ketchum KA, Kimmel BE, Kodira CD, Kraft C, Kravitz S, Kulp D, Lai Z, Lasko P, Lei Y, Levitsky AA, Li J, Li Z, Liang Y, Lin X, Liu X, Mattei B, McIntosh TC, McLeod

MP, McPherson D, Merkulov G, Milshina NV, Mobarry C, Morris J, Moshrefi A, Mount SM, Moy M, Murphy B, Murphy L, Muzny DM, Nelson DL, Nelson DR, Nelson KA, Nixon K, Nusskern DR, Pacleb JM, Palazzolo M, Pittman GS, Pan S, Pollard J, Puri V, Reese MG, Reinert K, Remington K, Saunders RD, Scheeler F, Shen H, Shue BC, Siden-Kiamos I, Simpson M, Skupski MP, Smith T, Spier E, Spradling AC, Stapleton M, Strong R, Sun E, Svirskas R, Tector C, Turner R, Venter E, Wang AH, Wang X, Wang ZY, Wassarman DA, Weinstock GM, Weissenbach J, Williams SM, Woodage T, Worley KC, Wu D, Yang S, Yao QA, Ye J, Yeh RF, Zaveri JS, Zhan M, Zhang G, Zhao Q, Zheng L, Zheng XH, Zhong FN, Zhong W, Zhou X, Zhu S, Zhu X, Smith HO, Gibbs RA, Myers EW, Rubin GM, Venter JC (2000) The genome sequence of Drosophila melanogaster. Science 287(5461):2185–2195

7. C. elegans Sequencing Consortium (1998) Genome sequence of the nematode C. elegans: a platform for investigating biology. Science 282(5396):2012–2018
8. Edgar R, Domrachev M, Lash AE (2002) Gene Expression Omnibus: NCBI gene expression and hybridization array data repository. Nucleic Acids Res 30(1):207–210
9. Microarray standards at last (2002). Nature 419(6905):323. doi:10.1038/419323a
10. Barrett T, Wilhite SE, Ledoux P, Evangelista C, Kim IF, Tomashevsky M, Marshall KA, Phillippy KH, Sherman PM, Holko M, Yefanov A, Lee H, Zhang N, Robertson CL, Serova N, Davis S, Soboleva A (2013) NCBI GEO: archive for functional genomics data sets—update. Nucleic Acids Res 41(Database issue): D991–D995. doi:10.1093/nar/gks1193
11. Brazma A, Hingamp P, Quackenbush J, Sherlock G, Spellman P, Stoeckert C, Aach J, Ansorge W, Ball CA, Causton HC, Gaasterland T, Glenisson P, Holstege FC, Kim IF, Markowitz V, Matese JC, Parkinson H, Robinson A, Sarkans U, Schulze-Kremer S, Stewart J, Taylor R, Vilo J, Vingron M (2001) Minimum information about a microarray experiment (MIAME)-toward standards for microarray data. Nat Genet 29(4):365–371
12. Gibney G, Baxevanis AD (2011) Searching NCBI databases using Entrez. Curr Protoc Hum Genet, Jonathan L Haines [et al] Chapter 6:Unit6 10. doi:10.1002/0471142905.hg0610s71
13. Barrett T, Suzek TO, Troup DB, Wilhite SE, Ngau WC, Ledoux P, Rudnev D, Lash AE, Fujibuchi W, Edgar R (2005) NCBI GEO: mining millions of expression profiles—database and tools. Nucleic Acids Res 33(Database issue):D562–D566. doi:10.1093/nar/gki022
14. Gentleman RC, Carey VJ, Bates DM, Bolstad B, Dettling M, Dudoit S, Ellis B, Gautier L, Ge Y, Gentry J, Hornik K, Hothorn T, Huber W, Iacus S, Irizarry R, Leisch F, Li C, Maechler M, Rossini AJ, Sawitzki G, Smith C, Smyth G, Tierney L, Yang JY, Zhang J (2004) Bioconductor: open software development for computational biology and bioinformatics. Genome Biol 5(10):R80. doi:10.1186/gb-2004-5-10-r80
15. Sean D, Meltzer PS (2007) GEOquery: a bridge between the Gene Expression Omnibus (GEO) and BioConductor. Bioinformatics 23 (14):1846–1847
16. Smyth GK (2004) Linear models and empirical Bayes methods for assessing differential expression in microarray experiments. Stat Appl Genet Mol Biol 3:Article3. doi:10.2202/1544-6115.1027
17. Benjamini Y, Hochberg Y (1995) Controlling the false discovery rate—a practical and powerful approach to multiple testing. J R Stat Soc B Methodol 57(1):289–300
18. Tejera E, Bernardes J, Rebelo I (2013) Co-expression network analysis and genetic algorithms for gene prioritization in preeclampsia. BMC Med Genomics 6:51. doi:10.1186/1755-8794-6-51
19. Ni M, Ye F, Zhu J, Li Z, Yang S, Yang B, Han L, Wu Y, Chen Y, Li F, Wang S, Bo X (2014) ExpTreeDB: web-based query and visualization of manually annotated gene expression profiling experiments of human and mouse from GEO. Bioinformatics 30 (23):3379–3386. doi:10.1093/bioinformatics/btu560
20. Meng J, Li P, Zhang Q, Yang Z, Fu S (2014) A radiosensitivity gene signature in predicting glioma prognostic via EMT pathway. Oncotarget 5(13):4683–4693
21. Li J, Zhang Y, Gao Y, Cui Y, Liu H, Li M, Tian Y (2014) Downregulation of HNF1 homeobox B is associated with drug resistance in ovarian cancer. Oncol Rep 32(3):979–988. doi:10.3892/or.2014.3297

Chapter 6

A Practical Guide to The Cancer Genome Atlas (TCGA)

Zhining Wang, Mark A. Jensen, and Jean Claude Zenklusen

Abstract

The Cancer Genome Atlas (TCGA) is one of the most ambitious and successful cancer genomics programs to date. The TCGA program has generated, analyzed, and made available genomic sequence, expression, methylation, and copy number variation data on over 11,000 individuals who represent over 30 different types of cancer. This chapter provides a brief overview of the TCGA program and detailed instructions and tips for investigators on how to find, access, and download this data.

Key words TCGA, The Cancer Genome Atlas, Cancer genomics, Next-generation sequencing, Mutation, Methylation, Proteomics, Transcriptome, Copy number, miRNA

1 Introduction to TCGA

Cancer is a complex and heterogeneous disease in which mutations and other genomic and epigenomic abnormalities play a role in both its initiation and progression. Accumulated research data implicate numerous somatic mutations and a more limited number of inherited mutations in carcinogenesis. Understanding cancer-specific somatic mutations can provide important clues regarding the molecular processes underlying the development and progression of certain tumors. Given cancer's complexity, it is generally believed that only a fraction of alterations that may be useful as characteristic markers of specific tumor types and/or potential molecular targets have been identified to date. Therefore, to be successful, comprehensive genomic analyses of cancer must overcome a broad range of challenges stemming from the biological complexity and heterogeneity of human tumors and subtypes. An important role in tumor heterogeneity is played by the genomic instability that is inherent to the progression of cancer. The dynamic changes in tumor genomes are influenced by the cellular and biological context, genetic characteristics of individual persons, and environmental factors. Certain similarities exist across tumor

Ewy Mathé and Sean Davis (eds.), *Statistical Genomics: Methods and Protocols*, Methods in Molecular Biology, vol. 1418, DOI 10.1007/978-1-4939-3578-9_6, © Springer Science+Business Media New York 2016

types, however any effort to characterize the genomes of tumors in a comprehensive, systematic manner must address the heterogeneity across distinct cancer types and subtypes.

The Cancer Genome Atlas (TCGA) is a large-scale, collaborative effort led by the National Cancer Institute (NCI) and the National Human Genome Research Institute (NHGRI) to map the genomic and epigenomic changes that occur in 32 types of human cancer, including nine rare tumors. Its goal is to support new discoveries through the generation of a catalog of somatic aberrations (part list) occurring in the different neoplasms, and accelerate the pace of research aimed at improving the diagnosis, treatment, and prevention of cancer.

TCGA is a community resource project [1], and as such, project data is released to the public immediately after it is judged to be technically correct through quality control metrics. The information generated by TCGA is centrally managed and entered into databases as it becomes available, making the data rapidly accessible to the entire research community. By January 2015, TCGA had generated about 1.7 petabytes of data for about 11,500 cases of tumor and matching normal tissue samples.

TCGA provides the cancer research community with a valuable resource. TCGA has:

1. Collected an unprecedented number of high-quality human cancer samples as well as their matching normal tissues, allowing researchers to identify important genomic changes that may play a role in the development of cancer.
2. Consistently characterized and analyzed each sample, yielding a deeper, more reliable, and broader perspective of the cancer genome compared to previous approaches of more limited scope.
3. Fostered collaborations across broad cross-sections of the cancer research community by making the data freely available in real time.

The TCGA program includes a broad cross-section of the cancer research community. The TCGA Research Network includes scientists, bioinformaticians, bioethicists, physicians, nurses, advocates, and many others working collaboratively to generate and analyze the data produced in the context of the project.

The TCGA Network consists of four basic components (Fig. 1).

TCGA data are available to the public and research community at large in two data repositories: the TCGA Data Portal (Subheading 2) and the Cancer Genomics Hub (Subheading 3). These two databases and the way to access the data stored within will be discussed at length in the following sections. The goal and purpose of this chapter is to facilitate the access to the TCGA data by those who can best explore its richness and help in making the discoveries that will alleviate suffering from cancer.

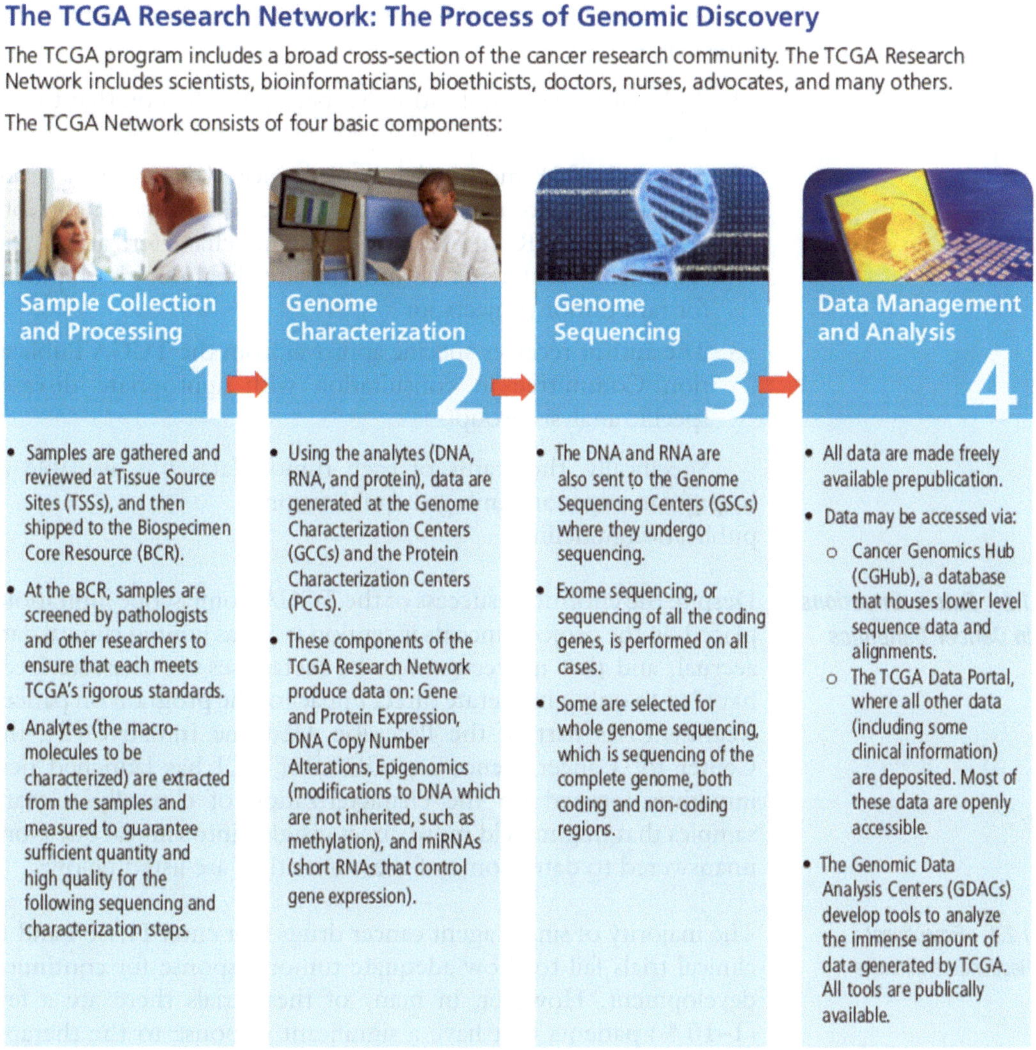

Fig. 1 The TCGA research network

1.1 Data Use Restrictions

TCGA data are made available rapidly after generation for community research use. To act in accord with the Fort Lauderdale principles and support the continued prompt public release of large-scale genomic data prior to publication, researchers who plan to prepare manuscripts containing descriptions of TCGA data that would be of comparable scope to an initial TCGA comprehensive, global analysis publication, and journal editors who receive such manuscripts, are encouraged to coordinate their independent reports with TCGA's publication schedule. The TCGA program has established the following policy to clarify freedom of TCGA and non-TCGA users to publish findings using TCGA data. There are no limitations on publications containing analyses using any TCGA data set if any one of the following three freedom-to-publish criteria holds:

1. A global analysis publication has been published on that tumor type; or
2. A period of 18 months has passed since 100 cases of a given tumor type have shipped from the TCGA Biospecimen Core Resource (BCR) to TCGA genomic characterization and sequencing centers or 18 months after final sample shipment for rare tumor projects; or
3. The author receives specific approval from the TCGA Publication Committee in consultation with appropriate disease-specific analysis group(s).

Specifically, the status of each tumor dataset is available at http://cancergenome.nih.gov/publications/publicationguidelines

1.2 Future Directions in Cancer Genomics

Despite the enormous success of the TCGA, some scope limitations placed on the project since its inception (such as limited clinical data accrual, and that no recurrences or metastases are characterized) have led to only a moderate direct impact of the program on patient treatment. As part of the Precision Medicine Initiative [2], the Center for Cancer Genomics (CCG) of NCI has launched new initiatives focused on the characterization of clinically relevant samples that might yield more direct insights into clinical questions unanswered to date. Some of these initiatives are listed below:

1.2.1 Exceptional Responders Initiative

The majority of single agent cancer drugs that enter Phase I and II clinical trials fail to show adequate tumor response for continued development. However, in many of these trials there are a few (1–10 %) patients that have a significant response to the therapy. These "failed" trials could be very informative with regard to molecular markers that predict a positive response to these single agent therapies in a small subset of patients, thus making these "inactive" agents useful. Even later phase clinical trials may have a few patients that experience an exceptional response, such as a complete response, that occurs in 10 % or less of patients. The Exceptional Responders initiative envisions that a genomic approach to understanding therapeutic responses may be broadly applicable in cancer. The ability of molecular technologies to stratify tumor types has resulted in many common cancers being separated into specific subtypes that respond to therapeutic agents in very different ways. Identifying additional molecular markers that are able to predict a clinical response in subsets of patients will render future cancer treatments more precise. The feasibility of the Exceptional Responders paradigm will be assessed in 100 patients with reliable outcome data, treated on clinical trials both with generic and targeted monotherapies or combinations. Tumor biopsy material will be obtained from these patients as well as germline tissue. The tissue will undergo next generation whole exome resequencing and, if practical, whole transcriptome resequencing. The success of the endeavor depends on having adequate tissue for analysis, robust analytical techniques/platforms, and reliable outcome data for patients who have been treated on defined and consistent drug regimens. This is now made possible thanks to the experience gained in the TCGA project, and the pipeline derived from it.

1.2.2 ALCHEMIST Clinical Trial

The Adjuvant Lung Cancer Enrichment Marker Identification and Sequencing Trials (ALCHEMIST) seek to capitalize on a number of events that have converged to create an opportunity for significant clinical advances in the treatment of subsets of adjuvant lung cancer and to take the next step in biological characterization of lung cancer on a national level. Recently, two of the NCI-supported US Cooperative Clinical Trials groups have brought study proposals forward for evaluation in the NCI Thoracic Malignancy Steering Committee. Each proposal would select resectable patients according to a biomarker validated in the metastatic setting and test for a large clinical effect. The two markers, EGFR mutation and ALK-positivity, each have an incidence of about 5–10 % in the general lung adenocarcinoma population, so each study design would require screening over 7000 patients. These two markers are mutually exclusive, so there is particular efficiency in screening one set of patients to identify those positive for one marker or the other. Surgical specimens will be available from this trial, providing adequate tissue for extensive analysis. Tissues from all patients screened for the Alchemist trial would be available for comprehensive genomic analyses, after appropriate patient consent. Patients testing negative for the EGFR and ALK biomarkers will be given standard treatment and followed as a prospective cohort, with repeat biopsies obtained upon tumor relapse.

1.2.3 Cancer Driver Discovery Project

The TCGA project characterized 500 cases per tumor for most tumor types. However, with 500 cases at a background mutation rate of ten mutations/Mb, only mutations that occur in at least 5 % of the cases can be identified. Analysis of more of tumors is needed to discover less common oncogenes and tumor suppressors [3]. The pilot phase of the Cancer Driver Discovery Project (CDDP), seeks to identify these lower frequency cancer drivers in lung, colorectal, and ovarian cancers. For the most part, CDDP cases will have the same qualifying metrics as TCGA. However in order to collect more cases, CDDP will accept cases with only FFPE tumor tissues in addition to cases with frozen tumor tissues. Since FFPE tumor tissues tend to be smaller, the minimal yield requirements for DNA and RNA in CDDP will be lower than has been typically been used in TCGA, creating both a challenge and an opportunity to advance the technical limits of the platforms used to data and opening the door to an expanded availability of cases for analysis.

In addition to these new characterization projects, the CCG has recognized that the data produced by several genomic scale projects sponsored by the NCI have been produced on different platforms that are not always compatible with each other and raise barriers to cross project analysis. To help solve this issue and to house all the data in a single site, the CCG has issued a contract with the University of Chicago to build the NCI Genomic Data Commons (GDC). The primary purpose of the GDC is to be a data service to the cancer research community. This service shall provide for the receipt, quality control, integration, storage, and redistribution of standardized cancer genomic data sets derived from multiple legacy and active CCG projects.

A signature characteristic of high quality genomic data and associated clinical annotations is that they can be combined and mined repeatedly to make additional discoveries, often by applying new algorithms and methods of analysis. The challenge for NCI is to ensure that a sophisticated and accessible repository of cancer genomics data, which is populated with current and future data from CCG-sponsored programs, is made available to all cancer researchers so that they can efficiently extract the maximum amount of knowledge from these resources. This will require:

1. That data from dissimilar sources can be aggregated and stored in a common format with appropriate security and access controls;
2. The provision of high capacity network connectivity so data can be deposited and accessed efficiently;
3. An extensible environment to enable analysis tools of increasing sophistication to be developed and run adjacent to the data by provisioning cloud-like resources next to the data, and making such tools available for broad use.

It is the NCI's hope that the combination of the GDC with the NCI Cloud Pilots (https://cbiit.nci.nih.gov/ncip/nci-cancer-genomics-cloud-pilots) will provide the users with harmonized genomic data that could be computed on in the cloud without the need of downloading the data or having a larger bioinformatics core at their research institutions.

2 Using the TCGA Data Coordinating Center (DCC) Portal

The TCGA DCC Data Portal is a one-stop shop for almost all data types for all TCGA studies. Only DNA and RNA sequence data reside outside the DCC (at CGHub, *see* Subheading 3), and even this data is cataloged by the DCC for the convenience of users. The basic data used in every TCGA Research Network publication was originally provided via the DCC, and an archived version of the data as used in each publication is preserved at the DCC.

Because the DCC has grown and evolved along with the 9-year TCGA project, and because part of its mission is to serve users with widely varying bioinformatics and data expertise, multiple methods for browsing, searching, and downloading data are available. This portion of the guide will introduce Data Portal applications and content to users who are new to TCGA data, and who have a basic understanding of common desktop applications. When more advanced tools and techniques exist, we will briefly mention these and provide links and resources that will help the interested user learn more.

2.1 Getting Help

The DCC maintains extensive public documentation on all its applications and processes. The first place to look for help is the TCGA Wiki. The wiki home page provides quick links to User Guides for each of the applications described below, as well as links for getting started with TCGA data. The TCGA Encyclopedia is an extensive resource containing over 200 short articles defining important TCGA and Data Portal data concepts.

The DCC team has wide current and historical project knowledge. They are pleased to answer questions on any aspect of the data, or provide users with the right resource to help. Questions may be directed to the DCC bioinformatics team by posting to the bioinformatics list: TCGA-DCC-BINF-L@lists.nih.gov. The team will acknowledge questions directly to the user's email within 24 h. The turnaround time for complete answers or referrals will depend on the question, but is typically 5–10 business days. The DCC also welcomes reports of data quality problems that are encountered by users; the bioinformatics list is the right place to report any such issues.

2.2 TCGA Data Access and Use

A guiding principle of the TCGA program since its inception has been to make clinical and genomic analysis data available to the public as widely, rapidly and reliably as possible, as outlined in Subheading 1. There are no inherent restrictions on the use or reuse of public TCGA data for biomedical research purposes, including commercial purposes. However, to protect personally identifiable information of TCGA enrollees, access restrictions are enforced on certain data types and levels that contain or may contain germline genetic variants. These data are described in Subheading 2.3, "Data Types and Data Levels".

Most data available from the DCC is classified as *open access*, available for download to anyone at any time. Data whose access is restricted is said to be *controlled access*. Users may apply for controlled access privileges. The Data Access Committee of the NCBI's Database for Genotypes and Phenotypes (dbGaP) is responsible for reviewing these applications and granting such privileges. The application process along with details about user commitments and links to application sites and materials can be found at https://tcga-data.nci.nih.gov/tcga/tcgaAccessTiers.jsp. The DCC itself does not participate in the process that grants controlled access to applicants.

Researchers are encouraged to report their analyses of TCGA data in scientific publications and presentations (*see* Subheading 1.1, "Data Use Restrictions"). The TCGA Publication Guidelines lay out the policies for use of TCGA data in publications and presentations, in research proposals and other communications. By downloading data from the DCC, the user acknowledges having read these policies and agrees to adhere to them.

2.3 Data Types and Data Levels

TCGA data available at the DCC can be classified into three broad types:

1. Biospecimen and clinical data,
2. Molecular analysis (genomic characterization) data, and
3. Analysis metadata.

Biospecimen data includes the standard TCGA identifiers that have been assigned to each biospecimen, as well as quality control information as provided to the DCC by TCGA's Biospecimen Core Resource (BCR). The BCR also collects clinical data on each participant. These data are submitted to the DCC and made available to users via the Data Portal. Molecular data is submitted to DCC by the Genomic Characterization Centers (GCCs) after centers have analyzed sample aliquots sent to them by the BCR. Genomic Sequencing Centers (GSCs) submit certain data, such as genomic variant calls, to the DCC for distribution to users. All centers provide additional metadata with their submissions that allow users to relate the data files back to the participant, sample, and aliquot from which the data were derived. Metadata is typically provided in the MAGE-TAB standard format, extended to support TCGA identifiers and multiple platforms.

The Data Types page on the TCGA Data Portal provides a comprehensive list of all data types from all platforms available at DCC, with detailed descriptions and examples. Not every TCGA study will have all data types, and not every participant within each study will necessarily have been analyzed on all platforms. The most important reason for this is that over the course of the TCGA program, genomic technologies transitioned from mostly microarray-based to mostly sequence-based platforms. The TCGA project generally has been an early adopter of novel high-throughput genomics technologies, and this is reflected in the data corpus. However, every participant in every study has or will have data for each of the following data dimensions:

1. Clinical data, including pathology reports and diagnostic images.
2. Whole exome sequence variation (tumor and normal samples compared).
3. mRNA Expression (tumor samples compared with a normal pool).
4. microRNA Expression (tumor samples compared with a normal pool).
5. Copy number variation (based on SNP6 arrays; tumor and normal samples compared).
6. Methylation (tumor and normal samples compared).

(For most solid tumors, the normal samples are taken from blood.) Many studies also possess complete reverse-phase protein array (RPPA) expression data, and a fraction of participants in all studies has associated whole genome sequence data.

Every data file at the Data Portal can be classified as either metadata ("Level 0") or one of three *data levels*. Descriptions of each level are shown in Table 1. The data levels 1, 2 and 3

Table 1
TCGA DCC data levels

Data level	Level type	Description	Example
1	Raw	Low-level or instrument data for single aliquot not normalized	Affymetrix CEL file
2	Processed	Normalized single aliquot data	Signal of a probe or probe set for a sample; VCF file; MAF file containing germline mutation calls
3	Segmented or interpreted	Aggregate of processed data from single sample; tumor/normal sample normalized differentials	Expression signal or differential expression of all genes; MAF file containing only somatic mutation calls

correspond roughly to those designations as typically used in microarray analyses.

Most Level 1 and Level 2 molecular data is controlled access. An important exception is Level 1 and 2 clinical data, which are all open access as provided by the DCC, since potentially identifying information has been obscured. For example, all dates in DCC clinical data are expressed in terms of days since diagnosis. The actual date of diagnosis does not appear anywhere in DCC data. Level 3 data, including variant calls in MAF files, is always open access.

2.4 TCGA Samples and Identifiers

To analyze TCGA molecular data, it must be associated to the clinical data of the source participant. Some basic facts about the structure of the TCGA data model and item identifiers are key to creating this relationship.

The provenance of biospecimens analyzed in TCGA studies can be organized into a source-product heirarchy (Fig. 2):

The hierarchy represents the material flow from acquisition of samples with associated clinical data, through processing at the BCR, to the final material (the "aliquot") that is shipped to the GCCs and GSCs for analysis. Each of these items is given a unique TCGA barcode as well as a universally unique identifier (UUID) that is assigned by the BCR and indexed and maintained by the DCC.

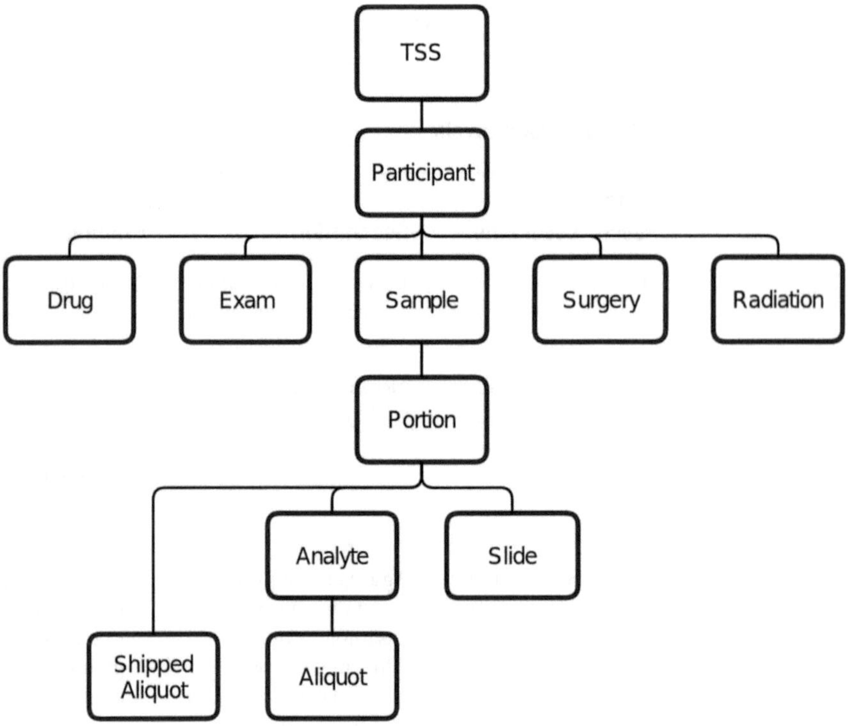

Fig. 2 The TCGA data hierarchy

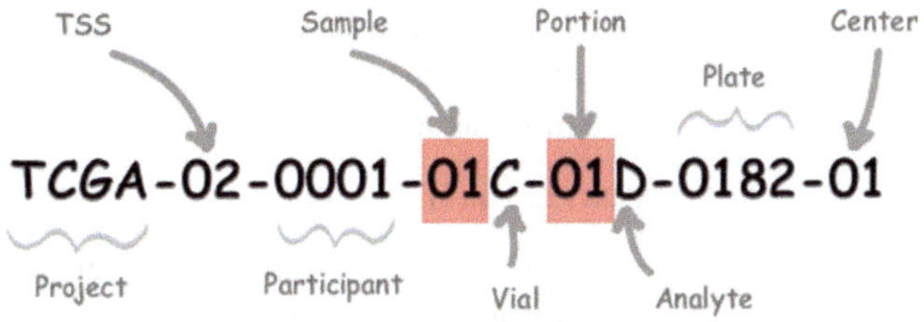

Fig. 3 TCGA aliquot barcode

The key items for the data user are the *participant*, the *sample*, and the *aliquot*. The participant identifiers are associated with clinical data. The sample identifiers can be used to determine whether biospecimen material comes from tumor tissue, normal blood (most normal samples), or tumor-adjacent tissue (some normal samples). The aliquot identifier is central to all analysis: every molecular analysis file is associated with a single source aliquot, and the aliquot identifier is used to relate that data back to the source participant and the clinical data.

The structure of the TCGA barcode allows the user to conveniently establish this key relationship. The barcodes are designed such that, for each item in the hierarchy, the barcodes for each higher-level source item are contained within it. Figure 3 describes the interpretation of an aliquot barcode:

For this example, the user would know that data files obtained for aliquot TCGA-02-0001-01C-01D-0182-01 are associated with participant TCGA-02-0001, and that clinical data obtained for this participant should be associated with these data files during analysis.

The metadata in TCGA barcodes are reliable, and indeed most TCGA Network papers are based on connecting data and participants in the manner just described. There do exist a few known anomalies however, and users are strongly encouraged to check data annotations (*see* Subheading 2.7.2, "Annotations Database") to determine if their data are affected. The UUID may always be used in the biospecimen metadata browser (*see* Subheading 2.7.3) to obtain the correct relationships and metadata. This is a good check to apply when examining outliers in your analysis.

2.5 Data Archives and Files

Data is stored at the DCC in the form of archives that group individual data files together in a single compressed file or under a single directory. An archive generally groups files of a single data level, for a single platform, submitted by a single center, over a group of participants in a single TCGA study. DCC archive names reflect this, according to the following template:

```
<center name>_<study>.<platform>.Level_<data level>.
<serial index>.<revision>.0.tar.gz
```

For example, a breast cancer archive of Level 2 DNA sequence data submitted by Washington University in St Louis might be named as follows:

```
genome.wustl.edu_BRCA.IlluminaGA_DNASeq.Level_2.2.0.0.tar.gz
```

Download methods that return archives use this convention for naming the archives returned to users. The archive is the primary unit of data organization at the DCC and will be encountered in many contexts. Complete descriptions of archives and archive types at the DCC are available via https://wiki.nci.nih.gov/display/TCGA/Archive.

The "serial index" and "revision" numbers deserve some discussion. A data submitting center may divide their data for a given study, platform, and data level into multiple archives for submission. This set of archives is called an *experiment*, and the serial index differentiates the archives in the experiment. Archives in an experiment must be mutually exclusive with respect to content; the same file may not appear in more than one archive in the series. Apart from these rules, centers are free to choose the value of the serial index. For BCR clinical and biospecimen archives, each archive contains participant files for one *batch* of participants (an arbitrary group of participants within a single study, whose samples are generally processed together). For these archives, the batch number is always used for the serial index.

Submitters may also update any currently deployed archive by submitting a revised archive. These revisions may update, add, or delete entire files or portions thereof. Revisions of an archive containing data for a given center, study, platform, data level and serial index (participant group) are differentiated by the revision number. For each revision, the revision number must increase by one sequentially, starting from zero. The *latest archive* is that archive containing the most updated revision of data for a given center, study, platform, data level, and serial index, and is always the revision with the highest revision number. Most DCC data applications search and download data *only* from latest archives. The latest archive (according to this definition) for any TCGA experiment may be taken as the best one for use in current research and analysis.

As noted above, the key step before proceeding to an analysis is to relate the data content of a given file to the source participant or aliquot. Most individual data files contain data for a single participant or aliquot. Biospecimen and clinical files include the TCGA participant identifier in the filename, as well as within the XML content of the files themselves. For other platforms (apart from DNAseq, *see* below), filename conventions are established by the submitting centers, and the most reliable way to connect identifiers to the data is via the sample-data relationship format file (SDRF), found in the associated MAGE-TAB archive. The SDRF is a tab-delimited text file containing extensive metadata regarding the

Table 2
Key SDRF columns for file identification

SDRF column	Example value	Description
Extract name	000d877f-8d03-44bc-8607-27b5ba84b5fe	TCGA aliquot UUID
Comment [TCGA barcode]	TCGA-E9-A1RD-11A-33R-A157-07	TCGA aliquot barcode
Derived data file REF	UNCID_1165062.....sorted_genome_alignments.bam	Level 1 data filename (reference—not available at DCC)
Comment [TCGA data level]	Level 1	Indicates data level of preceding file
Derived data file	unc.edu.....bt.exon_quantification.txt	Level 3 data filename (available in DCC archive)
Comment [TCGA include for analysis]	yes	Indicates passing QC at analysis center
Comment [TCGA data type]	exon_quantification	DCC data type of preceding file
Comment [TCGA data level]	Level 3	Indicates data level of preceding file
Comment [TCGA archive name]	unc.edu_BRCA.IlluminaHiSeq_RNASeqV2.Level_3.1.8.0	File contained in this DCC archive

analysis protocols and derived data products for each aliquot processed by a center. A single SDRF describes all data files for all levels of data for any study, center, and platform, so that it will contain metadata for files across several archives. The SDRF contains many columns, but the key relationship is found between the columns “Extract Name”, “Comment [TCGA Barcode]”, and the file name columns. Table 2 describes key columns in the SDRF that relate an aliquot identifier to data file names.

The main exceptions to the “one file, one aliquot” convention are the mutation annotation format (MAF) files and variant calling format (VCF) files, both of which contain mutation calls for all participants in a study. The latest TCGA specifications of MAF and VCF can be consulted to determine the associated aliquot IDs for each data record in these files.

2.6 Querying and Downloading TCGA Data

Four methods are available on the DCC Data Portal for querying and/or downloading data: the Data Matrix, File Search, Bulk Download, and the HTTP Directories. Detailed, step-by-step tutorials are available on the TCGA wiki for using each of these methods. Below, we provide links to the applications and tutorials, and describe typical usage scenarios for each application.

2.6.1 Data Matrix

Data Matrix Application/Tutorial

The Data Matrix (Fig. 4) is the most commonly used application on the TCGA Data Portal for querying and downloading data. It provides a simple search interface that allows users to filter data for a selected TCGA study according to several key parameters, including data type, analysis center, BCR participant batch number, TCGA sample idea, data level and data availability. The Data Matrix does not have cross-study query capability; a user must perform separate queries to obtain data from different studies.

Once the query is submitted, the matrix view (Fig. 5) screen appears. This is a tabular display of data availability, with data types and platforms in columns, and sample IDs in rows. The user builds a download archive by selecting columns, rows, and/or individual cells and submitting these selections. Many users find the matrix view useful for simply browsing available data on the basis of their queries.

When matrix selections are submitted by clicking "Build Archive", the user is presented with a "tree view" (Fig. 6) of the

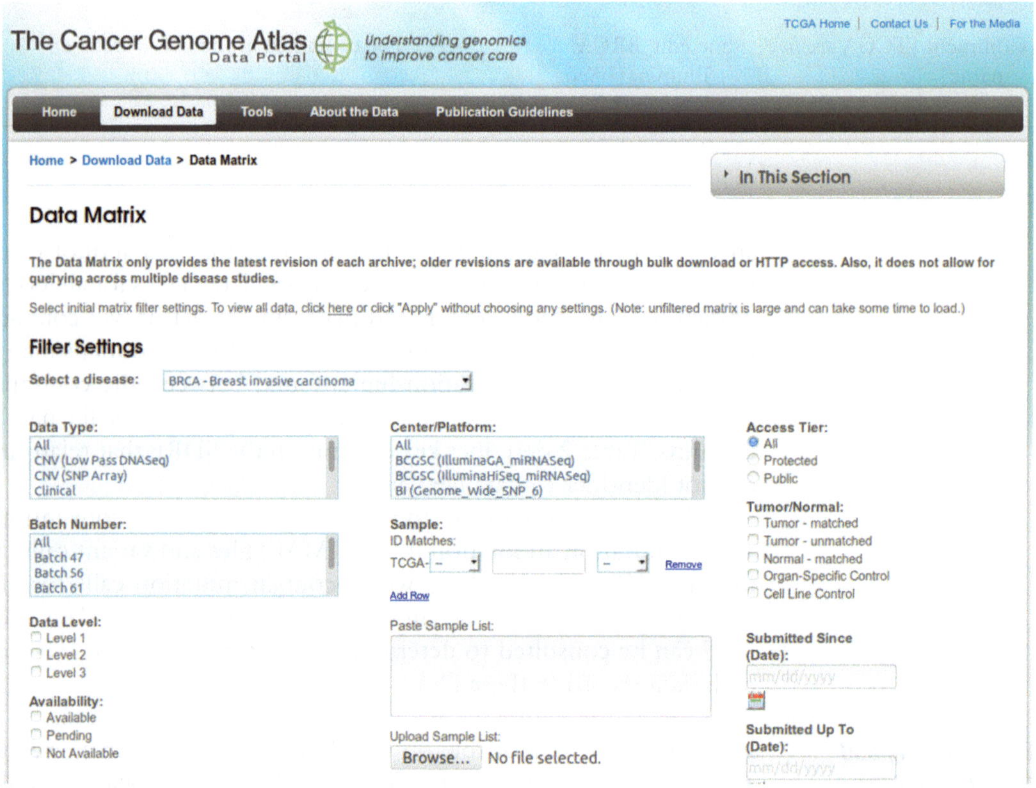

Fig. 4 Data Matrix query interface

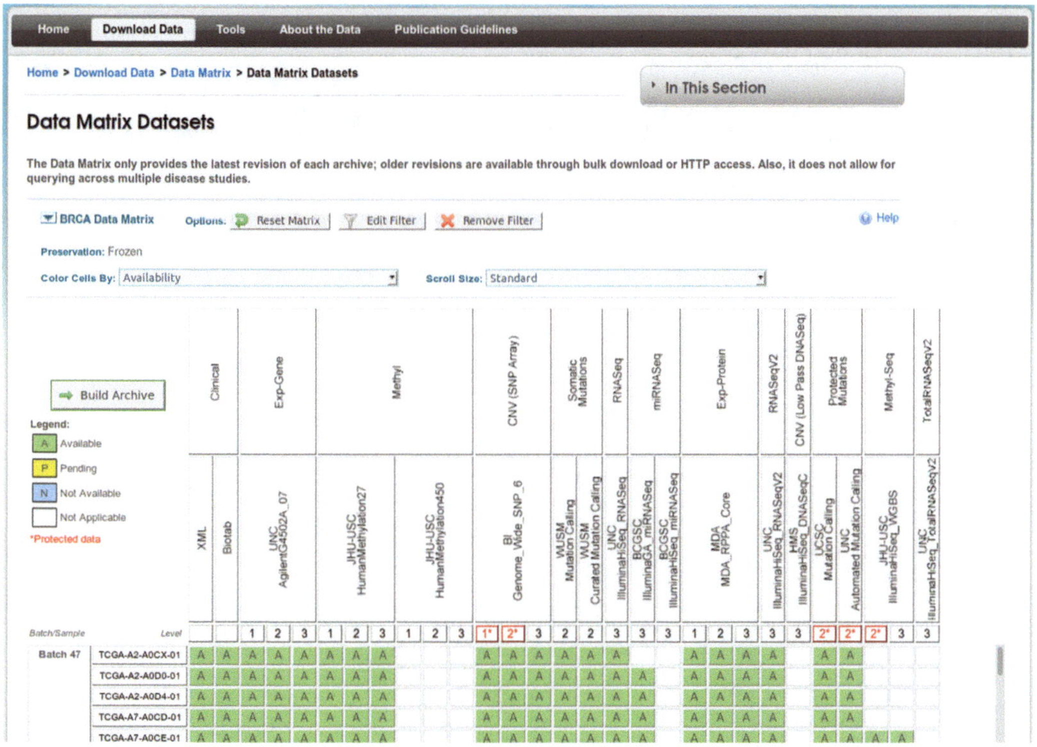

Fig. 5 Data Matrix—matrix view

files that will be incorporated in the download archives. The user may further modify the final archive by selecting or deselecting files.

To submit the final archive selections, the user first enters and confirms her email address and then clicks "Download". The DCC servers then build the archive. Depending on the size of the final archive, this process may take minutes to hours. The response time depends on the final size of the archive being built, the number of archive-building jobs running in the DCC server's queue, and whether compression is requested by the user. For large (1–70 GB) archives, if the user has the available space, it is better **not** to select compression, since the time to compress very large archives can be on the order of hours.

DCC will send a link to the archive by email. The archive remains available for download at this link for 24 hours. Note that if any controlled data are included in the archive, then the user will need use the username and password received upon con trolled access approval to download the archive from this link. The archive will contain not only the data requested, but also key metadata including relevant annotations (*see* Subheading 2.7.2) and any associated MAGE-TAB files (*see* Subheading 2.4).

Data Matrix archives contain only *latest data*, that is, only files contained in the latest revision of data archives that match the

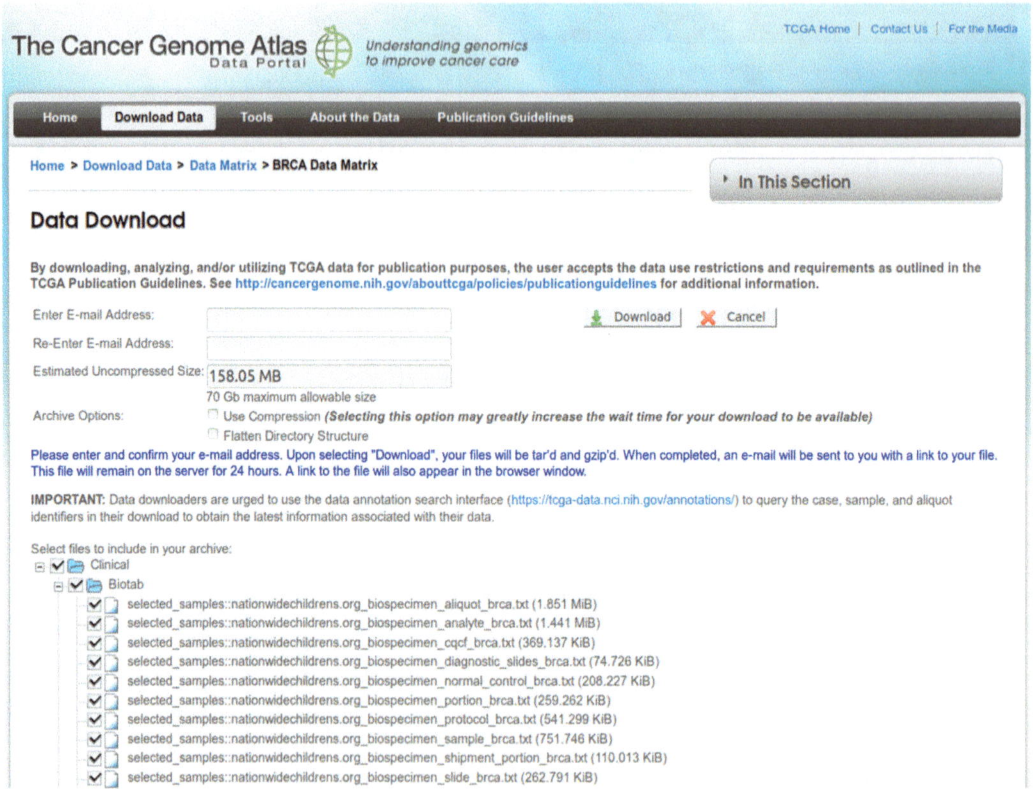

Fig. 6 Data Matrix—tree view

query. The Bulk Download and HTTP Directory methods may be used to retrieve files in older archive revisions.

The Data Matrix query and download process can be performed by user scripts or applications via the Data Matrix Web service interface, as described in its user guide.

2.6.2 File Search

File Search Application/Tutorial

The File Search application (Fig. 7) may be the most intuitive application for cancer biologists who wish to query and download the most up-to-date TCGA data. Users may query based on an easy-to-understand list of data categories, the disease or tumor type of interest, data level and access level. File Search permits searches across tumor types. File Search allows users to download files of interest without having to download other files in the containing archives.

File Search also allows users search based on TCGA Barcode or UUID. A simple text file containing a list of barcodes, one per line, may be uploaded to the application to obtain all relevant latest data files matching the other search criteria for just those items.

File Search uses the familiar shopping cart mechanism to allow the user to choose the desired files for download. After adding files to the cart, the user clicks "View/Download Cart". From the cart, the user clicks "Download All". The user then advances to a screen

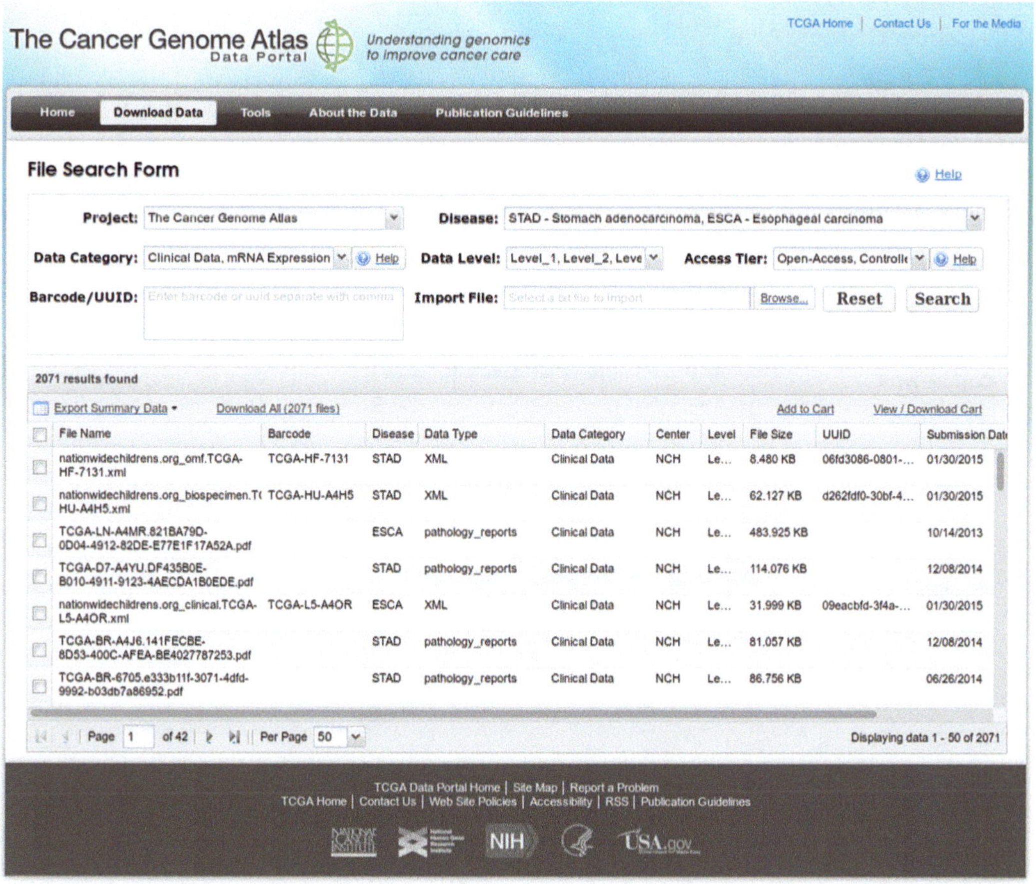

Fig. 7 File search interface

similar to the Data Matrix download screen. The user enters an email address and clicks "Download". Data is retrieved as for the Data Matrix.

2.6.3 Bulk Download

Bulk Download Application/Tutorial

The Bulk Download application (Fig. 8) is a facility for searching and downloading archives that contain specific files of interest across TCGA archives, including older, superseded archive revisions. The search dimensions are similar to those provided in the Data Matrix query interface.

The major differences from the Data Matrix interface here are the Submission Date and File Name controls. Submission Date can be used to concentration on a given date interval of data submission, for users who have interest in historical TCGA data. The File Name field can be very useful for find specific data for specific platforms. For example, all archives containing MAF (Mutation Annotation Format) files can be browsed and downloaded by entering "maf" in the File Name field.

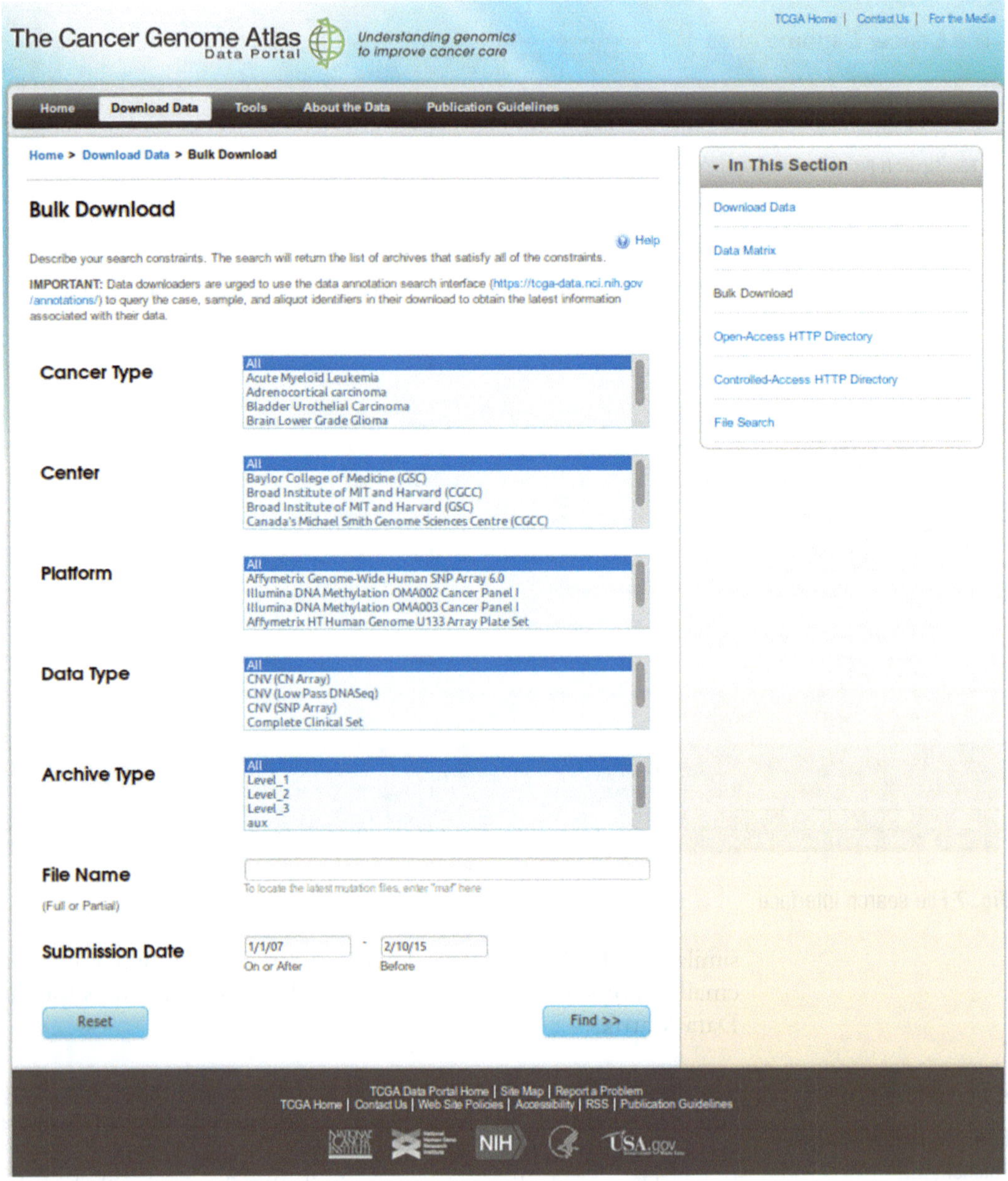

Fig. 8 Bulk download interface

From the Bulk Download archive view (Fig. 9), the filenames found in individual archives may be browsed, and archives may be selected for download by a process similar to that of the Data Matrix or File Search.

Using the Bulk Download application effectively requires some detailed understanding of DCC archives, their structure, and their file components. Much of this information may be found in the

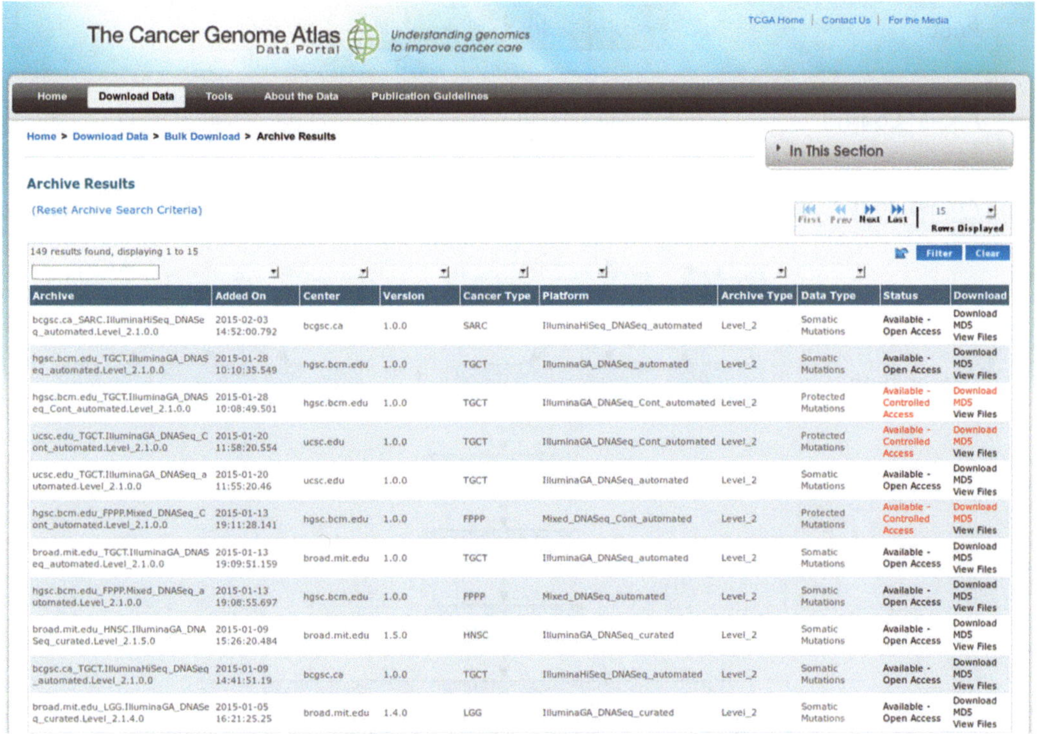

Fig. 9 Bulk download archive view

TCGA Encyclopedia and the Data Types and Data Levels Portal page. Brief descriptions of many of the codes and terms defined by and used in DCC naming conventions can be found at https://tcga-data.nci.nih.gov/datareports/codeTablesReport.htm.

2.6.4 HTTP Directories

Open Access Directories/Controlled Access Directories/Tutorial

The HTTP directories are window into the TCGA data storage of the DCC. All data search and download applications derive their delivered archives and files from this file system. The file system is divided into two branches, one for open access and the other for controlled access data. From this split, the directory structure organizes archives hierarchically according to TCGA study, center type, analysis center, and data platform, as depicted in Fig. 10.

Archives are available on the file system in both compressed (Unix tar plus gzip compression) and "exploded" form —i.e., as uncompressed directories that may be browsed in a typical Web browser. If a user knows exactly the files she is looking for, this can be the most convenient way to retrieve them. Keep in mind that older revisions of data archives are maintained on the file system. Care must be taken to identify the archive that represents the latest data (the archive having the highest revision number), if that is what is desired.

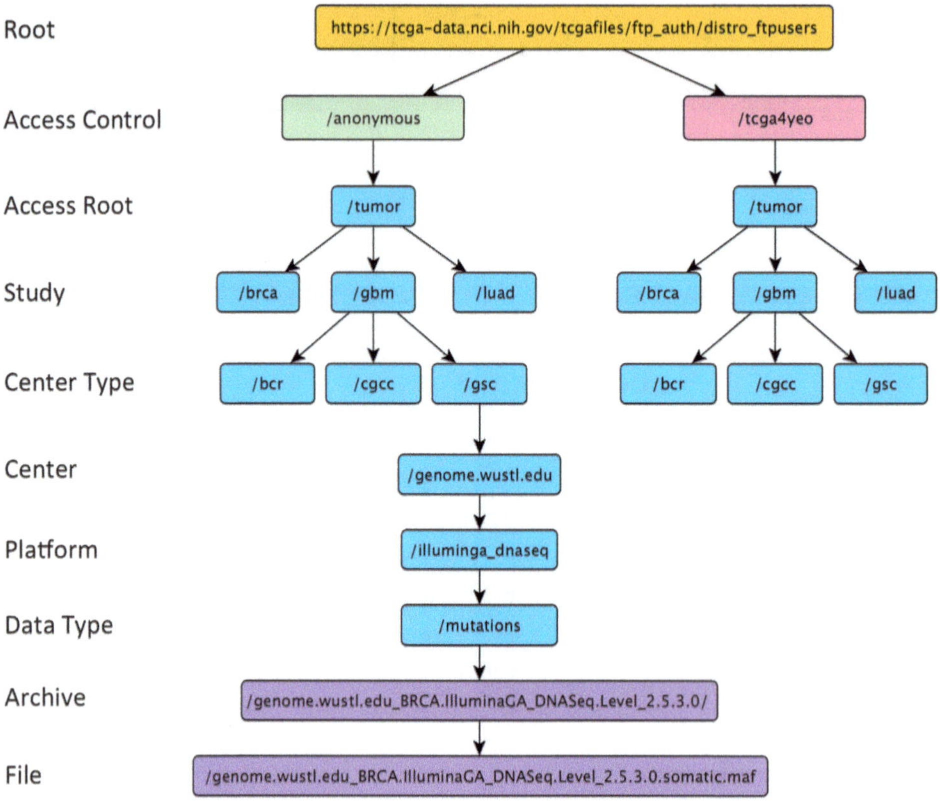

Fig. 10 DCC file system structure

Data may be downloaded via user scripts or applications by using command-line retrieval applications such as wget or curl.

2.7 Other Key Features and Resources

2.7.1 MAF Search Facility

The MAF Search (Fig. 11) provides an interactive view into the genomic variants identified and published by the TCGA Network in its series of comprehensive papers. The user can filter by study (disease), variant type (e.g., indel, SNP), mutation effect (e.g., missense, nonsense, frameshift, silent), and validation status. The results of the filter are displayed as a graphical overview and a table of results. The results may be downloaded in multiple formats, including in valid TCGA MAF format. A Users Guide is available.

2.7.2 Annotations Database

The DCC maintains a database of comments or "annotations" (not to be confused with genomic annotations) associated with TCGA participants, samples, and aliquots. Most annotations are made by the BCR and indicate special circumstances or information relevant to a given participant—for example, whether samples were included in an internal TCGA technical study, or whether a participant was later discovered to have received treatment prior to TCGA recruitment. Occasionally, participants are redacted from TCGA studies

Fig. 11 MAF search interface

for data quality reasons or at the request of the participants. These instances are also recorded in the annotations database.

The Annotations Manager (Fig. 12) enables users to search for annotations relevant to participants or samples in their data of interest. The TCGA Program Office encourages users to review annotations on all data to be used in analyses. Login to the Annotation Manager is not required in order to search and download annotations. Annotations may be exported to Excel or other formats for more convenient examination. The Users Guide provides details of this and other features.

2.7.3 Biospecimen Metadata Browser

The Biospecimen Metadata Browser (Fig. 13) enables users to identify the TCGA study, tissue type, sample type, tissue source site, and other metadata associated with the barcode or UUID of any TCGA entity, from participant to aliquot. The browser also can return sets of TCGA entities whose metadata matches a parameter filter entered by the user. This facility is used frequently by DCC bioinformaticians to track down data quality issues or anomalies observed by users, and by data submitters to verify the correctness and availability of metadata for the data items they submit. A User Guide is available.

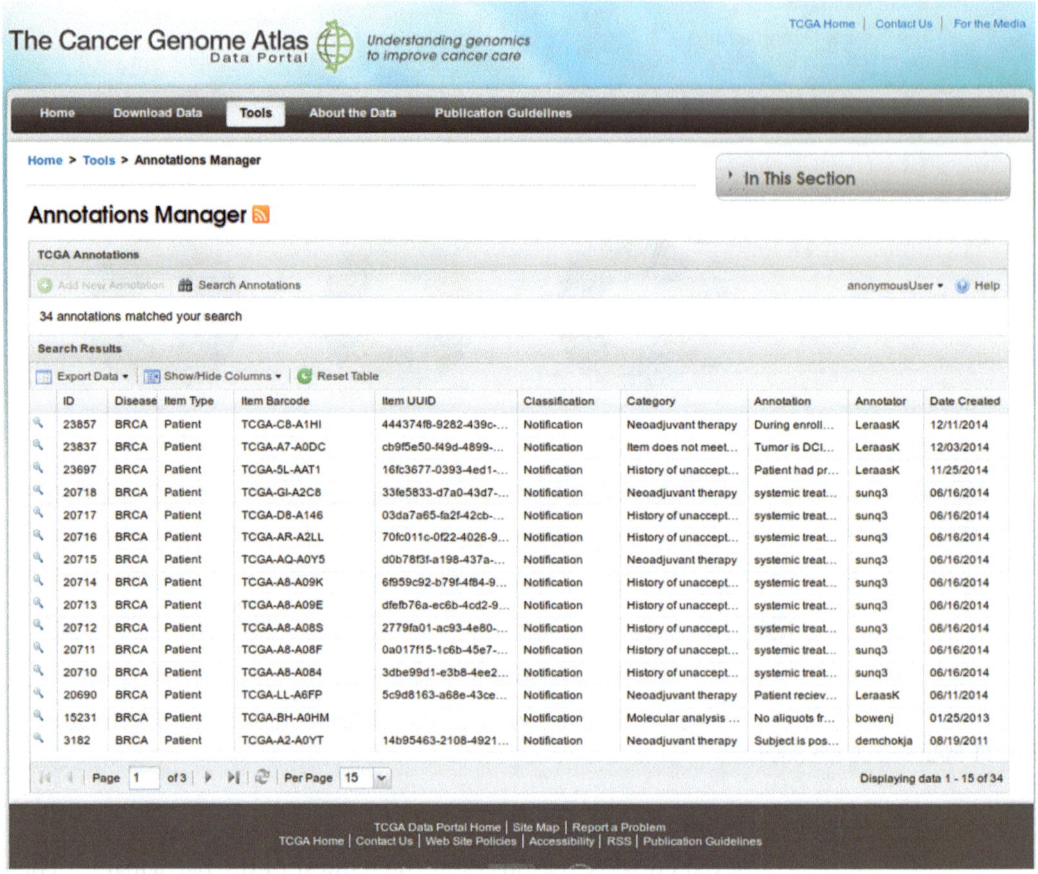

Fig. 12 Annotations manager interface

2.7.4 Clinical Data Dictionary

Nearly every clinical data element (CDE) in TCGA clinical data is formally defined as a term in the NCI Cancer Data Standards Registry and Repository (caDSR), and can be found using the CDE Browser. A convenient compilation of all active TCGA CDEs, with definitions and CDE Browser links, is available as the TCGA Clinical Data Dictionary.

2.7.5 Publication Pages

TCGA "marker papers" (initial scientific publication of comprehensive genomic analysis and interpretation) for each study have a companion webpage on the DCC Data Portal. These pages are collected at https://tcga-data.nci.nih.gov/docs/publications/. They contain links to final sample lists, static data archives, and supplementary materials associated with the publications. In general, all Level 3 and most Level 2 data as analyzed for the given publication are archived on the publication page. Lists of BAM sequence files used in the analyses are also usually available on these pages. These lists may be used to identify and download (with appropriate user access) the actual BAM files from CGHub.

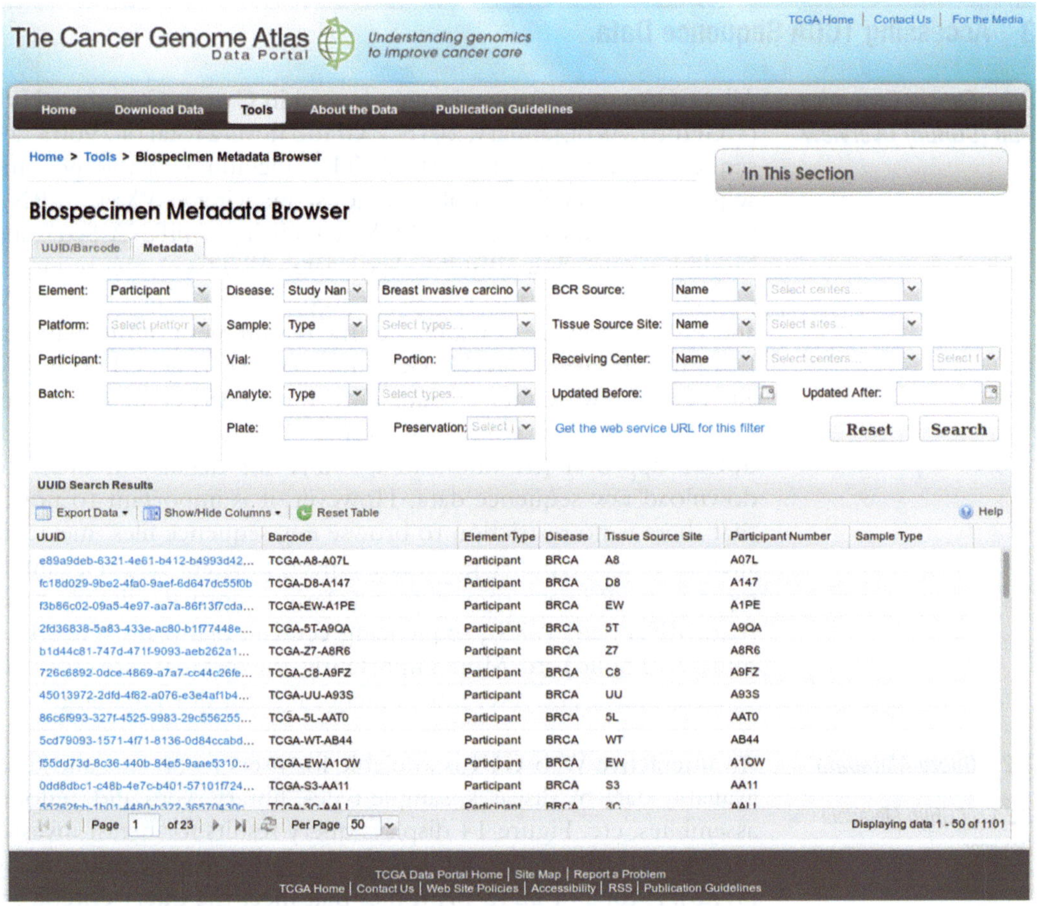

Fig. 13 Biospecimen metadata browser interface

2.7.6 Web Services

Several of the data access and metadata applications have associated Web services that may be used in user programs and scripts to query and download data automatically. While a full description of these is beyond the scope of this chapter, those interested can refer to the following User Guides for complete descriptions and examples:

1. Annotations.
2. Barcode to UUID Translator.
3. Biospecimen Metadata Browser.
4. Data Matrix.
5. Data Reports.

3 Accessing TCGA Sequence Data

3.1 Cancer Genomics Hub (CGHub) Overview

All TCGA raw sequence data are hosted in Cancer Genomics Hub (CGHub). As of January, 2015, CGHub hosts a total of 79603 raw sequence files (roughly 1092 TB) including whole-genome sequencing (WGS), whole-exome sequencing (WXS), mRNA sequencing (RNA-Seq), miRNA sequencing (miRNA-Seq) and bisulfite sequencing (Bisulfite-Seq) data. All TCGA raw sequence data are in BAM format except for RNA-Seq for which both BAM and FASTQ files are available. Per the requirement of the National Institutes of Health (NIH) Genomic Data Sharing (GDS) Policy, all TCGA raw sequence data (BAM or FASTQ files) are de-identified and are controlled access. Users need to apply for dbGaP approval (*see* Subheading 3.3.1 for details) in order to download raw sequence data. However, it is important to notice that there is no restriction to browse raw sequence files' metadata such as cancer types (e.g., breast cancer, prostate cancer), sample types (e.g., Primary tumor, blood derived normal), library types (e.g., RNA-Seq, WXS), sequencing centers, platforms (e.g., Illumina, ABI Solid) etc. More importantly, the metadata are critical in order to query and filter raw sequence files for downloading.

3.2 Query Metadata

3.2.1 CGHub Metadata Browser

An interactive Web GUI is available for users to browse and filter available data by diseases, sample types, library types and genome assemblies, etc. Figure 14 displays query results for breast invasive carcinoma (BRCA) of all sample types for exome sequencing (WXS). A total of 2242 BAM files that meet the query criteria are returned.

If a user has a list of barcodes or UUIDs, and wants to know whether their associated BAM or FASTQ files are available, it is most convenient to use the batch search tool at https://browser.cghub.ucsc.edu/search/batch/. Simply copy and paste a list of barcodes or UUID in the search box or upload a file that contains a list of barcodes or UUIDs, the batch search tool will return all matching objects. As shown in Fig. 15a, if we copy and paste two aliquot barcodes, TCGA-A7-A0DC-01B-04D-A22N-09 and TCGA-AN-A03X-10A-01W-A021-09, in the search box (Fig. 15a), it returns two matching BAM files (Fig. 15b).

3.2.2 cgquery

CGHub has also developed a command line tool to programmatically query and filter the available data. Once they have installed the GeneTorrent download client (*see* Subheading 3.3.3 for detail), users can run the following command to get the same results as Web GUI described in Subheading 3.2.1:

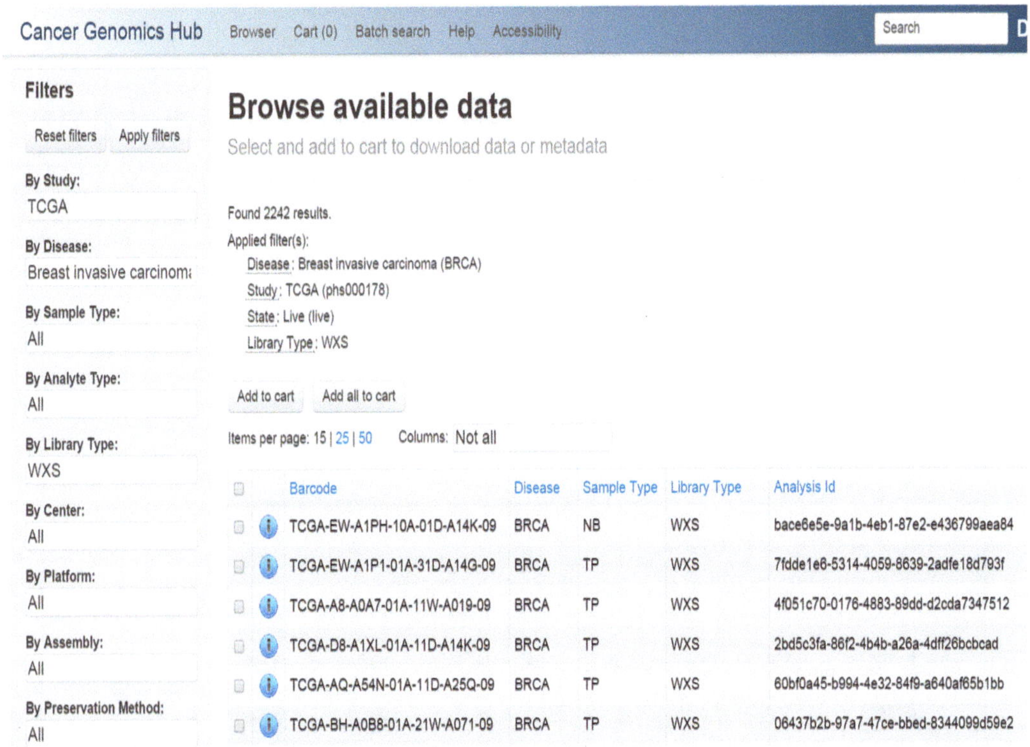

Fig. 14 CGHub metadata browser

```
cgquery "disease_abbr=BRCA & library_strategy=WXS & \
study=phs000178 & state=live"
```

As shown in Fig. 16, a total of 2242 objects are returned.

The above command will output results in screen. Users can redirect output to an XML file, results.xml, by running the following command:

```
cgquery -o results.xml \
"disease_abbr=BRCA & library_strategy=WXS & \
study=phs000178 & state=live"
```

3.3 Steps to Download BAM/FASTQ Files

3.3.1 Get a dbGAP Authorization

As mentioned earlier, raw sequence data (BAM or FASTQ files) are controlled access and are available only to approved users. It is important to notice that, like the DCC, CGHub itself does not have authority to approve an access request. The authorization process is handled by the Database of Genotypes and Phenotypes (dbGAP) at the National Center for Biotechnology Information (NCBI), National Institutes of Health (NIH).

To apply a dbGAP access, please follow the step by step guide described at https://wiki.nci.nih.gov/display/TCGA/Application+Process. Please read data use certificate (DUC) carefully and strictly follow the terms and conditions. Genomic data at CGHub

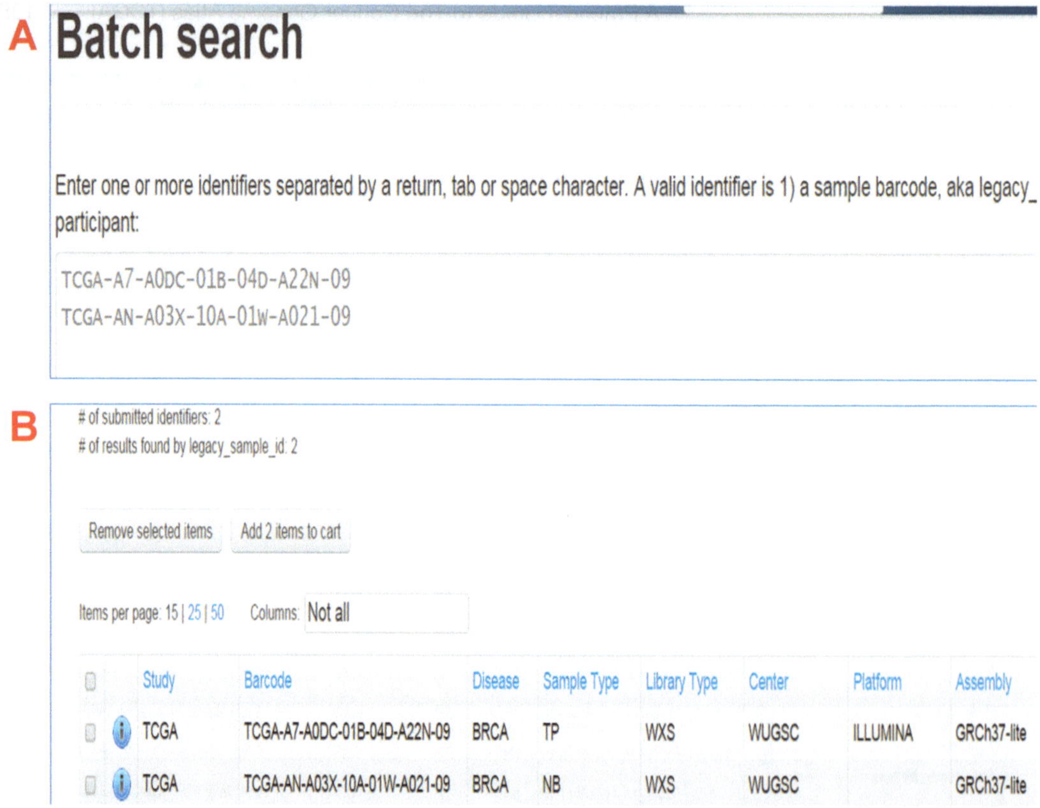

Fig. 15 CGHub batch search tool

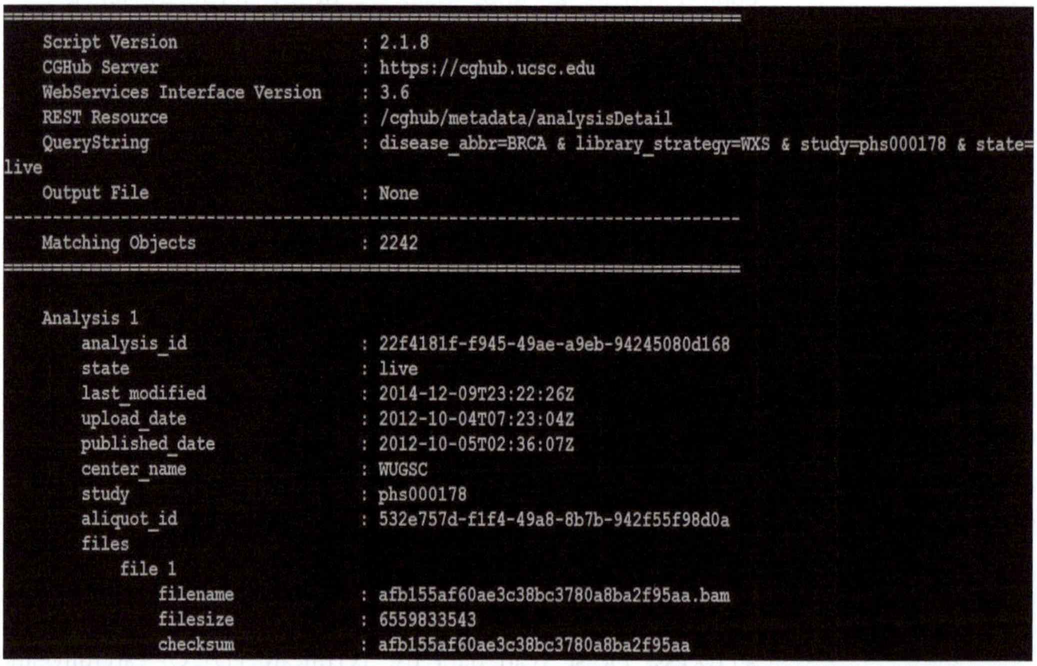

Fig. 16 Query metadata using cgquery

are all "de-identified", however, it is not theoretically impossible to "reidentify" a donor using genomic data. Users are not permitted to use TCGA data to "reidentify" the donor per data use agreement in order to protect donor's privacy. Any violation of the data use agreement could lead to serious penalties.

3.3.2 Get a CGhub Keyfile

Once a user gets a dbGaP approval and receives the user name and password, the user may then go to CGHub website at https://cghub.ucsc.edu/keyfile/keyfile.html to get the cghub.key file. As shown in Fig. 17, clicking "Returning user" (a) will bring the user to the NIH authentication site. If the user name and password is authenticated, a cghub.key file will be generated (b). The cghub.key file must be kept in a secure location and should never be shared with others. This file is specific to your user name and password, therefore, others could use your cghub.key file to download raw sequence data in your name in violation of your data use agreement. This file is valid for one year. When it expires, you will have to repeat the above process to generate a new cghub.key file. If you believe your cghub.key file is compromised, you must: (1) email the key file to CGHub so they can black list the file to prevent unauthorized access; (2) regenerate a new file using your user name and password.

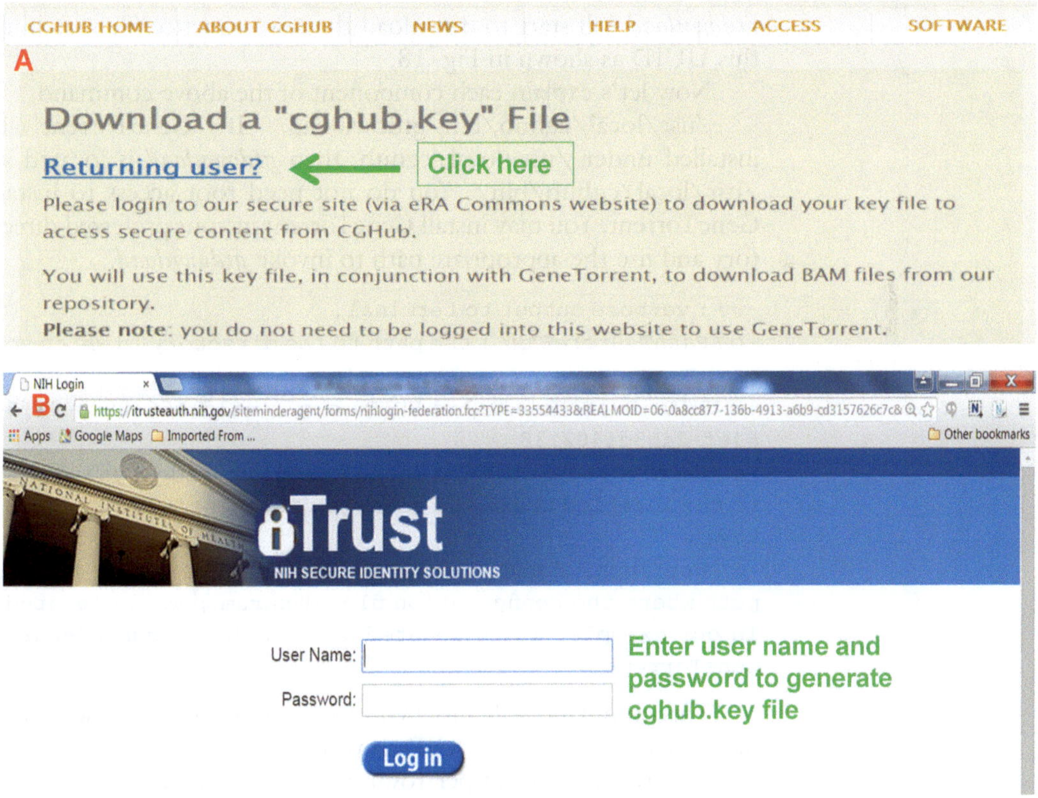

Fig. 17 Getting a CGHub keyfile

3.3.3 Get and Install GeneTorrent

Due to the large sizes of raw sequence data, CGHub developed a special tool, GeneTorrent, to facilitate secure and efficient transfer of raw sequence files. GeneTorrent is freely available and can be downloaded from https://cghub.ucsc.edu/software/downloads.html. GeneTorrent is distributed for multiple operating systems including Microsoft Windows, Mac OS, CentOS, Ubuntu, and Fedora Linux. It contains two main programs: *cgquery* and *gtdownload*. *cgquery* is used to query metadata, while *gtdownload* is used to download raw sequence files.

3.3.4 Retrieve BAM/FASTQ Files

Once you have GeneTorrent installed and the cghub.key file downloaded, you are ready to download raw sequence files. Here we introduce two basic scenarios to download raw sequence files. (In the following, Unix path formats are used; analogous commands in Windows may be run using the DOS path format.)

Scenario One

Assuming you know the UUID of a raw sequence file, you can download it by executing the following command:

```
/usr/local/cghub/bin/gtdownload -vv -p /PathToStoreBAM/
-d UUID \
  -c /PathToCredentialFile/cghub.key \
  -C /usr/local/cghub/share/GeneTorrent/
```

gtdownload will start to download the raw sequence file matching this UUID as shown in Fig. 18.

Now let's explain each component of the above command.

/usr/local/cghub/bin/gtdownload: If GeneTorrent is installed under /usr/local/cghub, then *gtdownload* is located at /usr/local/cghub/bin/. You do not need root access to install GeneTorrent. You may install GeneTorrent in your preferred directory and use the appropriate path to invoke *gtdownload*.

`-vv` : verbose output to terminal.

`-p /PathToStoreBAM/` : the path to the directory in which you want to store your raw sequence file.

`-d UUID`: Provide a valid UUID, e.g., cd3d1b66-08cb-4ff8-84ef-2ab9df402138

`-c /PathToCredentialFile/cghub.key` : Provide a valid path to the location where you store your credential file, cghub.key.

`-C /usr/local/cghub/share/GeneTorrent/` : Provide a valid path where the configuration file, dhparam.pem, is located. In our example, it is located at /usr/local/cghub/share/GeneTorrent/.

This command will allow you to download one raw sequence file matching that given UUID. If you have a file that contains a list of UUIDs (one UUID per row), you may write a script to loop

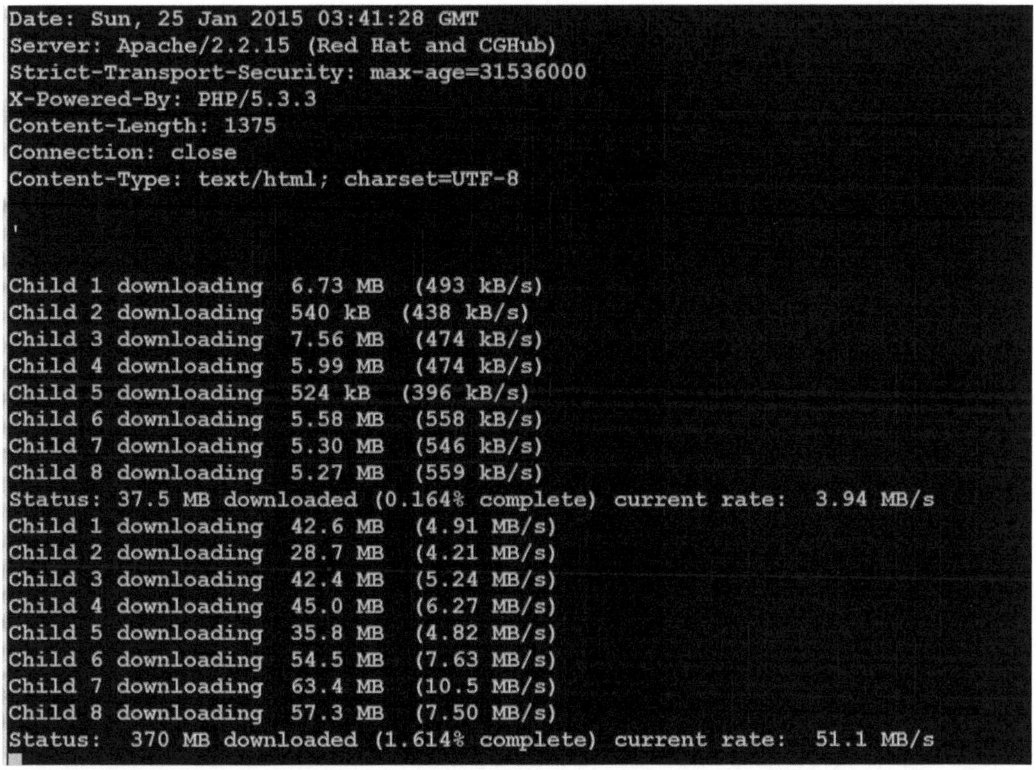

Fig. 18 GeneTorrent download

through these UUIDs to download all matching raw sequence files. The following is an example Perl script:

```
open $in, 'Your-UUID-File' or die "Unable to open the UUID
file"; while(<$in>)
{
  chomp;
  system <<CMD;
/usr/local/cghub/bin/gtdownload -vv -p /PathToStoreBAM/ \\
-d $_ -c /PathToCredentialFile/cghub.key \\
-C /usr/local/cghub/share/GeneTorrent/
CMD
  print "The raw sequence file, $_, has been downloaded\n";
}
close $in;
```

Note that if you have hundreds of UUIDs in your list, it could take several days or weeks to download. You may want to use *nohup* or *screen* commands in Unix.

Scenario Two

If you use CGHub Web GUI to query and filter your desired raw sequence files, after your query and filtering, you can click "Add all to cart" as shown in Fig. 19a. It will bring you to the next page as

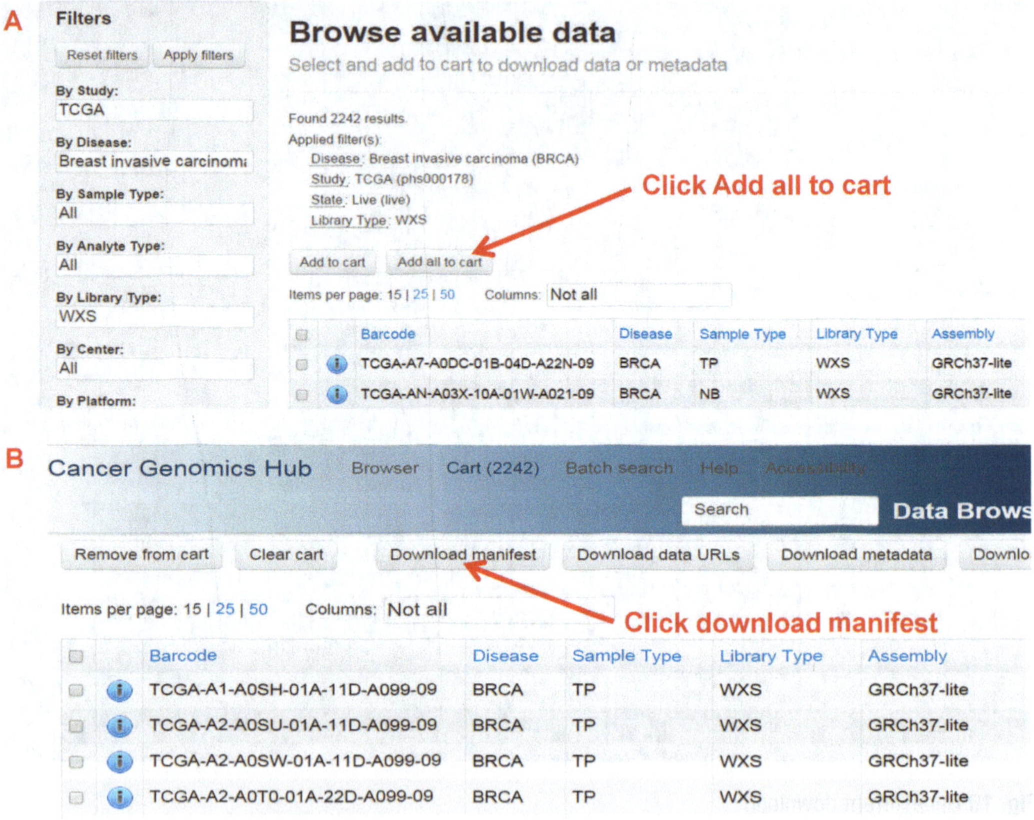

Fig. 19 Retrieving manifest using CGHub browser

shown in Fig. 19b. You can then click "download manifest" to generate the manifest.xml file. *gtdownload* is able to parse manifest.xml and download all raw sequence files listed in this manifest file. You may execute the following command to download all matching raw sequence files.

```
/usr/local/cghub/bin/gtdownload -vv-p /PathToStoreBAM/ \
-d manifest.xml -c /PathToCredentialFile/cghub.key \
-C /usr/local/cghub/share/GeneTorrent/
```

Notice that the only difference is the –d option, i.e., replacing a UUID with the manifest.xml file.

3.4 Safeguard Raw Sequence Files

3.4.1 Regarding Redistribution of Raw Sequence Files

Once you have your desired raw sequence files downloaded, you may only use these files as described in your data access request (DAR). You are ***not*** permitted to redistribute these files to others. Please note that you are also not permitted to redistribute Level 2 data such as VCF or germline MAF files that are derived from these raw sequence files.

3.4.2 Publication Moratorium

Before you decide to publish your results, please always check the TCGA publication moratorium page at http://cancergenome.nih.gov/publications/publicationguidelines.

References

1. (2003) Sharing data from large-scale biological research projects: a system of tripartite responsibility. report of a meeting organized by the Wellcome Trust and held on 14–15 January 2003 at Fort Lauderdale, USA
2. Collins FS, Varmus H (2015) A new initiative on precision medicine. N Eng J Med. doi:10.1056/NEJMp1500523
3. Lawrence MS, Stojanov P, Mermel CH et al (2014) Discovery and saturation analysis of cancer genes across 21 tumour types. Nature 505:495–501

Part III

Applications

Chapter 7

Working with Oligonucleotide Arrays

Benilton S. Carvalho

Abstract

Preprocessing microarray data consists of a number of statistical procedures that convert the observed intensities into quantities that represent biological events of interest, like gene expression and allele-specific abundances. Here, we present a summary of the theory behind microarray data preprocessing for expression, whole transcriptome and SNP designs and focus on the computational protocol used to obtain processed data that will be used on downstream analyses. We describe the main features of the *oligo* Bioconductor package, an application designed to support oligonucleotide microarrays using the R statistical environment and the infrastructure provided by Bioconductor, allowing the researcher to handle probe-level data and interface with advanced statistical tools under a simplified framework. We demonstrate the use of the package by preprocessing data originated from three different designs.

Key words Oligonucleotide microarrays, Preprocessing, Genotyping, Normalization, Probe-level data

1 Introduction

Microarrays are devices designed to collect information on thousands of genetic targets simultaneously. Manufacturers use short fragments of DNA (25–60 bases long) called oligonucleotides to serve as probes that capture data on the aforementioned targets, as Fig. 1 illustrates. These probes are placed on a solid surface, to which fluorescent-tagged target molecules are hybridized. Scanners are used to quantify the luminosity associated to the hybridized probes and the resulting image reflects the relative abundances of the targets.

Manufacturers often use multiple probes to query a single target. This strategy increases sensitivity and accuracy during the estimation procedures. The group of probes that share the same target is called probeset and it is the basic unit used to quantify the events of interest, like gene expression or allele-specific abundance. Figure 1 exemplifies one hypothetical probeset and the different physical location of the associated probes on the chip.

Ewy Mathé and Sean Davis (eds.), *Statistical Genomics: Methods and Protocols*, Methods in Molecular Biology, vol. 1418, DOI 10.1007/978-1-4939-3578-9_7, © Springer Science+Business Media New York 2016

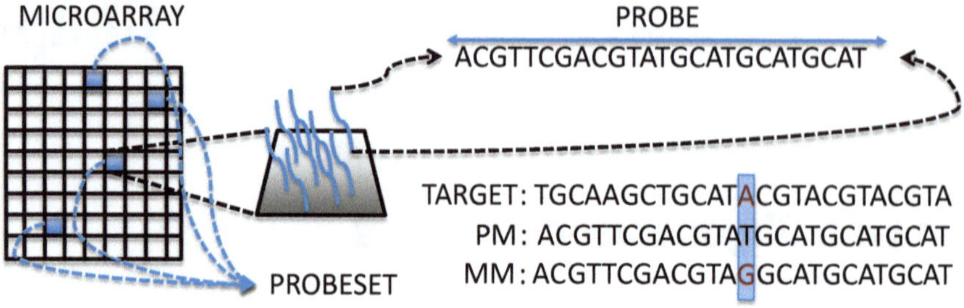

Fig. 1 The microarray is a solid surface that contains short oligonucleotides (25–60 bases long) called probes, which target the molecules of interest for hybridization, followed by image scanning. Probes are replicated on the chip to improve accuracy during estimation procedures. Earlier designs contained perfect-match (PM) probes and mismatch (MM). The base in the middle of the short sequence defined the difference between PM and MM probes. MM probes were thought to improve signal estimates, but researchers showed that PM-only algorithms outperformed methods that depended on MM probes

1.1 Array Types and Applications

For analyses involving gene expression, several products are currently available. Affymetrix microarrays that can be used for such tasks include the 3′ IVT arrays and the Whole Transcriptome (WT) chips, like the Exon ST and Gene ST arrays. The 3′ IVT designs estimate expression with probes located near to the 3′ end of the transcripts. The WT arrays provide a more comprehensive coverage of targets, with probes located more evenly across exons. The data analyst can choose one from three definitions of probesets for summarization to the transcript level:

1. Core Probesets: supported by RefSeq and full-length mRNA GenBank records;
2. Extended Probesets: supported by non-full-length mRNA GenBank records, EST sequences, ENSEMBL gene collections and others;
3. Full Probesets: supported by gene and exon prediction algorithms only.

Researchers use Single Nucleotide Polymorphism (SNP) arrays to perform analyses at the DNA level, inferring genotypes and obtaining copy number estimates. SNPs are single-base substitutions that account for a great amount of the genetic variation observed among individuals. At a SNP site, we often observe two possible alleles in the population. They are commonly referred to as alleles A and B and the microarray contains probes specific for each of them. The difference between the allele-specific quantifications is one of the most common statistics to classify individuals into genotype groups, like AA, AB, and BB. Genotyping arrays can query millions of markers in parallel and this characteristic transformed them into one of the standard tools for genome-wide association studies (GWAS), where the analyst seeks for association of specific markers with phenotypes of interest. A recent study [1] provides a catalog of identified associations between SNP markers and traits like diabetes, cancer, cardiovascular disease, and drug response.

Older microarray designs were comprised of perfect-match (PM) and mismatch (MM) probes. As presented on Fig. 1, the MM probes differ from the perfect-match ones by one nucleotide in the middle of the sequence and were thought to provide an improved way of estimating background noise and improving the estimates of true signal. Once probe sequences became available, their thermodynamic properties could be investigated in more details. This led researchers to observe that, although estimates were unbiased, the resulting variances were significantly large, seriously affecting the precision of the signal estimates.

Converting the fluorescence intensities into quantities that represent biological events is a series of complex statistical procedures called preprocessing. A group of researchers present a preprocessing algorithm that uses only the PM probes [2, 3]. This method, Robust Multiarray Average (RMA), provides remarkably better estimates, improving considerably the results of downstream analyses. These results stimulated manufacturers to produce designs containing only PM probes. Consequently, PM-only arrays cover more targets and can have more probes per probeset.

1.2 The Need for Flexible Software

Expression microarrays designed for human samples query between 10,000–20,000 targets, while current SNP arrays have probesets for millions of markers. These probesets are represented by multiple (between 4 and 20) probes to ensure statistical accuracy during the estimation procedures. This probe multiplicity within probesets increases significantly the volume of data to be processed, requiring efficient implementation of software for data preprocessing that can successfully interface with tools for downstream analyses.

In microarray data analysis, researchers often face difficulties in importing the raw information generated by the experiment and conducting data processing required for downstream analyses that include differential expression and genotype inference. Manufacturers provide software tools that perform some of these tasks, but it has been shown that alternative methods combined with software designed for appropriate data analysis outperform these proprietary solutions in a number of metrics [2, 3].

Bioconductor is an open source project [4] that fosters the development of efficient and transparent software based on the R programming language [5]. The tools distributed by the project offer solutions for the analysis of high-throughput genomic data. The *oligo* Bioconductor package [6] is designed to handle oligonucleotide microarray datasets from different manufacturers, allowing the user to import the files provided by the maker in their native formats, avoiding, for example, problems in format conversion and allowing detailed data analysis when desired. In addition to file parsers, the package provides alternative methods for the preprocessing steps involved in microarray data analysis. The package focuses on usability and flexibility, allowing the experienced user to apply its functionality on complex scenarios and the beginner to accomplish the task of preprocessing data with simple interactions with the Bioconductor environment.

The *oligo* package implements multiple containers for probe-level data, also called *FeatureSet* objects. Each array type is associated to a distinct *FeatureSet* category. This allows the customization of methods for different chips, making the software aware of nuances between array types and their specificities. It also enables the researcher to use complex array designs in a simplified way, allowing on to choose, for example, the level of summaries (exon or transcript) for whole transcriptome arrays, feature not available for 3′ IVT designs. The main types of *FeatureSet* objects are: *ExpressionFeatureSet*, designed for expression arrays; *GeneSTFeatureSet* and *ExonStFeatureSet*, for whole transcriptome designs; *SNPFeatureSet*, for SNP microarrays.

The *oligo* package requires design-specific information for every microarray dataset imported through its parsers. More specifically, certain tasks need the software to identify physical probe position on the chip or the short DNA sequences used on each probe; other tasks demand *oligo* to locate probes that measure the same target (i.e., probes that belong to the same probeset). This information is obtained from annotation packages created by the *pdInfoBuilder* Bioconductor package. These annotation packages are also distributed via Bioconductor. The *oligo* package is capable of detecting the need of downloading, installing, and loading such packages automatically as required.

1.3 Handling Oligonucleotide Microarrays Data

Regardless the application, the workflow for processing microarray samples with the *oligo* package can be summarized in three general steps: (1) data import; (2) preprocessing; (3) data export for downstream analyses. The software allows the researcher to conduct microarray data analysis starting from probe-level data, apply alternative algorithms for processing and explore options for data visualization and data exchange. The package enables the investigator to have a better understanding of the available data by simplifying the access to information in a powerful analytical environment.

1.3.1 Data Import and Manipulation of Probe-Level Data

During the data import step, the software reads the files containing the raw intensities associated to the experiment into R. In this phase, the system identifies the array type (expression, SNP, whole transcriptome), loads the appropriate annotation package and returns the respective *FeatureSet* object containing the probe-level data. The supported formats files containing probe-level information are binary and text CEL (Affymetrix) and text XYS (Nimblegen). The *oligo* package offers one parser for each format, made available to the user through the *read.celfiles* and *read.xysfiles* commands, respectively.

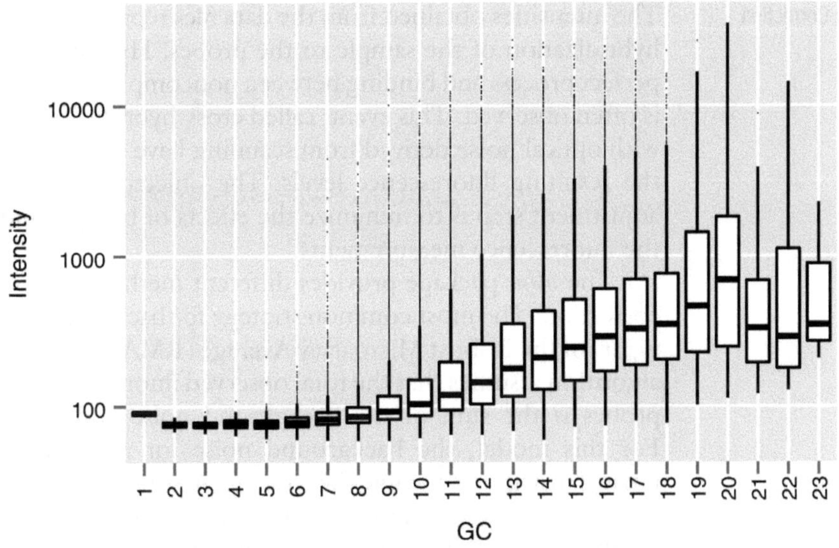

Fig. 2 The *oligo* package allows users to access the probe-level data, including probe sequences. This feature enables researchers to study and account for properties that were not considered important in older algorithms. This figure shows the strong association between GC contents and signal strength. Intensity levels present a systematic increase as GC content gets larger on probes

After the data import step, the software makes no distinction about the origin of the data, allowing the researcher to equalize strategies for data processing. The researcher can optionally study specific properties of the probe-level data, like assess the behavior of probe intensities obtained via the pm method as a function of the probe sequences, returned by the *pmSequence* method. The user can combine the resulting objects with analytic methods implemented in R to study specific associations between the variables of interest. Figure 2 illustrates the association between GC content in the probe sequence and probe intensities: the systematic increase in intensities as the number of nucleotides G or C increases (within a 25mer) is clear. Observing and modeling this behavior is only possible by having access to probe intensities and sequences, task simplified with the use of the *oligo* package.

Once the files with the raw data are loaded in the R environment, the user can investigate specific features of the probe-level data. This framework allows the user to combine microarray data accessors and statistical tools implemented in R to achieve excellence in biological data analysis.

1.3.2 Preprocessing

The probe-level data are a representation of the image obtained during scanning of the microarray. Converting these data into quantities that represent the biological events of interest (like expression levels or relative allelic abundance) is what we refer to as preprocessing. Preprocessing involves a series a complex statistical procedures that can be condensed in three major steps: background correction, normalization, and summarization.

Background Correction

The intensities obtained from the data files represent the amount of hybridization of the sample to the probes. Hybridization is not a perfect process and binding between noncomplementary sequences is often observed. This event, called cross-hybridization, combined with optical noise derived from scanning have an additive effect on the resulting fluorescence levels. The objective of a background adjustment step is to minimize the effects of background noise on the fluorescence measurements.

The *oligo* package provides different methods for background correction. The most common strategy for background adjustment is part of the Robust Microarray Average (RMA) methodology. The algorithm assumes that the total observed fluorescence level for PM probes is the sum of the background noise and specific signal. For this model, the background noise for all probes follows a normal distribution, while the true signal distribution is concordant with an exponential distribution that is a probeset-specific. This combination of distributions allows one to estimate the background noise parameters using closed-form equations. The resulting object contains adjusted intensities that better represent the amount of hybridization observed during the experiment.

Normalization

Background noise due to optical fluctuations and nonspecific binding are not the only sources of variation for microarray data. In addition to these, technical variables often present strong effects on the patterns of fluorescence. One common example of technical issues affecting the intensity levels is the comparison of sets of chips that were analyzed using different scanners.

Analysts apply normalization methods to microarray data to reduce the effects of technical factors that are not accounted for during experimental design. This technique allows the samples to be directly comparable to one another in a scale where only biological differences should be present.

One of the methods made available through the *oligo* Bioconductor package is the quantile normalization. It consists in a procedure of two steps: (1) estimate the average quantiles of the dataset and define this mean vector as the target empirical distribution; (2) for each sample, replace the observed quantiles by those from the target distribution. In summary, this strategy ensures that all samples share the same empirical distribution, providing means for direct comparisons between samples or their groups.

Summarization

Manufacturers use multiple probes for a single target to improve precision and sensibility. Collapsing these data points into one single statistic that represents the probeset in question is the strategy currently used to quantify the biological phenomenon of interest. On this step, for the i-th probeset, a linear model is the tool of preference of data analysts because the background adjusted, normalized, and log-transformed intensity for its j-th probe, I_{ij}, can be represented as

$$I_{ij} = \mu + \alpha_i + \beta_j + \varepsilon_{ij},$$

where $\mu + \alpha_i$ is the expression for the *i*-th probeset and β_j is the effect of the *j*-th probe in that probeset and ε_{ij} describes the associated measurement error. Different methodologies can be applied to fit the model above. The one used in RMA is the Tukey's Median Polish strategy due to its robustness and speed. Once data are summarized, researchers can compare different groups of samples to investigate the phenotypes of interest.

1.3.3 Preprocessing Different Array Types

There are subtle differences between preprocessing data from different microarray types, like expression, whole transcriptome, and SNP designs. Chaining the three aforementioned steps for standard expression chips is straightforward. For whole transcriptome arrays, the researcher needs to specify the desired target (exon or evidence level of the transcript) and, then, the process is carried on normally.

Due to the signal strength, the background correction step is not applied when preprocessing SNP microarray data. The SNPRMA method [7] performs quantile normalization by using a predefined target based on HapMap samples. This normalization strategy increases robustness for downstream procedures, like genotype calling [8], as the experiment-specific sources of variability do not affect the estimation of the empirical target distribution. The summarization step is then performed at the allelic-level for each SNP, starting from the normalized and log-transformed intensities, I_{ijA}, as shown below:

$$I_{ijA} = \mu_A + \alpha_{iA} + \beta_{jA} + \varepsilon_{ijA},$$

$$I_{ijB} = \mu_B + \alpha_{iB} + \beta_{jB} + \varepsilon_{ijB},$$

where $\theta_{iA} = \mu_A + \alpha_{iA}$ and $\theta_{iB} = \mu_B + \alpha_{iB}$ represent the allele-specific effects for the *i*-th SNP and $\beta_{j.}$ and $\varepsilon_{ij.}$ are, respectively, the probe effect and the associated measurement error. Using these summaries, one can infer genotypes, on diploid subjects, by studying the behavior of the log-ratio statistic, M_i, defined as the difference of the allele-specific effects. If all SNPs shared the same distributional properties across genotype clusters (i.e., there were no SNP-specific effects), one could use a simple rule, as described below, to infer genotype calls for one diploid individual at all SNP sites simultaneously:

$$M_i = \theta_{iA} - \theta_{iB} \begin{cases} \gg 0 & \therefore \quad \text{genotype} = AA \\ \approx 0 & \therefore \quad \text{genotype} = AB \\ \ll 0 & \therefore \quad \text{genotype} = BB \end{cases}.$$

With the assumption of diploid individuals, large values of M_i are associated to an excess of alleles A when compared to allele B, suggesting that the genotype is AA. The opposite effect occurs for genotype BB. For genotype AB, the quantifications for both alleles are comparable, causing the difference to be approximately zero.

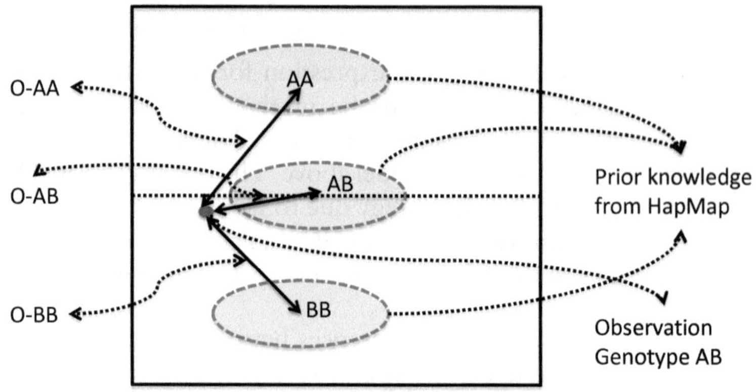

Fig. 3 The CRLMM method uses likelihood-based distances to assign genotype calls to observations. This genotyping algorithm was shown to outperform other solutions due to its robustness during the calling process, a direct effect of the preprocessing strategy designed for SNP microarray products, called SNPRMA, which uses quantile normalization to a fixed and known target and summarization by median-polish

However, different SNPs have different distributional properties. This requires the use of a more convoluted approach to make genotype calls. The specific model implemented in *oligo* to obtain genotypes from preprocessed data is the Corrected Robust Linear Model with Maximum-likelihood based distances (CRLMM) and it is described in [8]. CRLMM obtains the log-ratio estimates for every SNP on the chip using the normalization and summarization methods described above. At a given SNP site, the likelihood-based distances between sample log-ratio and the three (AA, AB, and BB) HapMap genotype clusters are assessed. For that SNP, a sample is called AA if the distance between the observation and the AA HapMap cluster (O-AA) is the minimum distance between the three possible distances (O-AA, O-AB, and O-BB), as shown on Fig. 3. Genotypes for other SNPs and samples are inferred using the same strategy.

The implementation of different, but internally connected, FeatureSet objects to represent distinct types of arrays enables the *oligo* package to detect the subtleties required by each design automatically. The software uses the appropriate methods conditionally on the array design, avoiding mistakes on data processing.

1.3.4 Using Preprocessed Data for Additional Analyses

The researcher uses preprocessed microarray to investigate the biological hypotheses that motivated the experiment in the first place. Despite their major importance, these downstream analyses are not the focus of the *oligo* package, which was developed to provide efficient strategies to access and preprocess probe-level data. Standard downstream analyses change depending on the type of microarray under consideration. Therefore, the results generated by the *oligo* package are distributed on standard Bioconductor data containers that can interface virtually with anything in R or be exported for manipulation via other software.

When working with gene expression designs, the researcher is often interested in assessing evidences of differential expression across two or more groups of samples. More specifically, investigators want to identify sets of probesets that show evidences of systematically different levels of expression between conditions and/or after adjusting for other variables. One common approach to assist the researcher to assess such hypotheses is combining linear models and the preprocessed data. In this case, probeset summaries obtained, for example, via RMA are used as dependent variables in models that use conditions and other variables as predictors. It is possible to integrate the features of the *oligo* package with those from the *limma* (Linear Models for Microarrays) Bioconductor package [9] and easily investigate such relationships. Other packages are distributed by the Bioconductor project to support analyses from different perspectives, like gene set enrichment analyses. The objects returned by *oligo* for expression designs are standard Bioconductor objects that can be easily used as input for other methods.

When the *oligo* package is used to preprocess data from SNP microarrays and call genotypes for samples present in a dataset, the resulting calls can be used on analyses that search for association between specific genomic sites and phenotypes [1]. These analyses often use generalized linear models (GLMs), like logistic regression, to quantify the association between markers and the trait of interest. The *snpStats* Bioconductor package [10] is one package that can be combined with *oligo* to perform such tasks.

2 Materials

The application of the protocols for preprocessing microarray data described here require the use of the following software and datasets:

1. R (version 3.1.2);
2. Bioconductor package: *oligo*;
3. R package: *R.utils*;
4. Datasets consisting of Affymetrix CEL files available from NCBI Gene Expression Omnibus (GEO) service:
   (a) Expression Arrays: GSE51808 [12].
   (b) Gene ST Arrays: GSE54229 [13].
   (c) SNP Arrays: GSE8271 [14].

3 Methods

On this section, we will load and preprocess data from three different experiments. They were generated from different array designs: (a) expression array; (g) gene ST array; (c) SNP array. The files associated to each of the experiments need to be downloaded and uncompressed. All the commands described below in italics need to be executed from within an R session. The *oligo* package can read uncompressed and gzip-compressed Affymetrix CEL files. If the data correspond to NimbleGen microarray designs, *oligo* can parse uncompressed and gzip-compressed XYS files. The protocols for Expression and Gene ST arrays presented below include simple examples on the assessment of differential expression.

3.1 Preprocessing Affymetrix Expression Arrays

1. Load R;
2. Load the *oligo* package;

```
library(oligo)
```

3. Identify the CEL files of interest using the *list.celfiles* command. The first argument to the command is the directory where the files are saved. The second argument specifies that R should return the full names (including full path) of the files. The third argument defines that compressed CEL files (with .CEL.gz extensions) should be listed;

```
cels = list.celfiles("GSE51808", full.names=TRUE,
listGzipped = TRUE)
```

4. Import the CEL files using the *read.celfiles* command. At this point, *oligo* identifies the required annotation package, which will be downloaded, installed, and loaded as necessary;

```
raw= read.celfiles(cels)
```

5. The object *raw* contains the probe-level data associated to the experiment. To perform preprocessing using the RMA algorithm (background correction via Normal/Exponential convolution; quantile normalization; and summarization via median-polish), the user can use the rma method directly on the raw data;

```
summ = rma(raw)
```

6. The object *summ* contains the preprocessed data and can be used on downstream procedures, like differential expression analyses. For this and other tasks using preprocessed data, the user needs the phenotypic data (created by the user) and the preprocessed data matrix (via the *exprs* method);

```
pdata = data.frame(trait=c(rep('dengue', 28), rep
('convalescent', 19), rep('control', 9)))
exprMat = exprs(summ)
```

7. Fit linear models using the *limma* package

```
library(limma)
fit = eBayes(lmFit(exprMat, model.matrix(~trait, data
=pdata)))
```

8. Obtain the list of candidates for differential expression for the conditions of interest. In this example, the "covalescent" and the "dengue" groups are associated to the coefficients 2 and 3, respectively;

```
deGenes2 = topTable(fit, coef=2)
deGenes3 = topTable(fit, coef=3)
```

Tables 1 and 2.

Table 1
This table shows the ten probesets with lowest adjusted *p*-values. These candidates were identified by comparing Covalescent Patient expression levels to those of Healthy Controls using linear models implemented in the *limma* Bioconductor package, which used data preprocessed by the *oligo* package as input

	logFC	Ave. expr.	*t*	*p*. value	Adj. *p*. val.	B
243796_PM_at	−0.941	5.868	−7.435	0.000	0.000	11.898
1554665_PM_at	−0.806	3.811	−6.426	0.000	0.001	8.485
236500_PM_at	−1.651	4.390	−6.306	0.000	0.001	8.082
239347_PM_at	−0.942	4.626	−6.215	0.000	0.001	7.774
239339_PM_at	−1.181	5.895	−5.594	0.000	0.007	5.703
232412_PM_at	−1.219	4.006	−5.546	0.000	0.007	5.544
232709_PM_at	1.097	7.501	5.484	0.000	0.007	5.337
200722_PM_s_at	0.609	7.391	5.425	0.000	0.007	5.145
223930_PM_at	−1.093	4.413	−5.423	0.000	0.007	5.137
217152_PM_at	−0.997	5.029	−5.319	0.000	0.008	4.798

Table 2
This table shows the ten probesets with lowest adjusted *p*-values. These candidates were identified by comparing Dengue Patient expression levels to those of Healthy Controls using linear models implemented in the *limma* Bioconductor package, which used data preprocessed by the *oligo* package as input

	logFC	Ave. expr.	*t*	*p*. value	Adj. *p*. val.	B
226959_PM_at	−2.011	8.901	−16.098	0.000	0.000	41.907
201761_PM_at	1.853	9.689	15.300	0.000	0.000	39.610
208805_PM_at	0.990	11.805	15.283	0.000	0.000	39.559
225834_PM_at	2.816	7.039	15.246	0.000	0.000	39.452
202779_PM_s_at	2.268	8.389	14.756	0.000	0.000	37.998
203728_PM_at	1.656	6.594	14.448	0.000	0.000	37.065
203396_PM_at	0.970	10.904	13.933	0.000	0.000	35.480
219545_PM_at	3.403	4.960	13.914	0.000	0.000	35.422
226382_PM_at	−2.038	9.228	−13.881	0.000	0.000	35.318
211475_PM_s_at	−2.332	10.702	−13.773	0.000	0.000	34.982

3.2 Preprocessing Affymetrix Gene ST Arrays

1. Load R;
2. Load the *oligo* package;

```
library(oligo)
```

3. Identify the CEL files of interest using the *list.celfiles* command. The first argument to the command is the directory where the files are saved. The second argument specifies that R should return the full names (including full path) of the files. The third argument defines that compressed CEL files (with .CEL.gz extensions) should be listed;

```
cels = list.celfiles("GSE54229", full.names=TRUE,
listGzipped=TRUE)
```

4. Import the CEL files using the *read.celfiles* command. At this point, *oligo* identifies the required annotation package, which will be downloaded, installed, and loaded as necessary;

```
raw= read.celfiles(cels)
```

5. The object *raw* contains the probe-level data associated to the experiment. To perform preprocessing using the RMA algorithm (background correction via Normal/Exponential convolution; quantile normalization; and summarization via median-polish), the user can use the *rma* method directly on the raw data. By default, the *rma* method uses only the core probesets to obtain the summaries. Gene ST arrays can be summarized to different levels: core or probeset. Exon ST arrays can be summarized to core, full, and extended probeset levels.

```
summCore= rma(raw, target='core')
```

6. The object *summCore* contains the preprocessed data and can be used on downstream procedures, like differential expression analyses. For this and other tasks using preprocessed data, the user needs the phenotypic data (created by the user) and the preprocessed data matrix (via the *exprs* method);

```
pdata = data.frame(trait=c(rep('Normothermia', 7),
rep('Hypothermia', 6)))
exprMat= exprs(summCore)
```

7. Fit linear models using the *limma* package

```
library(limma)
fit = eBayes(lmFit(exprMat, model.matrix(~trait,
data=pdata)))
```

8. Obtain the list of candidates for differential expression for the conditions of interest. In this example, the "*Normothermia*" and the "Hypothermia" groups are associated to the coefficients 2 and 3, respectively;

```
deGenes= topTable(fit)
```

Table 3

Table 3
This table shows the ten probesets with lowest adjusted *p*-values. We identified these candidates by comparing mouse embryonic fibroblasts exposed to normothermia to those exposed to hypothermia using linear models implemented in the *limma* Bioconductor package, which used data preprocessed by the *oligo* package as input

	logFC	Ave. expr.	*t*	*p*. value	Adj. *p*. val.	B
10420497	0.517	5.323	8.489	0.000	0.027	5.326
10509014	0.568	9.894	8.143	0.000	0.027	4.965
10561345	0.457	7.130	7.492	0.000	0.031	4.239
10543697	−0.299	8.241	−7.407	0.000	0.031	4.139
10482172	0.564	8.870	7.359	0.000	0.031	4.082
10516221	0.523	8.447	7.187	0.000	0.031	3.875
10583402	0.343	8.979	7.110	0.000	0.031	3.782
10598032	0.313	10.917	7.083	0.000	0.031	3.748
10339957	−0.228	10.263	−6.729	0.000	0.042	3.301
10377429	−0.759	8.294	−6.683	0.000	0.042	3.241

3.3 Preprocessing Affymetrix SNP Arrays

1. Load R;
2. Load the *oligo* package;

```
library(oligo)
```

3. Identify the CEL files of interest using the *list.celfiles* command. The first argument to the command is the directory where the files are saved. The second argument specifies that R should return the full names (including full path) of the files. The third argument defines that compressed CEL files (with .CEL.gz extensions) should be listed (if uncompressed CEL files are available, use *listGzipped=FALSE*);

```
cels = list.celfiles("GSE8271", full.names=TRUE,
listGzipped=TRUE)
```

4. (Optional) Identify files associated to the SNP 50K Xba Affymetrix array. If the directory contains only the CEL files of interest, this is not required;

```
celType = function(cels) sapply(cels, oligo:::getCel-
ChipType, TRUE)
types = celType(cels)
cels = names(grep('Xba', types, value=TRUE))
```

5. (Optional) Uncompress CEL files for better memory management during SNP data preprocessing. If the CEL files are already uncompressed, this is not needed;

```
library(R.utils)
cels = sapply(cels, gunzip)
```

5. Perform SNPRMA for SNP microarray data preprocessing combined with CRLMM for genotype calling. The *crlmm* method requires at least two arguments: the first is the set of CEL files and the second is the name of an output directory that will store the results generated by the methods;

```
crlmm(cels, 'outcrlmm')
```

6. Import the SNPRMA and CRLMM results back into R;

```
gtype = getCrlmmSummaries('outcrlmm')
```

7. Explore genotype calls and confidence scores. The second argument on the method *confs* defines that the confidence scores should be returned as computed by the *crlmm* method, i.e., without transformations;

```
head(calls(gtype))
head(confs(gtype, FALSE))
```

8. Methods that use genotype information to determine association between genomic markers and traits are used at this point. It should be noted that AA, AB, and BB genotypes are represented, respectively, by the integers 1, 2, and 3.

4 Notes

Among the genotyping arrays, designs like the Affymetrix SNP 5.0 and SNP 6.0 and many Illumina chips, became standard tools for large genome-wide association studies. These designs are denser and can target millions of SNPs simultaneously. They can be preprocessed and genotyped using the aforementioned approaches. The SNPRMA and CRLMM methods were improved, providing more accurate confidence levels, and reimplemented in the crlmm Bioconductor package [11] to accommodate the needs of larger studies.

References

1. Welter D, MacArthur J, Morales J et al (2014) The NHGRI GWAS Catalog, a curated resource of SNP-trait associations. Nucleic Acids Res 42(Database issue):D1001–D1006
2. Irizarry R, Hobbs B, Collin F et al (2003) Exploration, normalization, and summaries of high density oligonucleotide array probe level data. Biostatistics 4(2):249–264
3. Irizarry R, Bolstad BM, Collin F et al (2003) Summaries of Affymetrix GeneChip probe level data. Nucleic Acids Res 31(4), e15
4. Gentleman RC, Carey VJ, Bates DM et al (2004) Bioconductor: open software development for computational biology and bioinformatics. Genome Biol 5(10):R80
5. R Core Team (2014) R: a language and environment for statistical computing. R Foundation for Statistical Computing, Vienna, Austria, http://www.R-project.org/
6. Carvalho BS, Irizarry RA (2010) A framework for oligonucleotide microarray preprocessing. Bioinformatics 26(19):2363–2367
7. Carvalho B, Bengtsson H, Speed TP, Irizarry R (2007) Exploration, normalization, and genotype calls of high-density oligonucleotide SNP array data. Biostatistics 8(2):485–499

8. Zhang L, Yin S, Miclaus et al (2010) Assessment of variability in GWAS with CRLMM genotyping algorithm on WTCCC coronary artery disease. Pharmacogenomics J 10 (4):347–354
9. Ritchie ME, Phipson B, Wu D, Hu Y et al (2015) limma powers differential expression analyses for RNA-sequencing and microarray studies. Nucleic Acids Res 43(7):e47
10. Clayton D (2014) snpStats: SnpMatrix and XSnpMatrix classes and methods. R package version 1.16.0
11. Carvalho B, Louis TA, Irizarry RA (2010) Quantifying uncertainty in genotype calls. Bioinformatics 26(2):242–249
12. Kwissa M, Nakaya HI, Onlamoon N et al (2014) Dengue virus infection induces expansion of a CD14(+)CD16(+) monocyte population that stimulates plasmablast differentiation. Cell Host Microbe 16(1):115–127
13. Ilmjärv S, Hundahl CA, Reimets R et al (2014) Estimating differential expression from multiple indicators. Nucleic Acids Res 42(8), e72
14. Koeman JM, Russell RC, Tan M-H et al (2008) Somatic pairing of chromosome 19 in renal oncocytoma is associated with deregulated EGLN2-mediated [corrected] oxygen-sensing response. PLoS Genet 4(9): e1000176

Chapter 8

Meta-Analysis in Gene Expression Studies

Levi Waldron and Markus Riester

Abstract

This chapter introduces methods to synthesize experimental results from independent high-throughput genomic experiments, with a focus on adaptation of traditional methods from systematic review of clinical trials and epidemiological studies. First, it reviews methods for identifying, acquiring, and preparing individual patient data for meta-analysis. It then reviews methodology for synthesizing results across studies and assessing heterogeneity, first through outlining of methods and then through a step-by-step case study in identifying genes associated with survival in high-grade serous ovarian cancer.

Key words Biomarkers, Microarray analysis, Gene expression profiling, Ovarian neoplasms, Meta-analysis, Computational molecular biology

1 Introduction

This chapter introduces methods to synthesize experimental results from independent high-throughput genomic experiments, with a focus on adaptation of traditional methods from systematic review of clinical trials and epidemiological studies. We focus on differential gene expression because the public availability of data from gene expression microarrays far surpasses any other genomic assay; however, these methods are flexible and applicable to other genomic data types and study objectives.

The traditional systematic review and meta-analysis attempts to resolve inconsistency and uncertainty in a literature of, for example, clinical trials on effectiveness of a treatment or observational studies of association between a risk factor and health outcome. The analysis is of association between the outcome of interest and a single exposure or treatment. Great care must be taken in this situation to select and studies and exclude incomparable studies in order to avoid bias in the analysis, for example by adherence to the PRISMA guidelines [1]. Systematic review and meta-analysis has been well described in the primary literature, reviews [2, 3], and monographs [4, 5].

Ewy Mathé and Sean Davis (eds.), *Statistical Genomics: Methods and Protocols*, Methods in Molecular Biology, vol. 1418, DOI 10.1007/978-1-4939-3578-9_8, © Springer Science+Business Media New York 2016

Although the methodology we summarize in this chapter is substantially similar, the setting presents different challenges and priorities.

Unlike traditional studies of a single exposure or treatment, in this setting thousands of variables are observed—one per gene or transcript. Association measures, most commonly differential expression of each transcript, are almost never consistently reported, and instead must be calculated for each study using Individual Patient Data (IPD). It is hard to envision that synthesized results could be affected by inadvertent bias of the meta-analyst, but different challenges are present. The bioinformatic challenges can be substantial: locating and standardizing raw gene expression data and clinical data, handling large datasets, and repeating and interpreting thousands of meta-analyses. This chapter provides advice and recommends approaches for dealing with these challenges, and reviews relevant methods from traditional meta-analysis. Using straightforward and statistically well-established methods, meta-analysis makes it possible to overcome some of the limitations of high-dimensionality and batch effects that are inherent to high-throughput biology, and to develop extremely robust biomarkers.

2 Materials

2.1 Microarray Dataset Identification

If high coverage of available data is important, the most thorough approach to dataset identification is systematic literature review. Starting points for search terms of PubMed for numerous cancer types are provided in the GeneSigDB database [6]. However, the large majority of publicly available gene expression data are redistributed through the Gene Expression Omnibus (GEO) [7] or ArrayExpress [8], and data from these resources is significantly easier to access and is more stably available than data from the websites of authors or their institutions. Other alternatives providing greater curation but much lower coverage are InSilicoDB [9], Oncomine [10], and Bioconductor (BiocViews: ExperimentData, RNAExpressionData).

2.1.1 The Gene Expression Omnibus

Experiments in GEO are represented by Series (GSE codes), which are occasionally curated as Datasets (GDS codes). Series may be composed of a single platform (GPL) or multiple platform. Platforms (GPL) annotate platform-specific identifiers and usually provide maps to standard genes identifiers. However, it should be noted that the GPL annotations are generally author-provided and unstandardized, so different platforms provide different annotations or the same annotations based on different genome builds, and can even contain spreadsheet-introduced gene symbol errors [11]. When possible, it is safer to use annotations from Bioconductor [12] *.db* packages (BiocViews term *AnnotationData*) or from BioMart [13].

When manufacturer-specific annotations are not available, Bioconductor or Biomart can still be used to map stable identifiers such as Entrez Gene or Refseq to other annotations, rather than using unstable and potentially outdated identifiers, such as gene symbols, directly from the GPL annotations.

GEO is well-supported in Bioconductor by the GEOmetadb package [14] for searching meta-data and the GEOquery package [15] for downloading expression and platform data.

2.2 Dataset Preparation

Whereas meta-analysis for clinical trials and observational studies in epidemiology is often possible using published summary statistics and confidence intervals, thousands of rows of summary statistics are required in meta-analysis. Normally these must be calculated from Individual Patient Data (IPD) after appropriate standardizing steps. This section first describes steps to prepare datasets for the calculation of summary statistics appropriate for meta-analysis in ways that reduce the impact of unwanted technical variability between studies. However, we note that some amount of heterogeneity between datasets is inevitable, arising both from experimental settings and from differences in patient recruitment and treatment. Heterogeneity should not be viewed as an enemy of the meta-analyst. The existence of heterogeneity provides rationale for using meta-analysis to identify robust genomic signals present independently of the heterogeneity, and to investigate the impact of heterogeneity on identified genome/patient associations.

2.2.1 Curation

Individual-patient metadata must be standardized across studies, including variable names and the values they take. This process is error-prone and it can be difficult to catch mistakes later in high-throughput analysis, therefore a template-based syntax checking recommendable. For example the curatedOvarianData package [16] published an R script for template-based curation and *regular expression* checking in R, the InSilicoDB Web service [9] provides a graphical interface to curation.

2.2.2 Preprocessing

The application of different microarray preprocessing algorithms may introduce technical heterogeneity. Although in our experience (for example [17, 18]) even different microarray platforms do not necessarily contribute significant heterogeneity, when the analyst is in a position to preprocess raw data, certain pre-processing approaches will reduce the potential for heterogeneity. Common normalization and probe set summarization methods such as Robust Multi-array Average [19] use multiple arrays to estimate probe effects, and heterogeneity may be introduced by processing datasets separately and therefore using different estimates of probe effects for each dataset. One approach to avoiding these differences is to preprocess all datasets together, using a low-memory function such as justRMA from the Affymetrix Bioconductor [12] package. For supported platforms, the *frozen RMA* method [20] uses a frozen reference database of thousands of publicly available raw data files, and eliminates differences in estimated probe set effects across datasets. We emphasize, however, that it is also reasonable to use data already pre-processed by different algorithms, and assess the amount of heterogeneity post hoc using the I^2 or Cochrane's Q statistic.

When different technological platforms used prevent application of comparable pre-processing methods across all studies, a next-best approach is to scale the observations for each gene (or row) to *z*-scores, by subtracting the mean and dividing by the standard deviation. This should be done *after* ensuring that any dataset-wide variance-stabilizing transformation, such as the log-transform, has been applied consistently in all or none of the datasets. Scaling to each variable in each dataset to unit variance ensures that fold-change or other effect-size estimates are comparable across studies, which may be adequate when synthesizing results obtained from comparable but different measurement technologies.

2.2.3 Batch Effects

Anyone familiar with the problems that batch effects can cause for single studies [21, 22] will be concerned about the potential for batch effects to impact a meta-analysis. Using traditional methods of assessing heterogeneity, one can identify the extent to which heterogeneity between datasets is impacting a meta-analysis, and identify which datasets are most responsible for the heterogeneity. One can establish whether the amount of heterogeneity warrants the potentially large amount of effort required to identify and correct for batch effects in individual studies, and if so whether batch correction actually helps. One may be surprised to find that batch effects in some cases have limited practical effect on a meta-analysis.

2.2.4 Gene Collapsing

A basic requirement of in gene expression meta-analysis that each study contain overlapping sets of measurements. If all studies used the same technological platform, it is possible to perform meta-analysis either using manufacturer-specific probe set identifiers, or gene-level summaries. To synthesize analysis across different platforms, however, it is necessary to map and summarize manufacturer-specific probe set identifiers to standard identifiers such as Entrez Gene or gene symbols. Miller et al. [23] discuss and compare alternatives for merging probe set level data to gene level. We highlight one additional consideration for meta-analysis, that when using an approach that selects a single representative probe set per gene, it is preferable to select the same probe set for each study in the meta-analysis. For example, Ganzfried and Riester et al. [16] selected for each gene the representative probe set with maximum mean across all studies of a common platform. Approaches which do not use the dataset at hand for probe set selection, such as *Jetset* [24] or *BrainArray* [25], also avoid introducing heterogeneity that can arise from representing a gene by different probe sets in each dataset.

2.2.5 Pathway or Gene Set Collapsing

While pathway or gene set approaches are routinely used to test for enrichments in gene rankings, for example via cutoff-free methods such as GSEA [26] or via methods for analyzing gene lists such as implemented the DAVID Web service [27], the value of collapsing features to pathways is appreciated only recently. The idea of these approaches is to calculate for each sample and pathway (or gene set) a single pathway activation score, utilizing the expression values of all measured genes in this pathway. The gene-by-samples expression matrix is transformed into a pathway-by-samples expression matrix. Such a pathway activation score calculation is a potential noise reduction step and can be used to more robustly compare expression data across very different assays, for example even when the data was obtained from different species [28].

The first step is selecting appropriate gene sets for the problem at hand. A commonly used resource is MSigDB [26], which provides curated sets of genes in different categories, for example canonical pathways such as KEGG [29] or REACTOME [30, 31], downstream targets of gene regulators such as transcription factors or miRNAs, or genes associated with gene ontology (GO) terms [32]. The expected expression direction ("up" vs. "down") of genes within a set of genes representing an active pathway provides important information and high activity of both upregulated and downregulated genes may cancel each other out. It is thus recommended to split pathways into upregulated and downregulated gene sets when this information is available [33]. Final pathway scores can be calculated by subtracting the activation scores of down-regulated genes from the ones of the up-regulated activation scores.

Various methods for collapsing genes to gene sets have been proposed [34], most notably ssGSEA [35] or GSVA [33]. Since these methods need to distinguish, for all genes in the gene sets, activated or inhibited gene expression from normal expression levels, these methods work better the larger the dataset is [33]. This limitation could be in theory avoided when the future methods support platform-specific databases of gene expression ranges, as for example utilized in the fRMA approach [20] discussed in Subheading 2.2.2. In a meta-analysis setting, gene set collapsing methods have been used for example to robustly classify samples by subtype [36] or comparing in vivo, in vitro and murine data [37].

2.2.6 Duplicate Checking

The methods discussed here assume independence of studies and samples. This assumption can be violated by re-use of clinical tissue specimens by a research group in subsequent studies, or sharing of specimens between different research groups and in consortial studies. We developed the *doppelgangR* R package (https://github.com/lwaldron/doppelgangr) to facilitate identification of duplicates from gene expression profiles. Duplicates may also be identified by patient identifiers and inspection of published papers, and subsequent papers by the same research group deserve extra attention to duplicate-checking.

2.2.7 Gene Pre-filtering

Once the data for all studies has been preprocessed so that features (probe sets, genes, pathway activation scores etc.) are comparable across studies, it is further advisable to investigate whether the features indeed measure the same biological signal, especially when data was obtained from different platforms. The integrative correlation technique proposed by Parmigiani et al. [38] can be used to select "reproducible" genes. The basic idea behind this approach is that genes should be co-expressed with the same set of other genes across platforms and studies. Therefore, the correlation in expression of a given gene *G* is calculated between *G* and every other gene in a study, i.e., to identify the "neighborhood" of *G*. If this "neighborhood" is very different across datasets, the average correlation of correlation profiles across all pair of studies would be low; only if the average correlation of correlations exceeds a certain threshold, gene *G* is thus called reproducible and is included in the meta-analysis.

3 Methods

Although several methods have been proposed for meta-analysis of gene expression microarrays, traditional methods developed for synthesis of clinical trials and epidemiological studies remain highly relevant and are the most well-understood and implemented. Several implementations are available in the R environment, but our favorite for maturity and documentation is *Metafor* [39]. We refer to discussion therein for references to alternative software packages. The classic methods of fixed and random-effects synthesis are simply be applied for each gene or feature, with the main challenge being repetition of the methods for thousands of rows of a gene expression matrix, and summary and interpretation of thousands of results.

3.1 Fixed Effects Meta-analysis

Here we summarize the methods described by DerSimonian and Laird [2] for synthesis of effect size estimates across studies. Although their paper is concerned with risk ratio or risk difference in case-control studies, the methods are applicable to the synthesis of other statistics or *effect sizes* as long as they are accompanied by a standard error for each study. These methods are parametric in that a distribution of effect sizes across studies is assumed: constant in the fixed effects model, or normal in the random effects model. Let θ_i be a per-gene estimate of interest that is assumed to normally distributed, such as log fold-change for differential expression from a Limma analysis [40] or log hazard ratio in a univariate Cox proportional hazards model, where i indexes each independent study, $i = 1, \ldots, K$. It can represent the coefficient of a gene in a generalized linear model with non-normally distributed residuals and linear or non-linear link function, or a simple differential expression analysis. The coefficient can be corrected for clinical covariates in a multivariate regression model. What matters is that the method produces the estimates of interest, and accompanying standard errors, for each genomic feature in each study.

The fixed-effects model is developed under the assumption of one true effect size, with differences between studies attributed to individual-level sampling variation, i.e.,:

$$\theta_i = \theta_F \quad (1)$$

where θ_F is the common true effect size. Under a fixed-effects model, the synthesized estimate, $\hat{\theta}_F$, is a weighted average of estimates from each study:

$$\hat{\theta}_F = \frac{\sum_{i=1}^{K} w_i \hat{\theta}_i}{\sum_{i=1}^{K} w_i} \quad (2)$$

These weights are commonly taken to be the inverse squared standard error of the effect estimates from each study:

$$w_i = \frac{1}{\sigma i^2} \quad (3)$$

The standard error of the fixed-effects estimate is the inverse mean of the study-specific weights:

$$S.E.(\hat{\theta}_F) = \sqrt{\frac{1}{\sum_{i=1}^{K} w_i}} \quad (4)$$

Equations 2–4 are sufficient to calculate a fixed-effects meta-analytical estimate of log fold-change, log hazard ratio, etc. We note that some software and R functions produce confidence intervals rather than standard errors; these are converted to standard error assuming a normal sampling distribution, for example to convert a 95 % interval to Standard Error:

$$S.E. = \frac{C.I^{95\%}_{\text{upper}} - C.I.^{95\%}_{\text{lower}}}{2 \times 1.96} \tag{5}$$

Although this calculation can be performed by meta-analysis packages such as *metafor*, we present it to highlight the simplicity of the model used and to distinguish it from the random-effects model.

3.2 Random Effects Meta-analysis

Under the random-effects model, the synthesized estimate is given by Eq. 2 but with altered weights:

$$w_i^* = 1/\left(\hat{\tau}^2 + \hat{\sigma}_i^2\right) \tag{6}$$

where τ^2 is a heterogeneity parameter and standard output of meta-analysis software, $\hat{\tau}^2$ is estimated from the data, and $\widehat{\sigma_i^2}$ are the variances of each study's effect size estimate $\hat{\theta}_i$. As $\tau^2 \to 0$ (no heterogeneity) the random-effects solution converges to the fixed-effects solutions, and that as $\tau^2 \to \infty$ (very large heterogeneity) the random-effects solution is a simple average of the per-study effects, regardless of sample sizes or standard errors of each study. The null hypothesis of no heterogeneity $\left(\tau^2 = 0\right)$ can be tested by the *Cochrane's Q statistic*, which is just the summed product of study weights by squared residuals of per-study effect sizes from the fixed-effects estimate:

$$Q = \sum_{i=1}^{K} w_i \left(\hat{\theta}_i - \hat{\theta}_F\right)^2 \tag{7}$$

The Q statistic is $\chi^2{}_{K-1}$ distributed under the null hypothesis of no true between-study heterogeneity in effect sizes (e.g., as in Eq. 1), and is a standard test of heterogeneity. τ^2 can be understood as an estimate of total amount of heterogeneity present, in the same units as variance of the effect size. Another commonly reported measure of heterogeneity is I^2:

$$I^2 = \frac{\hat{\tau}^2}{\hat{\tau}^s + \hat{\sigma}^2_{\text{typical}}}, \tag{8}$$

In performing meta-analysis for thousands of gene expression features, these estimates are produced for *each* feature. A global picture of heterogeneity can be developed from histograms of I^2, τ^2, and Cochrane's Q-test p-values. Some features will exhibit more heterogeneity than others, and one can consider whether heterogeneity is likely technical or biological. For example, features can be ranked by evidence of heterogeneity, and gene set analysis can be performed on the basis of this ranking.

3.3 Fixed vs. Random Effects Meta-analysis

It is important to understand heterogeneity in the data. In the context of gene expression meta-analysis, a fixed-effects meta-analysis will identify the genes with strongest effect in the studies used as training data, while a random-effects model attempts to identify the genes with strongest average effect in a hypothetical population of studies. While the latter is theoretically preferable, it might not be optimal if, for example a source of heterogeneity is not expected to occur again in future data. For example, different microarray platforms might measure particular transcripts with different accuracy. If heterogeneity is due to technical issues in a single study or platform, then a random-effects analysis might remove genes with strong effect from a ranked list, because the random-effects model may unnecessarily down-weigh a useful predictor as a result of a problematic study or platform. In this context it is difficult to a priori decide whether fixed- or random-effects will work better. A sound approach is to try both, compare them, try to identify and understand the sources of heterogeneity, and choose the simpler approach if results are very similar.

3.4 Rank-Based Meta-analysis

The rank products method [41] was developed in the early days of microarray data analysis for identifying differentially expressed genes. Datasets in this era were typically small and noisy, which made a method free of distributional assumptions particularly useful. The method was soon extended for meta-analyses [42], and became a popular choice for microarray meta-analyses, mainly because of its simplicity, shown robustness [43], and for its more straightforward support for expression direction ("up" vs. "down"), compared to other simple and then commonly used methods such as synthesizing *p*-values via Fisher's or Stouffer's method. Since rank products weighs datasets by sample size, the results are in general expected to be more similar to a fixed-effects than to a random-effects meta-analysis [17]. A major practical disadvantage of the method, the computational cost of the permutation tests required to estimate the rank product statistic, was recently mitigated by the implementation of a fast approximation [44]. Rank products can give very different results compared to marginal tests for example when genes are highly correlated, clustered in so-called gene modules, since it breaks ties randomly. This will add random noise to the ranks of large gene modules, thus lowering their significance. This can be an advantage when final lists of differentially expressed genes comprise only highly correlated genes after adjusting for multiple testing.

3.5 Other Approaches to Synthesis

Several other methods have been frequently used in the literature, and we refer to the excellent review by Tseng et al. [45] for a comprehensive overview of those. These methods include "vote counting" approaches, probabilistic combining of *p*-values from different studies, and merging studies during data pre-processing.

Vote counting approaches, where genes are considered significant or validated when they reach a p-value cutoff in a minimum number of datasets, are statistically inefficient but popular due to their simplicity. Combining p-values requires only minimal information about studies and is frequently used for example when estimating standard errors of effect sizes is not possible. For an overview and recent improvements in this class of methods, see [46]. Direct merging of datasets will create batch effects, and any imbalance of the outcome of interest between datasets will result in confounding with these batch effects. This method should only be used when transcriptome differences between datasets is be expected to be larger than batch effects, and the meta-analysis approach described above is infeasible due to extreme imbalance between datasets, and confounding is unavoidable. When broad transcriptome modeling occurs between cases and controls, such as between oral squamous cell carcinoma and normal tissue [47], such merging of datasets can be acceptable even when datasets are unbalanced in case/control prevalence.

3.6 Case Study

In this section we work through a case study in identifying differentially expressed genes and assessing the extent and potential sources of heterogeneity. We focus on identifying genes associated with overall survival in ovarian cancer, using the curatedOvarianData Bioconductor package [16]. Full code and output for performing the analysis in this section are available from http://lwaldron.github.io/GeneExpressionMetaAnalysis/.

Although our original meta-analysis using the curatedOvarianData database was limited to late-stage, high-grade, serous ovarian cancer, here we include all patients for which overall survival is known, to demonstrate investigation of sources of heterogeneity. This analysis follows the basic steps:

1. Scale all genes to unit variance. In analyses where large numbers of samples are removed from a study, it is preferable to do scaling with the full number of samples. Scaling is critical when studies use different microarray platforms, so that coefficients have the same scale when synthesizing across studies.
2. Apply inclusion and exclusion criteria: microarray studies of primary tumors only, studies with at least 40 patients and 15 deaths, including only patients where censored overall survival is known.
3. Remove duplicate samples and studies that are subsets of another study. **Steps 2** and **3** reduce the database from 30 studies and 4411 samples to 15 studies and 2271 samples. With a curated database, these steps are performed automatically and can include Regular Expression filters on patient meta-data.

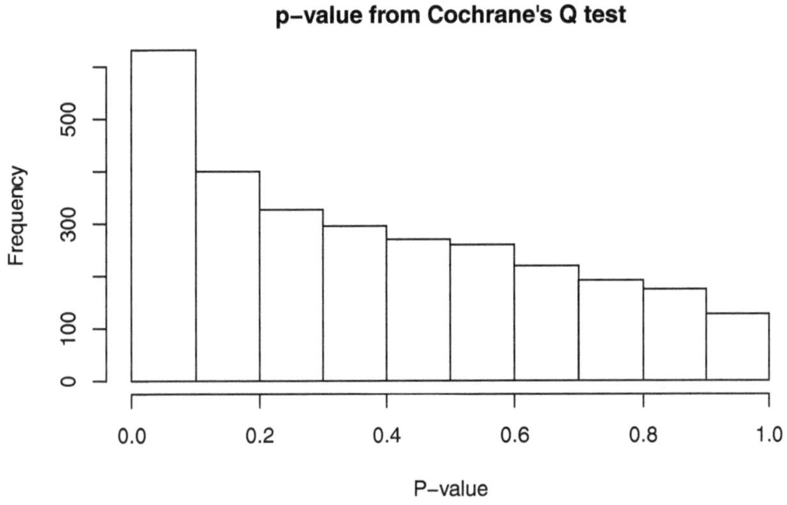

Fig. 1 Assessing heterogeneity of all genes by Cochrane's Q-test. The test was performed for all genes that present in every study, with *p*-values obtained from output of the rma.uni() function from the metafor library. In the absence of any heterogeneity a uniform distribution of p-values between 0 and 1 would be observed; this histogram indicates evidence of heterogeneity since small *p*-values are more frequent than larger *p*-values. This does not however imply that the magnitude of the heterogeneity is large, or that differentially-expressed genes identified are invalid

4. Restrict analysis to genes available on every platform. This is not necessary, but was a convenience for this analysis.
5. Perform meta-analysis for each gene, using both a random-effects and a fixed-effects model. In this example we fit a univariate Cox Proportional Hazards model and use the coefficient of the continuous gene expression variable for synthesis.

At this stage we can assess the presence of heterogeneity by plotting a histogram of *p*-values from Cochrane's Q-test (Fig. 1) for each gene. In the absence of any heterogeneity, these *p*-values would be uniformly distributed between 0 and 1. However, we observe that many genes exhibit some heterogeneity across studies in association with overall survival. We then identify NUAK1 as the gene with strongest evidence of prognostic association with overall survival (Fig. 2). This gene exhibits no evidence of heterogeneity, with a non-significant Cochrane's Q test and identical synthesized log hazard ratios by fixed or random-effects meta-analysis. We identify *KALRN* as the gene with greatest evidence for heterogeneity in log hazard ratio (Fig. 3), with the proportion of patients whose tumors were suboptimally debulked being one likely source of this heterogeneity (Fig. 4). Overall, although some global evidence of heterogeneity exists, its impact on the synthesized estimate even of the gene exhibiting greatest heterogeneity (*KALRN*) is not large, with the random-effects estimate differing from the fixed-effects mainly in having a somewhat larger confidence

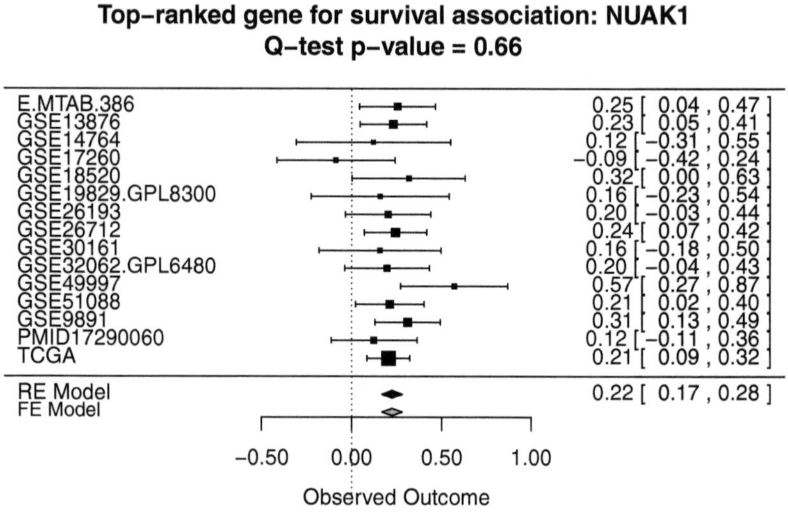

Fig. 2 Forest plot for the gene with the strongest evidence of association with overall survival in ovarian cancer. This gene, *NUAK1*, is the top-ranked gene according to synthesized *p*-values by both fixed and random-effects meta-analysis; in fact, it shows no evidence of heterogeneity and the synthesized estimates. *Fifteen rows with study names on the left* show the log Hazard Ratio of a Cox Proportional Hazards model for each study, with the point estimate and 95 % confidence interval, with box sizes proportional to the inverse square of the standard error of each study (or roughly to the number of deaths). *The right-hand column* provides the numeric point estimate and confidence interval. *Two diamonds* at the bottom show the point estimate and 95 % confidence interval of the synthesized log Hazard Ratio by fixed and random-effects estimate, which are identical. In other words, the study-to-study variation seen here is consistent with homogeneous studies and sampling variation only at the level of individual patients

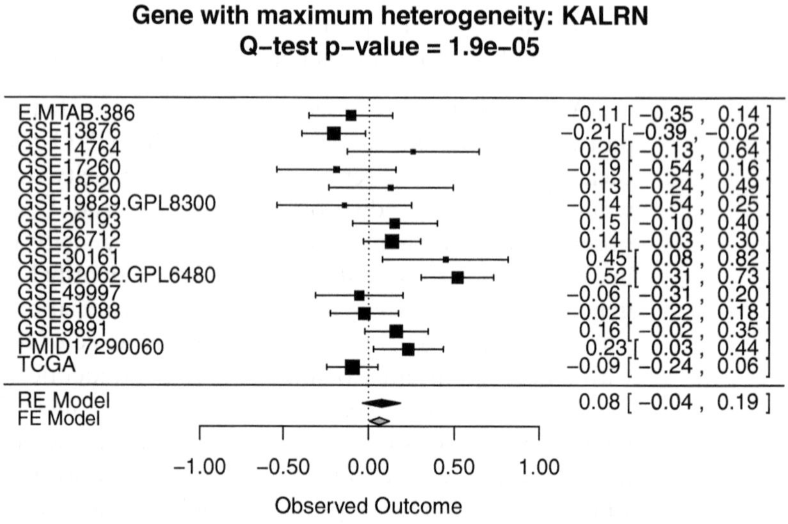

Fig. 3 Forest plot for *KALRN*, which demonstrates the strongest evidence of heterogeneity between studies. The *p*-value from Cochrane's Q-test is marginally significant after Bonferroni correction (FWER = 0.06). Heterogeneity is apparent as studies with both significantly positive and negative Cox coefficients are observed. Note however that the synthesized point estimates are similar by fixed and random effects, but the standard error and confidence interval are smaller in the fixed-effects model

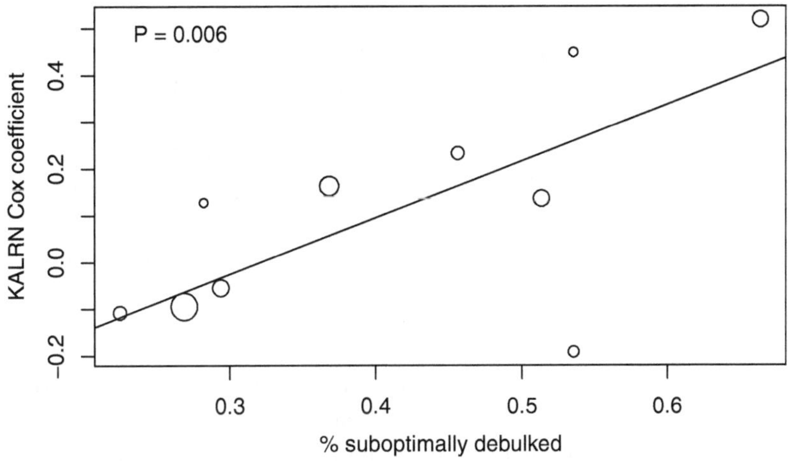

Fig. 4 Association between log Hazard Ratio and percentage of suboptimally debulked patients. Six covariates with prognostic relevance were considered as potential sources of heterogeneity between studies: suboptimal debulking of tumors, tumor histology, grade (high/low), stage (early/late), and age (greater or less than 70 years). For each of these covariates, a linear regression was performed between per-study covariate prevalence and per-study Cox coefficient (log Hazard Ratio), with points weighted by study size. Only the prevalence of suboptimal debulking was significantly associated with Cox coefficient ($P = 0.006$). The area of data points is proportional to study sample size and weighting in the linear regression. Higher proportions of suboptimally debulked patients are associated with greater association between *KARLN* expression and survival; in other words, KARLN is more strongly associated with bad prognosis in studies with more suboptimally debulked patients

interval. This is consistent with recent independent analysis using newly proposed methods for identifying homogeneous and heterogeneous variables in pooled cohort studies that found none of 15 genes with known prognostic significance exhibited significant heterogeneity in prognostic association in the curated Ovarian Data database [48].

3.7 Extensions to Predictive Modeling

The most likely objectives of meta-analysis are to identify candidate differentially expressed genes, or to develop and validate a predictive model. The former objective has been reviewed also by Ramasamy et al. [49]; the latter is a more recent application of genomic meta-analysis that has not to our knowledge been reviewed. Bernau et al. [50] recently proposed *leave-one-dataset-in cross-study validation* for validating of prediction models and comparing prediction algorithms using a collection of independent studies. In this approach, each dataset is used in turn for model training, and all other datasets for validation. The resulting matrix of independent validation statistics is analyzed for evidence of outlying studies, and for estimation of cross-study validation accuracy. Although this method is very useful for comparing algorithms and identifying outlying studies, a more accurate model can be developed using all available studies for training. Riester et al. [17] proposed a related leave-one-dataset-*out* cross-validation, whereby each study is used in turn for validation and the remaining $n - 1$ studies are used for training. Log fold-change or Cox regression coefficients are synthesized across the training studies by meta-analysis as described in this chapter, and synthesized coefficients are directly used as the coefficients of a linear prediction score.

4 Notes

Meta-analysis of genomic datasets, even using basic statistical methods adapted from unrelated fields, is a powerful tool for overcoming dimensionality and batch effects to develop robust biomarkers and prediction rules. Specialized statistical methodologies are still needed, for example, for aggregating evidence of heterogeneity across all gene expression measurements and appropriately re-weighting studies by their concordance with other studies, and for missing data imputation that leverages independent datasets. The most significant barriers to successful use of genomic meta-analysis are public availability of data and annotations, and the tedious work of standardizing datasets from disparate sources to enable straightforward analysis of individual patient data. As RNA sequencing catches up with and potentially overtakes microarray technology as the dominant method of transcriptome profiling, the necessity of patient privacy presents new challenges for the open sharing of data that makes meta-analysis possible. We hope that the methods outlined in this chapter, and the successes of recent gene expression meta-analyses, will help provide the necessary motivation for researchers, journal editors, reviewers, and funding agencies to continue the tradition of open data sharing that was developed for the microarray.

References

1. Moher D, Liberati A, Tetzlaff J et al (2010) Preferred reporting items for systematic reviews and meta-analyses: the PRISMA statement. Int J Surg 8:336–341
2. DerSimonian R, Laird N (1986) Meta-analysis in clinical trials. Control Clin Trials 7:177–188
3. Moher D, Olkin I (1995) Meta-analysis of randomized controlled trials: a concern for standards. JAMA 274:1962–1964
4. Lipsey MW, Wilson DB (2001) Practical meta-analysis. Sage, Thousand Oaks, CA
5. Borenstein M, Hedges LV, Higgins JPT, Rothstein HR (2011) Introduction to meta-analysis. John Wiley, New York, NY
6. Culhane AC, Schröder MS, Sultana R et al (2011) GeneSigDB: a manually curated database and resource for analysis of gene expression signatures. Nucleic Acids Res 40: D1060–D1066
7. Edgar R, Domrachev M, Lash AE (2002) Gene Expression Omnibus: NCBI gene expression and hybridization array data repository. Nucleic Acids Res 30:207–210
8. Kolesnikov N, Hastings E, Keays M et al (2015) ArrayExpress update-simplifying data submissions. Nucleic Acids Res 43: D1113–D1116
9. Taminau J, Steenhoff D, Coletta A et al (2011) inSilicoDb: an R/Bioconductor package for accessing human Affymetrix expert-curated datasets from GEO. Bioinformatics 27:3204–3205

10. Rhodes DR, Kalyana-Sundaram S, Mahavisno V et al (2007) Oncomine 3.0: genes, pathways, and networks in a collection of 18,000 cancer gene expression profiles. Neoplasia 9: 166–180
11. Zeeberg BR, Riss J, Kane DW et al (2004) Mistaken identifiers: gene name errors can be introduced inadvertently when using Excel in bioinformatics. BMC Bioinformatics 5:80
12. Gentleman RC, Carey VJ, Bates DM et al (2004) Bioconductor: open software development for computational biology and bioinformatics. Genome Biol 5:R80
13. Haider S, Ballester B, Smedley D et al (2009) BioMart Central Portal—unified access to biological data. Nucleic Acids Res 37: W23–W27
14. Zhu Y, Davis S, Stephens R et al (2008) GEOmetadb: powerful alternative search engine for the Gene Expression Omnibus. Bioinformatics 24:2798–2800
15. Davis S, Meltzer PS (2007) GEOquery: a bridge between the Gene Expression Omnibus (GEO) and BioConductor. Bioinformatics 23:1846–1847
16. Ganzfried BF, Riester M, Haibe-Kains B et al. (2013) curatedOvarianData: clinically annotated data for the ovarian cancer transcriptome. Database 2013: bat013
17. Riester M, Wei W, Waldron L et al (2014) Risk prediction for late-stage ovarian cancer by meta-analysis of 1525 patient samples. J Natl Cancer Inst. doi:10.1093/jnci/dju048
18. Waldron L, Haibe-Kains B, Culhane AC et al (2014) Comparative meta-analysis of prognostic gene signatures for late-stage ovarian cancer. J Natl Cancer Inst. doi:10.1093/jnci/dju049
19. Irizarry RA, Hobbs B, Collin F et al (2003) Exploration, normalization, and summaries of high density oligonucleotide array probe level data. Biostatistics 4:249–264
20. McCall MN, Bolstad BM, Irizarry RA (2010) Frozen robust multiarray analysis (fRMA). Biostatistics 11:242–253
21. Leek JT, Scharpf RB, Bravo HC et al (2010) Tackling the widespread and critical impact of batch effects in high-throughput data. Nat Rev Genet 11:733–739
22. Johnson WE, Li C, Rabinovic A (2006) Adjusting batch effects in microarray expression data using empirical Bayes methods. Biostatistics 8:118–127
23. Miller JA, Cai C, Langfelder P et al (2011) Strategies for aggregating gene expression data: the collapse Rows R function. BMC Bioinformatics 12:322
24. Li Q, Birkbak NJ, Gyorffy B et al (2011) Jetset: selecting the optimal microarray probe set to represent a gene. BMC Bioinformatics 12:474
25. Dai M, Wang P, Boyd AD et al (2005) Evolving gene/transcript definitions significantly alter the interpretation of GeneChip data. Nucleic Acids Res 33:e175
26. Subramanian A, Tamayo P, Mootha VK et al (2005) Gene set enrichment analysis: a knowledge-based approach for interpreting genome-wide expression profiles. Proc Natl Acad Sci U S A 102:15545–15550
27. Huang DW, Sherman BT, Tan Q et al (2007) The DAVID Gene Functional Classification Tool: a novel biological module-centric algorithm to functionally analyze large gene lists. Genome Biol 8:R183
28. Altschuler GM, Hofmann O, Kalatskaya I et al (2013) Pathprinting: an integrative approach to understand the functional basis of disease. Genome Med 5:68
29. Kanehisa M, Goto S (2000) KEGG: Kyoto encyclopedia of genes and genomes. Nucleic Acids Res 28:27–30
30. Croft D, Mundo AF, Haw R et al (2014) The Reactome pathway knowledgebase. Nucleic Acids Res 42:D472–D477
31. Milacic M, Haw R, Rothfels K et al (2012) Annotating cancer variants and anti-cancer therapeutics in reactome. Cancers (Basel) 4: 1180–1211
32. Gene Ontology Consortium (2004) The Gene Ontology (GO) database and informatics resource. Nucleic Acids Res 32:D258–D261
33. Hänzelmann S, Castelo R, Guinney J (2013) GSVA: gene set variation analysis for microarray and RNA-seq data. BMC Bioinformatics 14:7
34. Tarca AL, Bhatti G, Romero R (2013) A comparison of gene set analysis methods in terms of sensitivity, prioritization and specificity. PLoS One 8:e79217
35. Barbie DA, Tamayo P, Boehm JS et al (2009) Systematic RNA interference reveals that oncogenic KRAS-driven cancers require TBK1. Nature 462:108–112
36. Verhaak RGW, Tamayo P, Yang J-Y et al (2013) Prognostically relevant gene signatures of high-grade serous ovarian carcinoma. J Clin Invest 123:517–525
37. Ozawa T, Riester M, Cheng Y-K et al (2014) Most human non-GCIMP glioblastoma subtypes evolve from a common proneural-like precursor glioma. Cancer Cell 26:288–300
38. Parmigiani G, Garrett-Mayer ES, Anbazhagan R, Gabrielson E (2004) A cross-study comparison of gene expression studies for the

molecular classification of lung cancer. Clin Cancer Res 10:2922–2927

39. Viechtbauer W (2010) Conducting meta-analyses in R with the metafor package. J Stat Softw 36(3):1–48

40. Smyth GK (2004) Linear models and empirical Bayes methods for assessing differential expression in microarray experiments. Stat Appl Genet Mol Biol 3: Article 3

41. Breitling R, Armengaud P, Amtmann A, Herzyk P (2004) Rank products: a simple, yet powerful, new method to detect differentially regulated genes in replicated microarray experiments. FEBS Lett 573:83–92

42. Hong F, Breitling R, McEntee CW et al (2006) RankProd: a bioconductor package for detecting differentially expressed genes in meta-analysis. Bioinformatics 22:2825–2827

43. Hong F, Breitling R (2008) A comparison of meta-analysis methods for detecting differentially expressed genes in microarray experiments. Bioinformatics 24:374–382

44. Heskes T, Eisinga R, Breitling R (2014) A fast algorithm for determining bounds and accurate approximate p -values of the rank product statistic for replicate experiments. BMC Bioinformatics 15:367

45. Tseng GC, Ghosh D, Feingold E (2012) Comprehensive literature review and statistical considerations for microarray meta-analysis. Nucleic Acids Res 40:3785–3799

46. Li Y, Ghosh D (2014) Meta-analysis based on weighted ordered P-values for genomic data with heterogeneity. BMC Bioinformatics 15:226

47. Reis PP, Waldron L, Perez-Ordonez B et al (2011) A gene signature in histologically normal surgical margins is predictive of oral carcinoma recurrence. BMC Cancer 11:437

48. Cheng X, Lu W, Liu M (2015) Identification of homogeneous and heterogeneous variables in pooled cohort studies. Biometrics. doi:10.1111/biom.12285

49. Ramasamy A, Mondry A, Holmes CC, Altman DG (2008) Key issues in conducting a meta-analysis of gene expression microarray datasets. PLoS Med 5:e184

50. Bernau C, Riester M, Boulesteix A-L et al (2014) Cross-study validation for the assessment of prediction algorithms. Bioinformatics 30:i105–i112

Chapter 9

Practical Analysis of Genome Contact Interaction Experiments

Mark A. Carty and Olivier Elemento

Abstract

The three dimensional (3D) architecture of chromosomes is not random but instead tightly organized due to chromatin folding and chromatin interactions between genomically distant loci. By bringing genomically distant functional elements such as enhancers and promoters into close proximity, these interactions play a key role in regulating gene expression. Some of these interactions are dynamic, that is, they differ between cell types, conditions and can be induced by specific stimuli or differentiation events. Other interactions are more structural and stable, that is they are constitutionally present across several cell types. Genome contact interactions can occur via recruitment and physical interaction between chromatin-binding proteins and correlate with epigenetic marks such as histone modifications. Absence of a contact can occur due to presence of insulators, that is, chromatin-bound complexes that physically separate genomic loci. Understanding which contacts occur or do not occur in a given cell type is important since it can help explain how genes are regulated and which functional elements are involved in such regulation. The analysis of genome contact interactions has been greatly facilitated by the relatively recent development of chromosome conformation capture (3C). In an even more recent development, 3C was combined with next generation sequencing and led to Hi-C, a technique that in theory queries all possible pairwise interactions both within the same chromosome (intra) and between chromosomes (inter). Hi-C has now been used to study genome contact interactions in several human and mouse cell types as well as in animal models such as Drosophila and yeast. While it is fair to say that Hi-C has revolutionized the study of chromatin interactions, the computational analysis of Hi-C data is extremely challenging due to the presence of biases, artifacts, random polymer ligation and the huge number of potential pairwise interactions. In this chapter, we outline a strategy for analysis of genome contact experiments based on Hi-C using R and Bioconductor.

Key words Chromatin interactions, DNA looping, Hi-C, Enhancers, Promoters, Gene regulation

1 Introduction

Microscopic approaches have been developed to study the organization of genomes at a high resolution. Fluorescence in situ hybridization (FISH) was developed in the early 1980s to detect and localize the presence or absence of specific DNA sequences on chromosomes. Immunofluorescence with fluorescent antibodies in combination with RNA and DNA FISH has been used to study

Ewy Mathé and Sean Davis (eds.), *Statistical Genomics: Methods and Protocols*, Methods in Molecular Biology, vol. 1418, DOI 10.1007/978-1-4939-3578-9_9, © Springer Science+Business Media New York 2016

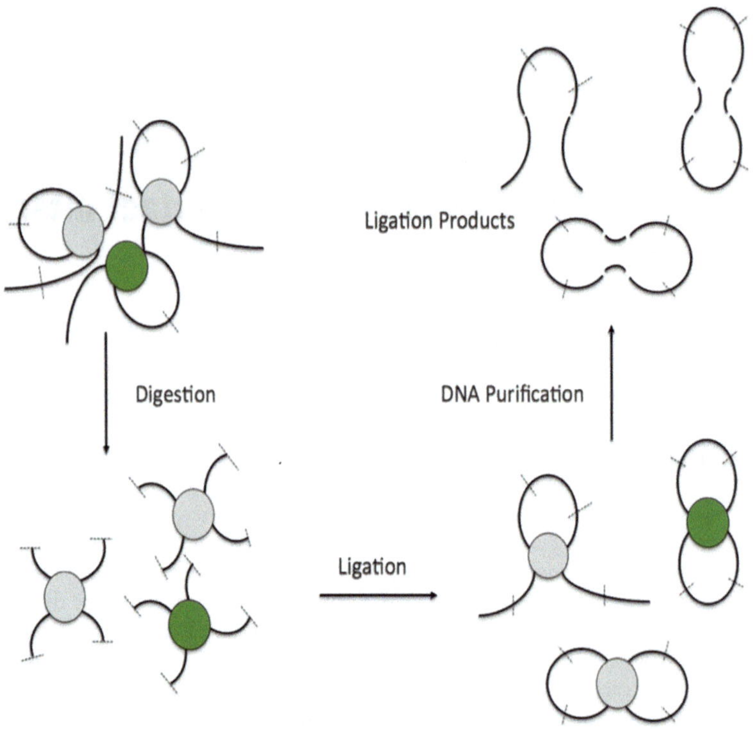

Fig. 1 All 3-C based protocols that start with treatment of cells with formaldehyde, leading to cross-linking of DNA segments in close proximity to one another and their eventual ligation

co-localization of loci. This technique has revealed that genomes are organized into specific chromosome territories in the nucleus [1]. However, FISH is limited to probing a few loci at a time, offers limited spatial resolution and can only detect distal interactions, typically involving genomic loci that are several megabases away on a chromosome or on distinct chromosomes. In the past decade, chromosome conformation capture (3C) technology was developed to interrogate the spatial structure of chromosomes in greater detail. 3C methods are based on cross-linking chromatin with formaldehyde, which provides us with a snapshot of physical interactions between pairs of loci. DNA is digested using restriction enzyme, and then ligated so that DNA fragments that are close in 3D are converted to a ligation product (Fig. 1), which usually is detected by PCR with locus-specific primers [2]. 3C methods have been adapted to use next-generation sequencing to enable the detection of chromatin interactions. Circular chromosome conformation capture with sequencing (4C-seq) is a variant of 3C technology that detects all chromatin interactions involving a specific locus. We refer to locus of interest as the "viewpoint" or "bait," while the other partner genomic regions are called "captures" [3].

In the 4C protocol, cells are treated with formaldehyde to crosslink DNA to DNA-associated proteins. Restriction fragments are produced after a restriction enzyme digest. The blunt ends of the fragments of the corresponding interacting loci are ligated to produce a unique ligation product. The ligation product is then heated, which results in the reversal of the cross-links. There is a second restriction digest, this time using a more frequent cutter that will produce smaller fragments. Circularization of DNA fragments follows easily since there are no bound proteins. Circular complex is PCR amplified, and then deep sequenced to produce a library of contacts between the viewpoint and all its captures. The 5C technique is an expansion of 3C and allows for the detection of interactions between selected loci. The ligation products are mixed with special primers that are designed to anneal at the very ends of restriction fragments, ones facing outward and the others inward, so that the end of each primer covers exactly a half of a restriction site [4]. In this way, the outward and inward primers abut tail-to-head across the ligation products. Hi-C is another 3C variant that allows for all-versus-all chromatin interaction analysis in cell populations. In Hi-C, cells are fixed with formaldehyde and then digested using a restriction enzyme. The fragment ends are filled in with biotinylated nucleotide and ligated. DNA is sheared using sonication and DNA fragments containing biotin-labeled ligation junctions are pulled down using streptavidin-coated beads. The fragments are sequenced to produce a library of chromatin contacts [5]. Tethered chromosome conformation capture is a variant of the Hi-C method, where the key reactions are carried out in solid phase instead of solution. This strategy increases the signal-to-noise ratio and enables stronger detection of chromatin interactions [6]. DNase Hi-C is another hybrid that uses DNase I for chromatin fragmentation instead of a restriction enzyme. This methodology overcomes the limitation of conventional Hi-C by increasing resolution. In situ Hi-C differs from conventional Hi-C by using a more frequent cutter (MboI), and in this modified approach, DNA remains in the nucleus during ligation, rather than being released into solution [7]. This approach reduces the potential for accidental binding of DNA fragments to other fragments that were not actually neighbors in the nucleus. This chapter primarily describes the analysis of Hi-C datasets. However the analyses introduced here are applicable to other 3C methods including 4C and 5C.

2 Materials

Materials to be used for analysis of Hi-C primarily include the R statistical software together with specific Bioconductor packages. Additional software frequently used for next-generation sequencing analysis such Bowtie2 [8] or BWA [9] is needed.

These programs typically run on Unix via the command line and need to be installed either as binary or from source code. However, installing and running such programs are not covered in this chapter.

3 Methods

3.1 Getting Started with Genome Contact Interaction Data Analysis

Hi-C data is typically generated using Illumina paired-end sequencing and consists of hundreds of millions of paired-end reads (50–150 bp long). It is generally good practice to perform a quality which are stored in the form of Fastq files check on the short reads in Fastq before any further processing of the data. There are several tools for assessing the quality of a sequence dataset, e.g., detecting abnormal GC content, low quality scores or untrimmed adapters. Frequently used tools include FastQC and FASTX.

In a Hi-C experiment, a read pair is formed as a result of a ligation junction joining two distinct DNA fragments originating from the genome. Since such fragments are obtained from restriction enzyme digestion, e.g., with HindIII, they are called restriction fragments. To identify the original regions in the genome, paired-end reads are mapped back to the reference genome using aligners such as Bowtie or BWA. If the biotinylated ligation junction of a fragment is located close to the center of ligated fragments, then the read pair will map correctly to two different genomic regions. However, if the junction lies close to either ends of the fragment, it may give rise to a chimeric read that is a composite of two genomic regions. When this occurs, the chimeric read is likely to map incorrectly to the reference genome. Additional artifacts can be introduced into the Hi-C protocol during the ligation phase. A successful ligation product involves two restriction enzyme fragments on opposite sides of the stem of chromatin loop. However, a fragment may ligate with itself or fail to ligate. A distance filter should be imposed to remove read pairs that are a result of self-ligation and dangling fragment ends (Fig. 2). Because a Hi-C experiment involves genomic regions that are in close proximity to restriction enzyme cutting site, a non-informative read pair occurs when either read is distal to the nearest restriction cut site (Fig. 2). Non-informative read pairs are eliminated from analysis by simply checking that either mate of the di-tags is within 500 bps of restriction enzyme cut site.

In what follows, we outline a procedure for preprocessing Hi-C reads using the Bioconductor package diffHic. The procedure involves aligning each of two reads that form the chimeric read pair separately to the reference genome. The first step is **Read Splitting**: The ligation junction sequence, which is obtained after ligation of sticky ends resulting from the restriction digest, is identified in each chimeric read, and the read is split into two fragments

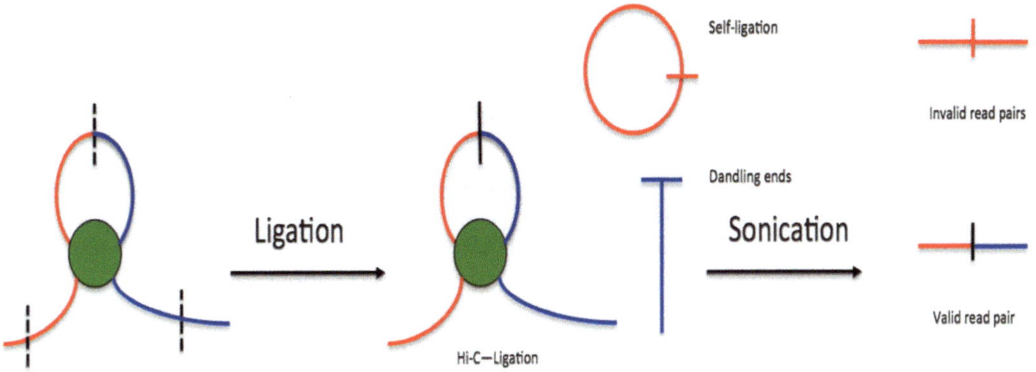

Fig. 2 Artifacts produced at different stages of the Hi-C protocol. Self-ligating fragments and dangling ends are artifacts from the ligation phase

at the center of the signature. For example, the HindIII enzyme cuts at AAGCTT leaving a 4 bp overhang. The ligation yields the signature AAGCTAGCTT. In the case of MboI that cuts at GATC, the resulting signature is GATCGATC. The next step is **Alignment**: Each of the split segments and unsplit segments are aligned separately to the corresponding reference genome using Bowtie2 [8]. The two files corresponding to each read of the read pairs are joined together in a single BAM file. Mapping by read splitting is performed using a Python script provided in the DiffHic package. This involves building a Bowtie2 index based on the reference genome.

```
bowtie2-build genome.fasta genome
```

where "genome.fasta" is the file with the reference genome sequence, eg. hg19, in fasta format. We then run the Python script on the two paired-end fastq files from the HiC data.

```
python presplit_map.py -G genome -1 reads_R1.fastq -
2 reads_R2.fastq -P 33 -sig "AAGCTAGCTT" -o output_bwt.bam
```

The script can be modified to use the BWA-mem aligner instead of Bowtie2. It is strongly recommended to mark duplicate read pairs in the resulting BAM file using the Picard software suite.

We first perform a diagnostic assessment in R using preparePairs function in the diffHic package. The preparePairs function matches the genomic position of each read to a restriction fragment in the reference genome, then inspects the number of chimeric tags, artifacts, and read pairs with low mapping quality.

```
> library(diffHic)
> library(BSgenome.Hsapiens.UCSC.hg19)
```

We first generate a restriction enzyme map using the cutGenome function in R.

```
> ms.frag <- cutGenome(BSgenome.Hsapiens.UCSC.hg19,
"GATC",4)
> chrom <- sprintf('chr%s',c(1:22,"M","X","Y"))
> idx <- unlist(sapply(chrom,function(x) which(seqnames
(ms.frag)==x)))
> ms.frag <- ms.frag[idx]
> ms.frag
GRanges object with 7127634 ranges and 0 metadata columns:
        seqnames          ranges strand
           <Rle>      <IRanges>  <Rle>
      [1]   chr1    [    1, 11163]      *
      [2]   chr1    [11160, 12414]      *
      [3]   chr1    [12411, 12464]      *
      [4]   chr1    [12461, 12689]      *
      [5]   chr1    [12686, 12832]      *
      ...    ...              ... ...
  [7127630]   chrY [59360039, 59360267]      *
  [7127631]   chrY [59360264, 59360317]      *
  [7127632]   chrY [59360314, 59360727]      *
  [7127633]   chrY [59360724, 59361569]      *
  [7127634]   chrY [59361566, 59373566]      *
  -------
  seqinfo: 93 sequences from hg19 genome

> ms.param <- pairParam(ms.frag)
> ms.param

Genome contains 7127634 restriction fragments across 25
chromosomes
No discard regions are specified
No limits on chromosomes for read extraction
No cap on the read pairs per pair of restriction fragments
```

We first then run the preparePairs function using the output of the Python script and the restriction enzyme map. Additional parameters include an output file for reads in HDF5 format, a parameter controlling PCR duplicate removal and filtering using a minimum mapping quality score to remove spurious alignments.

```
> diagnostic.bwt <- preparePairs(output_bwt.bam", ms.
param, file="output_bwt.h5",dedup=TRUE,minq=10)
> diagnostic_bwt
$pairs
   total  marked filtered   mapped
55817881       0  9399358 46418523

$same.id
  dangling self.circle
   1033143       24663

$singles
[1] 0
```

```
$chimeras
 total  mapped  multi invalid
21621256 17005090 11726950 4022451
```

As shown above, the preparePairs function reports the number of mapped reads as well as number of dangling self-circles and chimeras. The prunePairs function can then be used to remove read pairs that correspond to Hi-C artifacts.

3.2 Identifying Statistical Significant Hi-C Contacts Using a Model-Based Approach

Most statistical analyses assume that the majority of Hi-C contacts are the result of random inter-molecular ligations that result from random polymer diffusion and ligation. Under such conditions, the probability of a pair of restriction enzyme fragments are in physical contact with each other as a consequence of random polymer ligation can be estimated as:

$$p_{i,j} = 2 \times cvg_i \times cvg_j$$

where cvg_i is the relative coverage of a given genomic region i defined as

$$cvg_i = \frac{reads_i}{N}$$

where $reads_i$ is the total number of reads with a tag mapping to region i and N is the total of paired-end reads from the Hi-C experiment. Given N *and* $p_{i,j}$, the probability of observing $x_{i,j}$ reads or more between a pair of loci i and j can be estimated by

$$P(X \geq x_{i,j}) = 1 - \sum_{k=0}^{x_{i,j}-1} \binom{N}{k} p_{i,j}^{k} \left(1 - p_{i,j}\right)^{N-k}$$

Recall that our null hypothesis represents our belief that no interaction has occurred. A p-value is the probability of seeing a more extreme sample result given that the null hypothesis true. Suppose we set the significance level to 5 %, then a p-value below 5 % would suggest that the observed data is inconsistent with our assumption that the null hypothesis is true. We have stronger evidence that the observed result might be a real interaction.

We use here a dataset generated by Lieberman et al. and published in 2009 [10]. In this study, Hi-C experiments were conducted using a lymphoblastoid human cell line with two different restriction enzymes (HindIII and NcoI). We use the HiCDataLymphoblast Bioconductor package to obtain the data. We also use the GOTHiC Bioconductor package to assign a p-value to Hi-C interaction data.

```
> library(GOTHiC)
> library(HiCDataLymphoblast)
> dirPath = system.file("extdata", package="HiCData Lymphoblast")
> fileName1 = list.files(dirPath, full.names=TRUE)[1]
> fileName2 = list.files(dirPath, full.names=TRUE)[2]
> binom = GOTHiC(fileName1,fileName2, sampleName= "lymphoid",
  res=5e5, BSgenomeName="BSgenome.Hsapiens.UCSC.hg18",
  genome=BSgenome.Hsapiens.UCSC.hg18,
  restrictionSite= "A^AGCTT", enzyme="HindIII" ,cistrans="all", filterdist
                =10000, DUPLICATETHRESHOLD=1,
  fileType="Bowtie", parallel=FALSE, cores=NULL)
```

In the GOTHiC function, fileName1 is the file or files of the mapped reads of the left fragment ends, similarly fileName2 corresponds to file or files of the right fragment ends, res is the resolution of the contact map, restrictionSite is the enzyme's recognition site, cistrans indicates whether cis (intra) or trans (inter) or both interaction type should be considered, filterdist is a parameter for filtering out abnormally long fragments that are likely the result of incompletely digested DNA, and the parameter DUPLICATETHRESHOLD specifies the maximum duplication rate for PCR duplicates. The output of the GOTHiC function is a data.frame, where each row is an interaction and the columns represent location of the interaction, p-value, and adjusted p-value (calculated using the Benjamini–Hochberg method). A little extra work is needed to represent these interactions in a user-friendly way, as shown below for chr20 interactions.

```
> binom = as.data.table(binom)
> regionsI = GRanges('chr20',IRanges(start=binom$locus1, width = 5e5))
> regionsJ = GRanges('chr20',IRanges(start=binom$locus2, width = 5e5))
> binI = sprintf('%s-%d-%d','chr20',start(regionsI),end(regionsI))
> binJ = sprintf('%s-%d-%d','chr20',start(regionsJ),end(regionsJ))
> binsGR = c(regionsI,regionsJ)
> binsGR = binsGR[!duplicated(binsGR)]
> names(binsGR) = sprintf('%s-%d-%d',as.character(seqnames(binsGR)),
                        start(binsGR),end(binsGR))
> binom.modified = binom[,list(binI=binI,binJ=binJ,
                     counts=readCount,pvals=format(pvalue,scientific=TRUE),
                     adj.pvals=format(qvalue,scientific=TRUE))]
> binCenters = GRanges(seqnames='chr20',
                     IRanges(start=start(resize(binsGR[binom.modified$binI],
                     fix='center',width=1)), end=start(resize(binsGR[binom.
                     modified$binJ],
                     fix='center',width=1))))
> binom.modified = binom.modified[,list(binI=binI,binJ=binJ,counts=counts,
                        D = width(binCenters)-1,pvals=pvals,adj.pvals)]
>  binom.modified[1:10,list(binI=binI,binJ=binJ,counts=counts,D=D,adj.pvals
                     =adj.pvals)]
```

```
         binI              binJ counts          D  adj.pvals
 1: chr20-0-499999 chr20-500000-999999     12  500000 1.695880e-13
 2: chr20-0-499999 chr20-1000000-1499999    7 1000000 8.724342e-05
 3: chr20-0-499999 chr20-1500000-1999999    3 1500000 2.874491e-01
 4: chr20-0-499999 chr20-2000000-2499999    2 2000000 1.000000e+00
 5: chr20-0-499999 chr20-2500000-2999999    4 2500000 3.224721e-02
 6: chr20-0-499999 chr20-3500000-3999999    1 3500000 1.000000e+00
 7: chr20-0-499999 chr20-4000000-4499999    1 4000000 1.000000e+00
 8: chr20-0-499999 chr20-4500000-4999999    4 4500000 5.934340e-02
 9: chr20-0-499999 chr20-5000000-5499999    1 5000000 1.000000e+00
10: chr20-0-499999 chr20-5500000-5999999    2 5500000 1.000000e+00
```

To visualize the interaction, we use two methods: Sushi plots and contact maps. To implement such visualizations, we have written two R functions: contactMap and SushiPlot that utilizes the packages aqfig, data.table, GenomicRanges, Matrix, and Sushi to construct these two types of contact matrices. These packages are available at https://bitbucket.org/leslielab/hic-dc and must be downloaded and installed in order to use the two functions. Sushi plots shows loops between interacting loci where the height of loops indicates interaction significance (–log10 (*p*-value)), and degree of transparency is from high to low statistical significances. Here is the R code to create a Sushi plot for a region of chr20:

```
> source('ContactMap.R')
> source('SushiPlot.R')
> binom.modified$pvals <- as.numeric(binom.modified
$pvals)
> binom.modified$adj.pvals <- as.numeric(binom.modified
$adj.pvals)
> chromstart = 0; chromend = 6.2e7
> SushiPlot(binom.modified, chromstart, chromend, binsGR)
```

The resulting Sushi plot is shown in Fig. 3. Contact maps either show raw read counts or nominal p-values after -log10 transformation. To plot contact maps for the chr20 region used above, use the following code.

```
> contactMap(binom.modified,chromstart,chromend,binsGR)
```

The resulting contact maps are shown in Fig. 4a, b.

3.3 Detecting Genome Contact Interaction Differences between Two Conditions

To identify interaction differences between two conditions, we must first summarize a Hi-C library into counts for each interaction, then normalize counts for each interaction between samples using scaling methods like Trimmed Mean of M-values (TMM) normalization [11], and finally call for differential contact frequencies using the statistical method edgeR adapted from RNA-Seq analysis [12]. We compare here contact frequencies for EBV-transformed lymphoblastoid cells to K562 erythroleukemia cells from the Lieberman et al., 2009 Hi-C dataset [10]. We assume

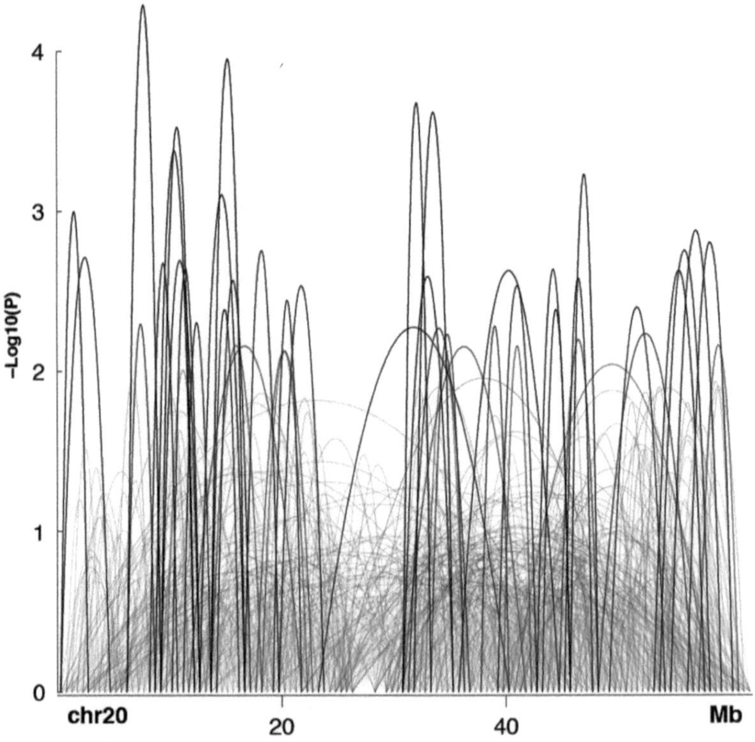

Fig. 3 A Sushi plot of all Hi-C contacts bins from 1 bp to 62 Mb on chr20. The *height of loops* indicates −log10 (*p*-value), and degree of transparency is from high to low statistical significances

here that the corresponding files were processed using the DiffHiC package and available in the working directory as GM1_lieberman.h5, GM2_lieberman.h5, K5621_lieberman.h5, and K5622_lieberman.h5. Counting of read pairs between bin pairs is performed for multiple libraries using the squareCounts function in DiffHiC. The latter function uses a pairParam object as input, typically from a restriction enzyme site map.

```
> library(edgeR)
> require(diffHic)
> input <- list.files(".", pattern = "lieberman.h5")
> input <- sprintf(".../data/%s",input)
> bin.size <- 1e6
> data <- squareCounts(input, ms.param, bin.size, filter = 1)
> group <- factor(c(rep("GM",2),rep("K562",2)))
> d <- asDGEList(data,group=group)
```

Only interactions with minimum read counts are used further.

```
> keep <- rowSums(cpm(d) > 5) >= 2
> d <- d[keep,keep.lib.sizes=FALSE]
> new.data <- data[keep,]
> dim(d)
[1] 15897 4
```

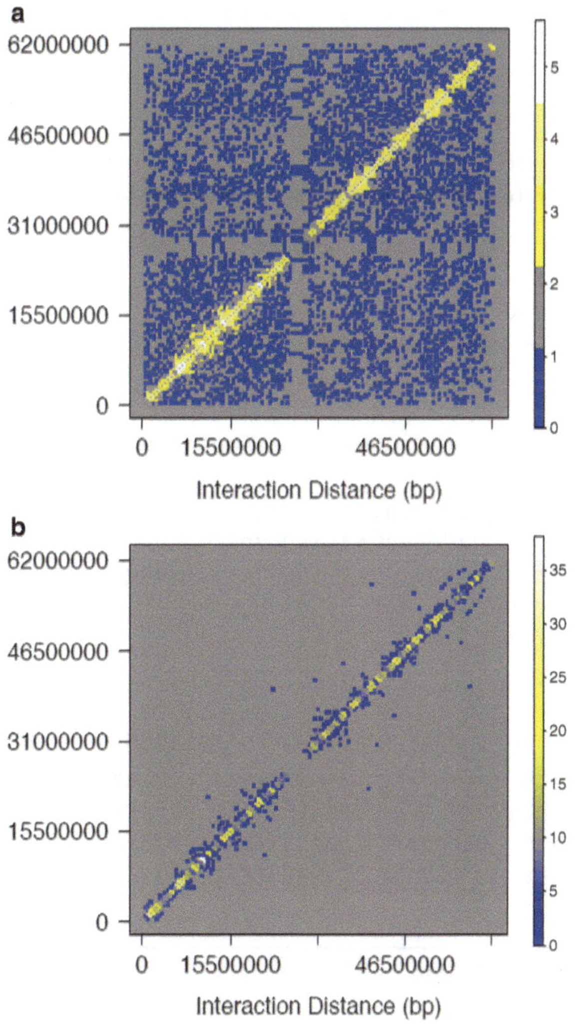

Fig. 4 (**a**) raw contact map of Hi-C (**b**) contact map with nominal p-values produced by GOTHIC. Contact matrices of all Hi-C contacts bins from 1 bp to 62 Mb on chr20. The contact matrix at the top is a matrix of raw Hi-C bin counts, and the contact matrix below is –log10(*p*-values) of Hi-C contact bins

A Generalized Linear Model is then used to estimate significance of counts between replicate pairs.

```
> design <- model.matrix(~group)
> d <- calcNormFactors(d,method = "RLE")
> d <- estimateGLMCommonDisp(d,design)
> d <- estimateGLMTrendedDisp(d,design)
> fit <- glmQLFit(d,design = design,robust = TRUE)
> result <- glmQLFTest(fit)
```

The number of differential interactions between GM and K562 can then be summarized as follows:

```
> summary(de <- decideTestsDGE(result, p=0.01, adjust ="BH"))
[,1]
-1 280
0 15590
1 27
```

Individual interactions can be visualized as follows:

```
> adj.p <- p.adjust(result$table$PValue, method="BH")
> ax <- anchors(new.data)
> tx <- targets(new.data)
> final <- as.data.table(final)
> final <- final[,list(anchor.chr=anchor.chr,anchor.start=anchor.start,
                       anchor.end=anchor.end,target.start=target.start,
                       target.end=target.end,logFC=logFC,FDR=FDR)]
> final
anchor.chr anchor.start anchor.end target.start target.end logFC FDR
1: chr1 1        1000707  1 1000707 3.68596978 0.01716097
2: chr1 1000704 1997191  1 1000707 3.85660173 0.01432674
3: chr1 1000704 1997191  1 000704 1997191 4.09673507 0.01943975
4: chr1 1997188 2997222  1 1000707 4.07016333 0.01748925
5: chr1 1997188 2997222  1 000704 1997191 4.35917769 0.01009404
—
15893: chrX  152000583 152998989 152000583 152998989  0.02836132 0.96686269
15894: chrX  152998986 154000896 152000583 152998989 -0.10639135 0.89985118
15895: chrX  152998986 154000896 152998986 154000896 -0.08071166 0.91691954
15896: chrX  154000893 154996366 152998986 154000896 -1.58248341 0.06887743
15897: chrX  154000893 154996366 154000893 154996366 -1.68881726 0.01296032
```

4 Notes/Conclusion

Statistical analysis of Hi-C and other 3C data is a complex and fast evolving topic. Bioconductor packages such as DiffHiC and GOTHiC implement a broad array of preprocessing procedure and relatively simple statistical analyses. Due to space limitations, several important aspects are not discussed in this book chapter, such as discovery interaction domain and integration with epigenetic marks and gene expression. Analyses of specific technologies such ChIA-PET or 4C are not covered either but many of the concepts described in this chapter can be adapted and reused to analyze datasets generated with these techniques. Likewise, improvement the 3C protocols technical improvements are frequently made, enabling increased resolution and more accurate detection of interact. It is for example possible to perform Hi-C on single cells [13]. It is likely that more sophisticated statistical modeling techniques and corresponding R/Bioconductor packages will be made available to analyze these novel and exciting datasets.

References

1. Williamson I et al (2014) Spatial genome organization: contrasting views from chromosome conformation capture and fluorescence in situ hybridization. Genes Dev 28(24):2778–2791
2. Gibcus JH, Dekker J (2013) The hierarchy of the 3D genome. Mol Cell 49(5):773–782
3. van de Werken HJ et al (2012) Robust 4C-seq data analysis to screen for regulatory DNA interactions. Nat Methods 9(10):969–972
4. Gavrilov A et al (2009) Chromosome conformation capture (from 3C to 5C) and its ChIP-based modification. Methods Mol Biol 567:171–188
5. Belton JM et al (2012) Hi-C: a comprehensive technique to capture the conformation of genomes. Methods 58(3):268–276
6. Kalhor R et al (2012) Genome architectures revealed by tethered chromosome conformation capture and population-based modeling. Nat Biotechnol 30(1):90–98
7. Rao SS et al (2014) A 3D map of the human genome at kilobase resolution reveals principles of chromatin looping. Cell 159(7):1665–1680
8. Langmead B, Salzberg SL (2012) Fast gapped-read alignment with Bowtie 2. Nat Methods 9 (4):357–359
9. Li H, Durbin R (2009) Fast and accurate short read alignment with Burrows-Wheeler transform. Bioinformatics 25(14):1754–1760
10. Lieberman-Aiden E et al (2009) Comprehensive mapping of long-range interactions reveals folding principles of the human genome. Science 326(5950):289–293
11. Robinson MD, Oshlack A (2010) A scaling normalization method for differential expression analysis of RNA-seq data. Genome Biol 11 (3):R25
12. Robinson MD, McCarthy DJ, Smyth GK (2010) edgeR: a Bioconductor package for differential expression analysis of digital gene expression data. Bioinformatics 26 (1):139–140
13. Nagano T et al (2013) Single-cell Hi-C reveals cell-to-cell variability in chromosome structure. Nature 502(7469):59–64

Chapter 10

Quantitative Comparison of Large-Scale DNA Enrichment Sequencing Data

Matthias Lienhard and Lukas Chavez

Abstract

DNA enrichment followed by sequencing (DNA-IP seq) is a versatile tool in molecular biology with a wide variety of applications. Computational analysis of differential DNA enrichment between conditions is important for identifying epigenetic alterations in disease compared to healthy controls and for revealing dynamic epigenetic modifications throughout normal and distorted cell differentiation and development. We present a protocol for genome-wide comparative analysis of DNA-IP sequencing data to identify statistically significant differential sequencing coverage between two conditions by considering variation across replicates. The protocol provides a detailed description for the comparative analysis of DNA-IP sequencing data including basic data processing, quality controls, and identification of differential enrichment using the Bioconductor package "MEDIPS".

Key words ChIP-seq, DNA methylation, Sequencing, MeDIP-seq, Epigenetics, DNA enrichment, Computational biology, Bioconductor, Quality control

1 Introduction

DNA enrichment methods are widely used for genome-wide identification of many different kinds of epigenetic marks. These techniques include chromatin-immunoprecipitation for localizing transcription factor binding sites or for revealing the genomic distributions of diverse types of histone modifications. To profile chromatin dynamics of scarce in vivo cell populations, an indexing-first chromatin IP approach (iChIP) has recently been developed [1]. *Me*thylated *D*NA *I*mmuno-*P*recipitation (MeDIP) [2] and methyl-CpG binding domain (MBD) protein capture [3] are similar techniques, but target the enrichment of DNA fragments containing methylated cytosines. Similarly, 5-hydroxymethylcytosines can be detected by antiserum specific to cytosine-5-hydroxymethylenesulfate (CMS) after treatment with sodium bisulfite [4]. To identify genomic loci enriched for DNA fragments in the DNA-IP sample compared to a negative control (e.g., Input-DNA), many different

Ewy Mathé and Sean Davis (eds.), *Statistical Genomics: Methods and Protocols*, Methods in Molecular Biology, vol. 1418, DOI 10.1007/978-1-4939-3578-9_10, © Springer Science+Business Media New York 2016

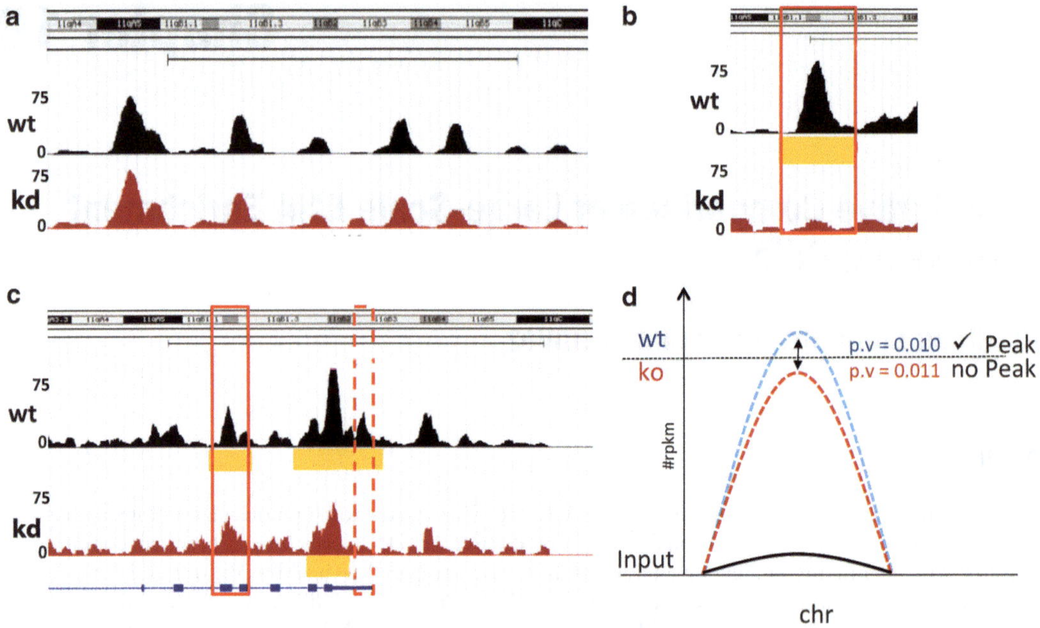

Fig. 1 Critical assessment of methods that solely rely on peaks for identifying differential enrichment between conditions. (**a**) Reproducible 5hmC enrichment in v6.5 wild type (*wt*) and Tet2 knock down (*kd*) mouse embryonic stem cells [12]. (**b**) In an ideal situation, differential enrichment is defined by the presence of a peak in one condition and the absence of enrichment in the other condition. (**c**) A peak identified in the *wt* condition has not been detected in the *kd* condition although the enrichment appears not to be depleted (*red box*). A peak detected in *wt* extends over a region with a highly variable enrichment profile throughout. Depletion of DNA enrichment in *kd* occurs in only a fraction of the peak identified in *wt* (*dashed red box*). (**d**) Schematic representation of the peak calling threshold problem that causes false positive differential enrichment. To define differential DNA-IP enrichment between conditions, MEDIPS evaluates the difference of DNA enrichment strength at distinct genomic loci instead of considering the presence or absence of previously calculated peaks ("peak free approach")

tools are available that can be categorized into the group of peak-callers [5, 6]. While these tools are capable of identifying DNA enrichments per sample, it is a common pitfall to rely on the identified sets of peaks for concluding sample specific and differential DNA enrichment comparing conditions (Fig. 1c–d). This procedure can cause false results due to several reasons. First, identification of peaks depends on statistically derived thresholds for elevated local sequencing coverage over background. This can lead to situations where a peak has been detected in one sample but not in the other although the sequencing coverage differs only slightly between samples (Fig. 1c–d). Second, peaks can be detected in both conditions without considering that there exist strong imbalances of DNA enrichment towards increased enrichment in one condition compared to the other.

Third, peaks often extend over long DNA regions incorporating several sites of epigenetic marks with distinct dynamics.

Figure 1c shows an example where the genomic coordinates of the peak identified in a control (*wt*) condition partly overlaps with a peak identified in a treatment condition (*kd*), indicating epigenetic alterations in only a fraction of the peak region. Appropriate segregation of such peaks into small fractions and subsequent differential enrichment analysis of each sub-region is mandatory to identify crucial epigenetic alterations between conditions. Finally, peak callers do not consider information on biological or technical variance across replicates. We therefore propose a method that calculates DNA-IP sequencing coverage separately for all samples at small genome wide windows [7]. Modeling the window coverage using negative binomial distribution allows for incorporation of biological variation between replicates. Based on this model, we apply a statistical test [8] to identify differentially enriched genomic regions between conditions. Afterwards, neighboring genomic regions with significant differential coverage into the same direction (i.e., either loss or gain of DNA enrichment) can be merged to define distinct loci of epigenetic alterations.

In this protocol we describe the analysis of large-scale ChIP-seq data obtained by profiling the histone modification H3K4me2 in naive and T_H2 memory (or CCR4 positive, respectively) T cells [9]. We demonstrate how to process and compare data of approx. forty distinct ChIP-seq assays per condition to ultimately identify cell type-specific enhancers marked by statistically significant differential H3K4me2 enrichment.

2 Materials

The MEDIPS analysis pipeline is implemented as an *R/Bioconductor* package, and therefore, runs on a wide variety of UNIX platforms, MS Windows and MacOS. We recommend installing the latest version of R freely available at www.r-project.org. The hardware requirement for the analysis depends on the genome size, the targeted resolution (window size), and the number of samples. We observe that processing of 84 human samples at distinct genomic windows of length 500bp requires approx. 40 GB of memory.

2.1 Installation of MEDIPS

To install the latest version of MEDIPS, it is recommended to first download and install the latest R version available at www.r-project.org. Consequently, the latest version of the MEDIPS package will be installed when employing the Bioconductor installer *biocLite*. To install MEDIPS start R and write:

```
R> source("http://bioconductor.org/biocLite.R")
R> biocLite("MEDIPS")
```

For this protocol we have employed R version 3.1.2, which is linked to Bioconductor version 3.0 (www.bioconductor.org)

containing MEDIPS version 1.17.2. To start a MEDIPS workflow, the MEDIPS library needs to be loaded into the R environment by writing:

```
R> library(MEDIPS)
```

2.2 Preprocessing of the Sequencing Data

The typical outputs of IP-sequencing experiments are short sequencing reads together with quality scores in *fastq* format. The H3K4me2-ChIPSeq data processed in this protocol is color-space sequencing data produced by SOLiD sequencing and the raw data (*csfastq* and *qual* files) but also *fastq* files are available at GeneExpressionOmnibus under accession id GSE53646 (http://www.ncbi.nlm.nih.gov/geo/query/acc.cgi?acc=GSE53646). To reproduce the analysis steps described here, ChIP-seq data for naïve and T_H2 samples have to be downloaded (these are the samples GSM1297924 to GSM1298001). Subheading 3.6 shows an R script that accomplishes this task by employing the Bioconductor package *SRAdb*. In order to apply the MEDIPS analysis, the reads first have to be aligned to a reference genome using an alignment tool, such as bowtie [10]. For the example data, the following alignment parameters have been applied:

```
> bowtie --sam -m 1 -C hg19 sample_A_runX.fastq sample_A_runX.sam
```

The option "–sam" determines the output format, -m 1 ensures that only unique hits in the genome are reported, and -C specifies that the reads are SOLiD colorspace encoded. The script for aligning the sequencing data is given in Subheading 3.7. Each sample has been multiplexed and sequenced across several sequencing lanes and runs. Therefore, data across different lanes and runs are merged to a single sam file per sample using *samtools* (http://samtools.sourceforge.net/). In order to save memory and to facilitate fast random access to the alignments, we recommend to sort the sam files by genomic position, to convert the text based sam files into a binary and compressed bam format and to create according index files using *samtools*:

```
> samtools view -Sb sample_A_runX.sam | samtools sort - sample_A_runX
> samtools merge sample_A_runX.bam sample_A_runY.bam [...] sample_A
> samtools index sample_A.bam
```

3 Methods

In this section, we describe a MEDIPS workflow for quality control and quantitative comparison of ChIP-seq data and explain all relevant options and parameters of the given examples. The according R script that reproduces the H3K4me2 ChIP-seq data analysis of naïve and T_H2 T-cell subtypes [9] is given in Subheading 3.6. First, we import the table containing all sample information into R (this table is generated in Subheading 3.6):

```
> samples=read.table("Seumois_NatImmu2014_sample_table.txt",
header=T, sep="\t", stringsAsFactors=FALSE)
> head(samples)
  accession            name
1 GSM1297951 CCR4Neg-Donor19-rep1
2 GSM1297950 CCR4Neg-Donor18-rep2
3 GSM1297957 CCR4Neg-Donor22-rep2
4 GSM1297969  CCR4pos-Donor7-rep1
5 GSM1297977 CCR4pos-Donor11-rep2
6 GSM1297993 CCR4pos-Donor19-rep2
                              filename celltype
1 bam/GSM1297951_H3K4me2_ChIPSeq_CCR4Neg-Donor19-rep1.bam CCR4Neg
2 bam/GSM1297950_H3K4me2_ChIPSeq_CCR4Neg-Donor18-rep2.bam CCR4Neg
3 bam/GSM1297957_H3K4me2_ChIPSeq_CCR4Neg-Donor22-rep2.bam CCR4Neg
4  bam/GSM1297969_H3K4me2_ChIPSeq_CCR4pos-Donor7-rep1.bam CCR4pos
5 bam/GSM1297977_H3K4me2_ChIPSeq_CCR4pos-Donor11-rep2.bam CCR4pos
6 bam/GSM1297993_H3K4me2_ChIPSeq_CCR4pos-Donor19-rep2.bam CCR4pos
```

In addition we have to install, load, and specify a reference BSgenome package. Here, we install and load the human reference build hg19/GRCh37:

```
> biocLite("BSgenome.Hsapiens.UCSC.hg19")
> library("BSgenome.Hsapiens.UCSC.hg19")
> reference="BSgenome.Hsapiens.UCSC.hg19"
```

3.1 Parameter Settings and Quality Control

3.1.1 Stacked Reads

Due to PCR amplification applied during experimental processing of ChIP-seq assays, library complexity can become low. In this case the same DNA fragments are sequenced multiple times what can lead to elevated amounts of "stacked" reads. Such sequencing reads will have the same sequence and will align to the same genomic position. High abundance of stacked reads is indicative for over-amplification and for libraries constructed from low quantities of input DNA. To avoid false positive differential enrichment between conditions due to stacked reads, MEDIPS offers the option to substitute each stack by only one representative by setting the *uniq* parameter of the *MEDIPS.createSet()* function to TRUE (*uniq=T*). The fraction of reads that build such stacks should be monitored by checking the standard output of MEDIPS ("*Total number of imported short reads*" vs. "*Number of unique short reads*"), and by visualizing the bam files in appropriate genome browsers. Low complexity samples should be excluded from further analyses. Please note that the concept of stacked reads is different to the concept of unique and multiple mapping reads. Multiple mappers cannot be assigned to a unique position of the reference genome whereas unique mappers can be assigned to a unique position of the reference genome. We typically consider only unique mappers for differential enrichment analysis although it is in principle possible to consider multiple mapping positions per read. However, both multiple and unique mappers can pile up to stacked reads and will be replaced by only one representative when setting the *uniq* parameter of the *MEDIPS.createSet()* function accordingly.

3.1.2 Extend Reads

The ChIP-seq protocol applied by Seumois et al. [9] generates and selects DNA fragments of length 100–250 base pairs (bp) for sequencing. By single-end sequencing, only one end of these DNA fragments is sequenced and the exact length of the individual fragments is unknown. Because the sequenced DNA fragments are typically longer than the actual sequencing reads, MEDIPS offers the option to extend the reads to the estimated average DNA fragment length by specifying a concrete value for the *extend* parameter of the *MEDIPS.createSet()* function. Given the DNA fragment size distribution of 100–250 bp we decided to extend all sequencing reads to 120 bases (*extend* =120).

3.1.3 Window Size

The fine resolution of DNA-IP enrichment experiments for narrowing down the exact location of the epigenetic mark, or the transcription factor binding site of interest, is restricted by the length of sonicated and immunoprecipitated DNA fragments. Consequently, an optimal window size for tiling the genome into nonoverlapping consecutive genomic regions ("*windows*") is influenced by the estimated average DNA fragment length and also by the expected distribution of the analyzed epigenetic mark across the genome (broad enrichment vs. sharp peaks). When mapping histone modifications, the maximal resolution of the ChIP-seq data is limited by the 146 bp DNA sequence that is wrapped around a nucleosome. However, the higher the targeted resolution of the results, the deeper sequencing is required. Decreasing the targeted resolution of the results might compensate lower sequencing coverage, but local effects might be overseen due to data smoothening. MEDIPS allows for controlling the targeted resolution of the results by the *window_size* parameter of the *MEDIPS.createSet()* function.

3.1.4 Coverage Saturation Analysis

An import quality control is to ensure that the sequence complexity of the libraries and depth of sequencing coverage is sufficient for subsequent data analysis. In order to assess the effect of the actual sequencing depth of a given sample as well as the impact of the *window_size* parameter on the reproducibility of the results, MEDIPS offers a coverage saturation analysis that randomly splits the given sequencing data into distinct subsets of increasing size and iteratively calculates correlations between such artificially created technical replicates. Please note that high abundance of stacked reads can have an undesired positive impact on the Pearson correlation calculated between two data sets (or data subsets, respectively). Therefore, it is recommended to either use the rank based Spearman correlation or to remove stacked reads (*uniq* = T). As a rule of thumb, an estimated correlation of around 0.9 should be observed for a sufficient sequencing coverage. If the correlation is lesser than 0.8, either additional sequencing or a larger window sizes should be considered. The actual number of reads necessary for obtaining a sufficient sequencing coverage depends on the genome size and on the genome wide abundance of the epigenetic mark of interest or on the number of transcription factor binding sites, respectively.

The coverage saturation analysis can be applied to individual samples using the function *MEDIPS.saturation()*. As input, the function requires the alignment file of a sample, the reference genome, and concrete settings for the parameters *window_size*, *extend*, and *uniq* as described above. Typically it is sufficient to apply the saturation analysis to only one selected chromosome what avoids importing the entire alignment file. The results are stored in an R list object that can be plotted by the *MEDIPS.plotSaturation()* function. The following code compares the coverage saturation of an arbitrary sample for window sizes 150 bp and 500 bp, respectively:

```
> sat150=MEDIPS.saturation(file=samples$filename[1],
reference,
    uniq=TRUE, extend=120, window_size=150,
    chr.select="chr22")
> sat500=MEDIPS.saturation(file=samples$filename[1],
reference,
    uniq=TRUE, extend=120, window_size=500,
    chr.select="chr22")
> par(mfrow=c(1,2))
> MEDIPS.plotSaturation(sat150, main=
    paste(samples$name[1], "Saturation analysis", "\nwin-
dow size 150"))
> MEDIPS.plotSaturation(sat500, main=
    paste(samples$name[1], "Saturation analysis", "\nwin-
dow size 500"))
```

As depicted in Fig. 2, the maximal estimated saturation is 0.8 at a window size of 150 bases, and improves to 0.85 at 500 bases. Based on these results, we chose a window size of 500 bp for further analyses.

3.2 Importing Alignment Data

When having determined the required parameters for *window_size*, *uniq*, and *extend*, the alignment data can be imported into MEDIPS by applying the function *MEDIPS.createSet()* to each of the samples. For each genomic window, the function counts the number of overlapping reads regardless of the fraction of the overlap.

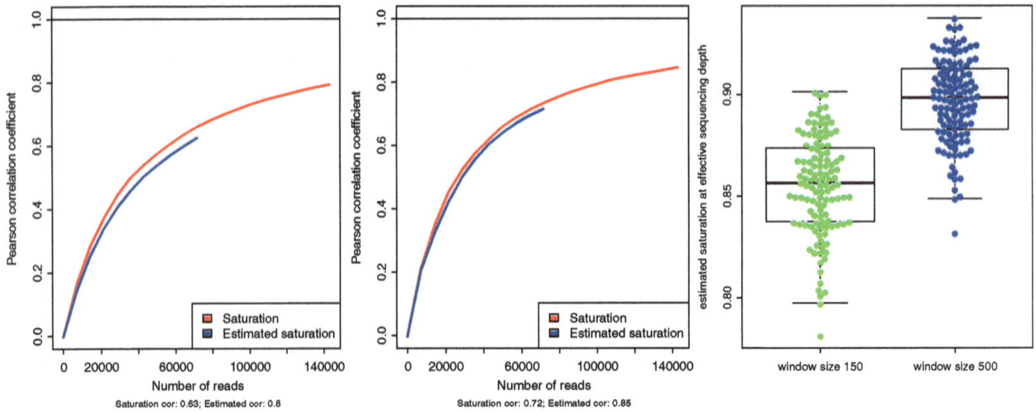

Fig. 2 Saturation analysis using window sizes of 150 bp (*left*) and 500bp (*middle*). Compared to 150 bp windows, the lower resolution of 500 bp windows leads to higher correlations (*y*-axis) between distinct random subsets throughout the tested range of sequencing depth (*x-axis*). The *x*-axis shows the increasing number of reads randomly assigned to the artificial replicates in each of the iterations. The maximally obtained estimated saturation increases from 0.8 for 150 bp windows (*left*) to 0.85 for 500 bp windows (*right*) when the entire data set is considered (maximum of the *red curve*). When applying the saturation analysis to all of the samples individually, we observe that the window size of 500 bp generally results in a higher maximal estimated saturation compared to the smaller window size of 150bp (*right, box*-and-*whisker plot*)

In addition to the three described parameters, the *MEDIPS.createSet()* function requires the alignment file (typically a bam or a bed file) and the name of the reference genome *BSgenome* package. Moreover, it is possible to restrict the analysis to a set of chromosomes by specifying the *chr.select* parameter.

```
> CCR4pos1=MEDIPS.createSet(samples$filename[1],
BSgenome=reference,
    uniq=FALSE, extend=120, window_size=500,
    chr.select=c(paste0("chr", 1:22), "chrX", "chrY") )
```

We have observed only a moderate level of stacked reads and therefore, decided to set *uniq*=FALSE for this data set (please note that such a moderate level of stacked reads is rather unusual and removal of stacked reads is typically recommended). The resulting *MSet* objects of several replicates can be concatenated into an R list, one for each experimental group. Please note, concatenating of *MSet* object does not merge the data and all replicates will be treated separately in the downstream analysis. In this study, this procedure results in two lists of *MSets*, one for naive and one for T_H2 T-cells (for the actual code please *see* Subheading 3.8). In principle, the code for three replicates per condition will look similar like this (please note that the MSets *CCR4pos2*, *CCR4pos3*, and *Naive1*, *Naive2*, *Naive3* have not been created in this example):

```
> MSet=list(
    CCR4pos=c(CCR4pos1, CCR4pos2, CCR4pos3),
    Naive=c(Naive1, Naive2, Naive3))
```

3.3 Differential Coverage Analysis

To identify genomic regions that are significantly differentially enriched between groups of samples, we will have to (1) normalize the genome wide count data for different library sizes of the samples, (2) find a minimal coverage across samples to avoid extensive unnecessary tests, (3) model variation across replicates, and (4) apply a statistical test suitable for DNA sequencing derived count data. An obvious choice for modeling the distribution of count data is the *Poisson* distribution. However, our samples are derived from different individuals with inherent biological variation among these biological replicates. This variation causes an overdispersed *Poisson* distribution, which can be described by the *negative binomial distribution*. Please note, to estimate the biological variation, replicates are required. Without replicates, biological variation will be set to a fixed value; however, this approach will always be an unsatisfactory approximation of unknown technical and biological variability. To accomplish the described tasks, MEDIPS employs the Bioconductor package *edgeR* [8] and its methods for library size normalization (*TMM*) and estimation of common and tag-wise dispersion. The function *MEDIPS.meth()* accepts two lists of *MSet* objects, each containing an arbitrary number of previously imported replicates. Moreover, the function requires further parameter settings for determining the type of statistical test applied (*diff.method*="edgeR"), the minimal number of read counts per window across replicates required for a window to be tested for differential coverage (*minRowSum*), and others (*see* also the MEDIPS vignette at http://www.bioconductor.org/packages/release/bioc/html/MEDIPS.html).

```
> resNvPos=MEDIPS.meth(MSet1=MSet[["CCR4pos"]], MSet2=MSet[["Naive"]],
diff.method="edgeR", MeDIP=F, minRowSum=10)
```

As a result, the *MEDIPS.meth()* function returns a table with test statistics for all windows. In order to depict the differences between the sets of samples, an M-vs-A-plot can be created from this table. The M-vs-A-plot contrasts the log ratios (M-value, y-axis) of the column with column name *edgeR.logFC* and the average log coverage (A-value, x-axis) of the column with column name *edgeR.logCPM* (*see* Fig. 3).

While the *MEDIPS.meth()* function returns a data table containing genome wide data, the function *MEDIPS.selectSig()* can be applied to reduce the results to those windows that fulfill the specified statistical requirements. Available parameters include a *p*-value threshold (*p.value*), the option to apply this threshold to either the raw (*adj=F*) or the adjusted *p*-value (*adj=T*), and a threshold for the coverage ratio between the two conditions (*ratio*). We term windows with significant differential DNA-IP enrichment at an adjusted *p*-value of ≤ 0.01 as differentially enriched regions (*DERs*). For visualization purpose only (*see* the

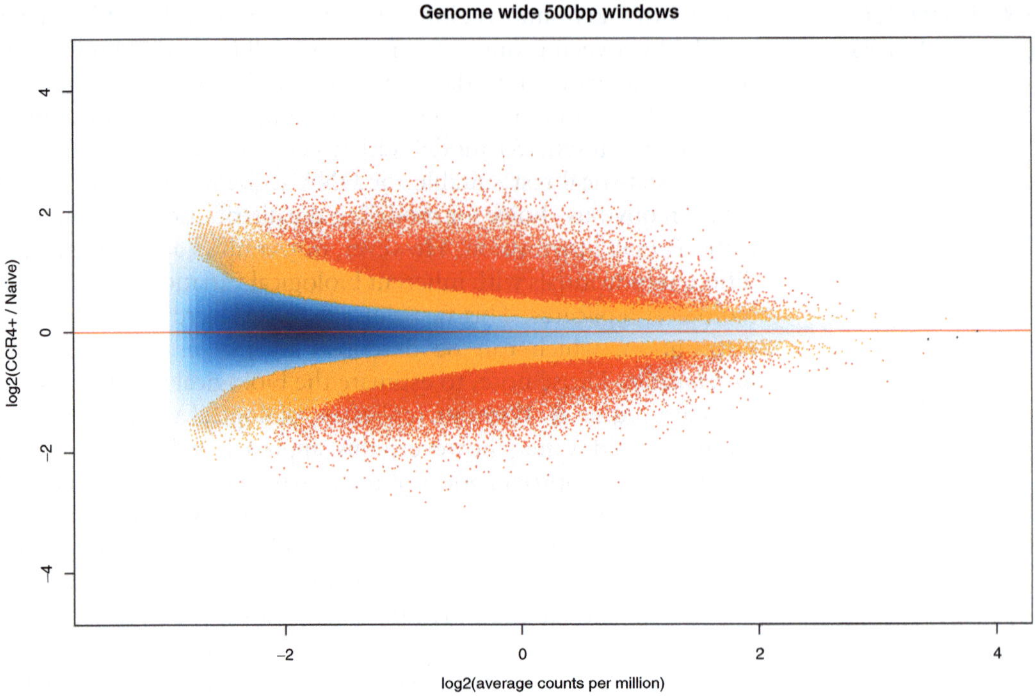

Fig. 3 "Minus-average" (MA) plots for genomic regions with differences in H3K4me2 enrichment (DERs) for comparing naïve and T_H2 T-cells. *Red* and *orange dots* indicate DERs with an adjusted $P < 0.05$ and raw $P < 0.005$, respectively. The MA plot has been generated by processing the entire ChIP-seq data set of naïve and T_H2 memory T-cells [9]

M-vs-A plot in Fig. 3), we also select windows that fulfill the relaxed threshold of a raw *p*-value ≤ 0.005:

```
> sigNvPos= MEDIPS.selectSig(resNvPos, p.value = 0.01,
adj = T)
> trendNvPos= MEDIPS.selectSig(resNvPos, p.value=0.005,
adj=F)
```

3.4 Merging Neighboring DERs and Annotation

DERs obtained by the *MEDIPS.selectSig()* function can be either focal events with no other immediate epigenetic differences in the vicinity, or they can be located in longer stretches of potential regulatory regions like for example super enhancers [11]. Adjacent DERs with imbalanced enrichment towards the same condition can be merged into extended and distinct loci by applying the function *MEDIPS.mergeFrames()*. Here, we are first dividing the DERs into those that gain H3K4me2 in T_H2 cells and those that lose H3K4me2 in T_H2 cells compared to naïve T-cells. Please note, log2 ratios are reported as log2(MSet1/MSet2):

```
> sigNvPosGain = sigNvPos[which(sigNvPos$edgeR.logFC>0),]
> sigNvPosLoss = sigNvPos[which(sigNvPos$edgeR.logFC<0),]
```

Afterwards, we can merge neighboring DERs:

```
sigNvPosGainMerged = MEDIPS.mergeFrames(sigNvPosGain)
sigNvPosLossMerged = MEDIPS.mergeFrames(sigNvPosLoss)
```

To further investigate potential functions of the identified DERs, MEDIPS provides functions to access the ENSEMBL database for gene annotation and to annotate the DERs by their proximity to known genes. The function *MEDIPS.getAnnotation()* connects to the ENSEMBL database and receives exon, gene or transcription start site information. The DERs can then be annotated by their proximity to the obtained annotations by applying the function *MEDIPS.setAnnotation()*. Here, we annotate merged DERs by ENSEMBL transcript names, if they are located in a promoter region (-1kb to +0.5kb around the transcription start sites):

```
> tss_ens = MEDIPS.getAnnotation(host="www.biomart.org",
dataset="hsapiens_gene_ensembl",
annotation="TSS",tssSz=c(-1000,500))
```

```
sigNvPosGainMerged = MEDIPS.setAnnotation(sigNvPosGain-
Merged, tss_ens)
sigNvPosLossMerged = MEDIPS.setAnnotation(sigNvPosLoss-
Merged, tss_ens)
```

3.5 Notes

We have described a peak-free approach for identifying statistical significant differential enrichment of DNA-IP sequencing data at distinct genome wide small regions. The described method, implemented in the Bioconductor package MEDIPS, enables the quantitative comparison of two conditions across an arbitrary number of biological replicates per group. In addition, we have demonstrated how to apply these methods for the identification of cell-type specific enhancers by comparing large scale H4K4me2 ChIP-seq data between naive and T_H2 memory T cells [9]. Although the ChIP-seq data processed in this protocol is single-end data, MEDIPS provides the functionality to process paired-end sequencing data (see the parameter *paired* of the *MEDIPS.createSet()* function).

Further functionalities of the MEDIPS package, not demonstrated in this protocol, include incorporation of Input-DNA sequencing data. In case such control data is available and genomic backgrounds vary between the samples (e.g., due to copy number variations in cancer samples compared to healthy controls), the parameter *CNV* of the function *MEDIPS.meth()* can be enabled to calculate potential CNVs based on the given Input-seq data. Subsequently, the fold change of the IP-DNA sequencing data can be corrected by the fold change of the Input-DNA sequencing data to exclude DERs that can be explained by differences in the genomic data alone. Excluding such DERs is controlled by the *ratio* and *CNV* parameters of the *MEDIPS.selectSig()* function.

The high number of statistical tests applied in such genome wide screens at small genomic windows can cause a severe multiple testing problem in case the number of replicates is small. While the number of biological replicates available in the discussed study [9] is sufficiently high for applying stringent significance thresholds to the corrected *p*-values, experimental designs with much smaller sample numbers will likely result in almost no or only a small number of sufficiently significant adjusted *p*-values. To overcome this problem, the number of applied tests can be reduced by two different ways. First, a minimal number of reads that fall into a genomic window across all of the samples can be required prior to testing. By setting the *minRowSum* of the *MEDIPS.meth()* function, the user can specify the number of reads that must fall into a genomic window in all of the samples together, otherwise this window will not be tested for differential enrichment. We generally recommend a *minRowSum* value of at least 10 to avoid massive amounts of tests at almost non-covered genomic regions. However, this value depends on the sequencing depth of the individual samples and an optimal *minRowSum* value can be estimated by investigating the count distribution of the DNA-IP sequencing samples or of Input-DNA sequencing samples, if available. Please note, there is currently no method implemented in MEDIPS that estimates an optimal *minRowSum* parameter. Second, instead of testing genome wide genomic windows, MEDIPS can be applied to any set of predefined regions of interest. Instead of a *window_size* parameter, the function *MEDIPS.createROIset()* accepts the genomic coordinates of any set of regions of interest as input for its *ROI* parameter (see also the man page of this function available by writing *?MEDIPS.createROIset* in R).

Besides applying the statistical test implemented in edgeR [8], MEDIPS allows for deriving *p*-values by applying the t-test to each of the genomic windows, if at least three replicates are available per condition (see the *diff.method* and *type* parameters of the *MEDIPS.meth()* function).

Varying enrichment efficiencies of DNA-IP sequencing experiments can have a negative impact on proper differential enrichment analysis when comparing DNA-IP assays from different batches. We have previously shown that systematic ChIP-seq batch effects can introduce likely false positive differential enrichment between conditions [9]. To approach this issue, MEDIPS version $\geq$1.18.0 enables Quantile normalization of the read counts across all samples prior to testing for differential enrichment between conditions.

The MEDIPS package provides further functionalities for DNA-IP sequencing data processing, including export of Wiggle files for visualization of DNA-IP sequencing data in genome browsers (see *MEDIPS.exportWig()*), merging of data sets in R to avoid unnecessary merging of bam files (see *MEDIPS.mergeSets()*), and others. Moreover, MEDIPS maintains its functionalities specific for MeDIP-seq experiments, including CpG density normalization and calculation of relative methylation scores. The MeDIP-seq specific functionalities can be enabled by the *MeDIP* parameter of the *MEDIPS.meth()* function. A complete list of functions is available in the reference manual at http://www.bioconductor.org/packages/release/bioc/html/MEDIPS.html.

3.6 R Script to Download the H3K4me2 ChIP-seq Data from SRA

```
#This script demonstrates how to download the entire human ChIP-seq data (SRA id: SRP034717) presented by Seumois et al. [9] into the designated working directory. The entire data set consists of 1789 fastq files across several multiplexed lanes and runs that can be assigned to 120 distinct ChIP-seq samples.

#Install the SRAdb package
source('http://bioconductor.org/biocLite.R')
biocLite('SRAdb')

#Load the SRAdb library and download the latest database content
library(SRAdb)
srafile = getSRAdbFile() #download of a SRA database snapshot (~800 mb)
sra_con = dbConnect(SQLite(),srafile)

#Fetch a table with the relevant run and sample information
runs = getSRA( "SRP034717", out_types="sra", sra_con)
runs = unlist(strsplit(paste0(runs$run,": ",runs$experiment_title),"[:;] "))
runs = as.data.frame(matrix(runs,ncol=5, byrow=T))

#For this protocol we restrict the analysis to only a small subset of the samples.
#Here, we select three TH2 and three naïve T-cell samples:
subselection=c("GSM1297960", "GSM1297962", "GSM1297964", "GSM1298002",
"GSM1298005", "GSM1298007")
runs=runs[(runs[,2]%in%subselection),] #remove this line, if the entire data
set should be processed

#Create a fastq table that contains the necessary information to link the individual fastq files to their respective samples (in this example there are 70 fastq files that can be assigned to the six selected sample):

write.table(runs, "Seumois_NatImmu2014_run_table.txt", col.names=F, row.
names=F, sep="\t", quote=F)

#Create a sample table that lists the sample names, their GO id and the anticipated location of the bam files in a separate bam sub-folder. The bam files will be generated based on the downloaded fastq files as shown in Subheading 3.7.
samples = unique(runs[,2:3])
samples[,2] = sub("H3K4me2_ChIPSeq_","",samples[,2])
names(samples) = c("accession", "name")
samples$filename = paste0("bam/", samples$accession, "_H3K4me2_ChIPSeq_",sam-
ples$
          name,".bam")
samples$celltype=sub("-Donor.+$","",samples[,2])
```

```
write.table(samples, "Seumois_NatImmu2014_sample_table.txt",
   col.names=T, row.names=F, sep="\t", quote=F)
#Download the fastq files into a fastq sub-folder:
#Note that the reduced set of 6 samples is about 7.5 gb, the complete dataset is
                  about 152 gb
dir.create("fastq")
for(acc in samples$accession){
 sra=runs[runs[,2]==acc,1]
 getSRAfile(in_acc=sra, sra_con=sra_con, fileType="fastq", destDir="fastq")
}
```

3.7 Shell Script for the Alignment of the H3K4me2 ChIP-seq Data (Fastq Files)

```
#!/bin/bash
#Set the path to your bowtie index (here this is a hg19 color-
space index):
hg19=/path/to/bowtie/genome/reference/hg19_CS
#specify number of cores
n_cores=30

#Set the variable dir to the designated working directory which is supposed to be
the same working directory as for the R script in Subheading 3.6. The working
directory is supposed to contain the fastq sub-folder containing the downloaded
fastq files:
dir='pwd'
#create the sub-folder that will contain the resulting alignment bam files
mkdir bam

#Assign the information of the fastq table (created in Subheading 3.6) to three
separate list variables:
run=($(cut -f1 Seumois_NatImmu2014_run_table.txt)) #Unique IDs for the 70 indi-
vidual fastq files
sampleAcc=($(cut -f2 Seumois_NatImmu2014_run_table.txt)) #GEO sample IDs
sampleName=($(cut -f3 Seumois_NatImmu2014_run_table.txt)) #Sample names

#Assuming that your PATH variable contains the path to bowtie and samtools bin-
aries and assuming that your compute server has at least 30 cores, the fastq files
can now be aligned and sorted into sample specific bam files as follows:
#The computation of the alignment takes about 30 minutes per sample on 30 cores
for s in $(printf '%s\n' "${sampleAcc[@]}"|sort|uniq)
do
list=""
skip="no"
nrruns=0
#get all fastq files for sample $s
for (( i=0; i<${#run[@]}; i++ ))
do
if [ ${sampleAcc[$i]} = $s ]; then
fq=${dir}/fastq/${run[$i]}.fastq.gz
if [ -e $fq ]; then
```

```
list="$list $fq"
nrruns=$((nrruns+1))
sName=${sampleName[$i]}
else
echo "warning: [sample $s] $fq not found"
skip="yes"
fi
fi
done
if [ $nrruns -ge 1 -a $skip != "yes" ]; then
echo "aligning reads for $s from $nrruns runs"
#
cat $list | gunzip |bowtie -sam -n 2 -m 1 -C -p 30 $hg19 - | \
   samtools view -Sb - | \
   samtools sort - ${dir}/bam/${s}_${sName}
samtools index ${dir}/bam/${s}_${sName}.bam &
fi
done

#This script also creates bam index files for each of the six resulting bam files.
```

3.8 R Script for Quality Control and Differential Enrichment Analysis of the Aligned H3K4me2 ChIP-seq Data Comparing T_H2 and Naïve T-Cells

```
#Go to the designated working directory as for Subheadings
3.6 and 3.7 and start R. This working directory is supposed
to contain a bam sub-folder containing the bam files created
in Subheading 3.7.
```

```
library(MEDIPS)
library(BSgenome.Hsapiens.UCSC.hg19)
samples=read.table("Seumois_NatImmu2014_sample_table.txt",
   header=T, sep="\t", stringsAsFactors=FALSE)
dir.create("plots")
#Coverage saturation analysis for all samples and for two different window sizes
                    each:
reference="BSgenome.Hsapiens.UCSC.hg19"
saturation=data.frame(ws150=numeric(), ws500=numeric())
#The saturation analysis should take < 1 minute per sample (depending on speed of
                    HDD)
for (i in 1:nrow(samples)){
sat150=MEDIPS.saturation(file=samples$filename[i],reference,
      uniq=TRUE, extend=120, window_size=150,
      chr.select="chr22")
sat500=MEDIPS.saturation(file=samples$filename[i],reference,
      uniq=TRUE, extend=120, window_size=500,
      chr.select="chr22")
```

```
saturation[samples$name[i],] =
   c(sat150$maxEstCor[2],sat500$maxEstCor[2])
png(file=paste0("plots/saturation_",samples$accession[i],".png"),
   width=800, height=400 )
par(mfrow=c(1,2))
MEDIPS.plotSaturation(sat150, main=
paste(samples$name[i], "Saturation analysis","\nwindow size 150"))
MEDIPS.plotSaturation(sat500, main=
paste(samples$name[i], "Saturation analysis","\nwindow size 500"))
dev.off()
}

#Import of bam files: for each sample an MSet object is created and the MSet objects are concatenated into a list of two separate MSet list objects. The individual MSets can be assigned to their respective cell types by their list names. We restrict the imported mapping data to the major chromosomes:
#reading the alignment files and computing the coverage takes about 10 minutes per sample (depending on speed of HDD)
MSet=list()
for (i in 1:nrow(samples) ){
MSet[[samples$celltype[i]]]=c(MSet[[samples$celltype[i]]],
  MEDIPS.createSet(samples$filename[i],
BSgenome=reference,
    uniq=FALSE, extend=120, window_size=500,
    chr.select=c(paste0("chr", 1:22), "chrX", "chrY")
))
}

#Differential Enrichment Analysis (takes about 30 minutes)
resNvPos=MEDIPS.meth(MSet1=MSet[["CCR4pos"]], MSet2=MSet[["Naive"]], diff.method="edgeR", MeDIP=F, minRowSum=10)
#Selecting a set of highly significant DERs by applying a significance threshold to the p-values adjusted for multiple testing:
sigNvPos=MEDIPS.selectSig(resNvPos, p.value=0.01, adj=T)
#Selecting a second set of less significant windows by applying a significant threshold to the unadjusted p-values (for visualization purpose only, see the MvA plot below):
trendNvPos= MEDIPS.selectSig(resNvPos, p.value=0.005, adj=F)
#Dividing the DERs into those that gain H3K4me2 in TH2 cells and those that lose H3K4me2 in TH2 cells compared to naïve T-cells. Please note, log2 ratios are reported as log2(MSet1/MSet2).
sigNvPosGain = sigNvPos[which(sigNvPos$edgeR.logFC>0),]
sigNvPosLoss = sigNvPos[which(sigNvPos$edgeR.logFC<0),]
#Merging neighboring DERs:
sigNvPosGainMerged = MEDIPS.mergeFrames(sigNvPosGain)
sigNvPosLossMerged = MEDIPS.mergeFrames(sigNvPosLoss)
```

```
#Annotating DERs by Ensembl transcript names, if the DERs are locate in a promoter
region (-1kb to +0.5kb around the transcription start sites):
tss_ens = MEDIPS.getAnnotation(host="www.biomart.org",
   dataset="hsapiens_gene_ensembl",
   annotation="TSS",tssSz=c(-1000,500))
sigNvPosGainMerged = MEDIPS.setAnnotation(sigNvPosGainMerged, tss_ens)
sigNvPosLossMerged = MEDIPS.setAnnotation(sigNvPosLossMerged, tss_ens)

#Create an MvA plot. This section creates the plot shown in Fig. 3, if the complete
data set has been processed (and not only the three example samples per
condition):
png("plots/MA_Comparison_Naive_vs_CCR4Pos.png",
   width=800, height=500)
smoothScatter(x=resNvPos$edgeR.logCPM,
   y=resNvPos$edgeR.logFC,
   xlim=c(-3.5, 4), ylim=c(-4.5, 4.5),
   pch=".", main="Genome wide 500bp windows",
   xlab="log2(average counts per million)",
   ylab="log2(CCR4+ / Naive)")
points(x=trendNvPos$edgeR.logCPM, y=trendNvPos$edgeR.logFC,
   pch=".", col="orange")
points(x=sigNvPos$edgeR.logCPM, y=sigNvPos$edgeR.logFC,
   pch=".", col="red")
abline(h=0, col="red")
dev.off()
```

Acknowledgements

This work is supported by the German Federal Ministry of Education and Research with the grant EPITREAT (No. 0316190A) and by the Max Planck Society with its International Research School program (IMPRS-CBSC) (to M.L.). L.C. is the recipient of a Feodor Lynen postdoctoral Research Fellowship granted by the Alexander von Humboldt Foundation.

References

1. Lara-Astiaso D, Weiner A, Lorenzo-Vivas E, Zaretsky I, Jaitin DA, David E et al (2014) Immunogenetics. Chromatin state dynamics during blood formation. Science 345(6199):943–949
2. Weber M, Davies JJ, Wittig D, Oakeley EJ, Haase M, Lam WL et al (2005) Chromosome-wide and promoter-specific analyses identify sites of differential DNA methylation in normal and transformed human cells. Nat Genet 37(8):853–862
3. Serre D, Lee BH, Ting AH (2010) MBD-isolated Genome Sequencing provides a high-throughput and comprehensive survey of DNA methylation in the human genome. Nucleic Acids Res 38(2):391–399
4. Pastor WA, Pape UJ, Huang Y, Henderson HR, Lister R, Ko M et al (2011) Genome-wide mapping of 5-hydroxymethylcytosine in embryonic stem cells. Nature 473 (7347):394–397

5. Zhang Y, Liu T, Meyer CA, Eeckhoute J, Johnson DS, Bernstein BE et al (2008) Model-based analysis of ChIP-Seq (MACS). Genome Biol 9(9):R137
6. Zang C, Schones DE, Zeng C, Cui K, Zhao K, Peng W (2009) A clustering approach for identification of enriched domains from histone modification ChIP-Seq data. Bioinformatics 25(15):1952–1958
7. Lienhard M, Grimm C, Morkel M, Herwig R, Chavez L (2014) MEDIPS: genome-wide differential coverage analysis of sequencing data derived from DNA enrichment experiments. Bioinformatics 30(2):284–286
8. Robinson MD, McCarthy DJ, Smyth GK (2010) edgeR: a Bioconductor package for differential expression analysis of digital gene expression data. Bioinformatics 26 (1):139–140
9. Seumois G, Chavez L, Gerasimova A, Lienhard M, Omran N, Kalinke L et al (2014) Epigenomic analysis of primary human T cells reveals enhancers associated with TH2 memory cell differentiation and asthma susceptibility. Nat Immunol 15(8):777–788
10. Langmead B, Trapnell C, Pop M, Salzberg SL (2009) Ultrafast and memory-efficient alignment of short DNA sequences to the human genome. Genome Biol 10(3):R25
11. Hnisz D, Abraham BJ, Lee TI, Lau A, Saint-Andre V, Sigova AA et al (2013) Super-enhancers in the control of cell identity and disease. Cell 155(4):934–947
12. Huang Y, Chavez L, Chang X, Wang X, Pastor WA, Kang J et al (2014) Distinct roles of the methylcytosine oxidases Tet1 and Tet2 in mouse embryonic stem cells. Proc Natl Acad Sci U S A 111(4):1361–1366

Chapter 11

Variant Calling From Next Generation Sequence Data

Nancy F. Hansen

Abstract

The use of next generation nucleotide sequencing to discover and genotype small sequence variants has led to numerous insights into the molecular causes of various diseases. This chapter describes the use of freely available software to align next generation sequencing reads to a reference and then to use the resulting alignments to call, annotate, view, and filter small sequence variants. The suggested variant calling workflow includes read alignment with novoalign, the removal of polymerase chain reaction duplicate sequences with samtools or bamUtils, and the detection of variants with Freebayes or bam2mpg software. ANNOVAR is then used to annotate the predicted variants using gene models, population frequencies, and predicted mutation severity, producing variant files which can be viewed and filtered with the variant display tool VarSifter.

Key words Variant detection, Next generation sequencing, Genotyping, Whole exome, Annotation

1 Introduction

In the years following the completion of the human genome reference sequence [1], a small set of new technologies became available to researchers hoping to extend DNA sequencing to a large number of human genomes, as well as to other unsequenced organisms. These "next generation" technologies were not notable for reading long stretches of DNA, and in fact, the sequence reads they produced were far shorter (anywhere from 25 to 150 bases) than the read lengths obtained using the capillary electrophoresis-based methods used for the Human Genome Project. What made these new sequencing technologies revolutionary was their ability to produce millions of reads in a single machine run, with lower cost and effort than was previously needed to produce thousands [2].

As a result, the development of next generation DNA sequencing has dramatically altered the process of discovering single nucleotide variants (SNVs) and small deletion and insertion variants (DIVs) from nucleotide sequencing data. By allowing large-scale, accurate interrogation of short single-molecule sequences from

Ewy Mathé and Sean Davis (eds.), *Statistical Genomics: Methods and Protocols*, Methods in Molecular Biology, vol. 1418, DOI 10.1007/978-1-4939-3578-9_11, © Springer Science+Business Media New York 2016

large numbers of samples, next generation sequencing has replaced older, more expensive methods involving subcloning or polymerase chain reaction (PCR) amplification followed by gel electrophoresis. As a result, there has been unprecedented characterization of genetic variation in both healthy and diseased tissues from humans and other species [3, 4].

1.1 Statistical Methods in Variant Analysis

The conversion of many millions of short segments of DNA sequence, each on the order of 150 nucleotide bases in length and accompanied by per-base probabilistic quality scores, into accurate chromosomal sequence differences from a reference genome involves large-scale data processing using complex algorithms [5]. Since the sequencing process itself involves random sampling of DNA molecules, the analysis of the data will inevitably make use of statistical methods, both to determine the original genomic locus of each of the bases that have been read by the sequencer (mapping and alignment) and to infer the alleles that were actually present in the sample (genotyping) as well as whether those alleles differ from a user's selected reference sequence (variant calling).

The most widely used read alignment programs use a "seed and extend" approach for the first stage of alignment to the reference. Short, highly similar matches ("seeds") are determined between a single read and various reference locations, using highly efficient algorithms which make use of suffix arrays, suffix trees, hashes, or other types of indexes. One of these types of indexes, created using the "Burrows-Wheeler Transform," is employed by several popular aligners for next generation reads [6–8]. Once seeds are determined, they are generally extended into full read alignments using a dynamic programming algorithm like the Smith–Waterman or Needleman–-Wunsch algorithm [9, 10], and then, if possible, the paired read is aligned in the same vicinity of the reference as the first read's match.

Variant calling generally employs a Bayesian approach, considering each locus of the genome one at a time. The program Freebayes groups short-range haplotypes into single loci [11], while bam2mpg considers each single base, insertion, or deletion variant within the read alignments separately [12]. At each locus, a Bayes model specifying the probability of the observed allele counts within the sequencing reads (i.e., the likelihood of the data) is incorporated into a maximum likelihood approach or the application of Bayes' theorem to predict the presence or absence of a variant and the exact genotype at that locus. This probabilistic framework also allows for the assignment of meaningful phred-scaled variant and genotype scores [13].

1.2 Choice of Analysis Software

There are numerous software packages available for nearly every step in calling small variants from next generation sequencing data, and each has its own particular advantages and drawbacks. Perhaps the best-known software pipeline for small variant calling is the Broad Institute's "best practices" pipeline, which consists of the BWA aligner [6] and the Genome Analysis Toolkit (GATK) [14]. The GATK pipeline produces variant calls that have been shown to be highly accurate, and offers versatility with respect to handling a variety of experimental designs, including pooling samples without identifying barcodes before sequencing, as well as sequencing samples to only a low depth of coverage. It also employs sophisticated filtering methods by training models to recognize features characteristic of false positives. Unfortunately, running the GATK pipeline requires extensive skill in software installation and configuration, a substantial investment of computational time and memory, and experimentation with the user's datasets in order to make optimal choices for software parameters and settings. In addition, the filtering steps require access to large sets of known variants, which are not always available for the species or genome build of interest.

Alternatively, if the sequence reads to be analyzed are from single samples, rather than unidentified pools, and these samples have been sequenced to sufficient depth of coverage to yield high quality variant calls (generally 25× or greater), then simpler software pipelines exist which are easier to implement and faster to run, without a demonstrable decrease in accuracy. One such alternative pipeline for the detection of small variants from next generation sequencing data is described here. For the alignment of reads to the reference sequence, novoalign is used, and for calling variants, the user can choose one of two Bayesian genotypers: Freebayes [11] or bam2mpg [12]. Variants are then annotated with the program ANNOVAR, filtered using quality scores and genomic location, and viewed with the VarSifter variant viewer.

1.3 Variant Filtering and Accuracy

Any method for calling variants will have reduced accuracy in regions of the genome that are prone to misalignment, due either to repetitive sequences or inaccuracies in the genome reference build, but appropriate use of filters can lead to highly accurate variant calls. A recent study [15] found that for a variety of variant calling pipelines combined with the application of a small handful of filters, variant call sets from high-coverage sequencing data can have less than one error in 100,000 bases, on average, and that the different software pipelines show high agreement in these call sets after filtering. The filters applied included the removal of variant calls overlapping low-complexity regions (LCRs) and the removal of variants in regions with significantly large read depth. In addition, the study found that the inclusion of "decoy" sequences representing regions missing from the genome reference prevented misalignments that result in false variant calls. This provides additional evidence that the use of quality score and genomic location filters can effectively replace GATK's approach using machine learning models to recognize features of true variants.

1.4 Software Pipelines and Reproducibility

Once software tools are chosen, the user needs to run them all in the appropriate sequence, sometimes merging multiple datasets from the same sample or merging multiple samples' variant sets for comparison. Using "pipeline" software to simplify sample processing enhances the reproducibility of analyses, and allows bioinformaticians to track and check job outcomes, log software versions, and avoid errors through automation.

Several publicly available pipelines exist for this purpose. The Galaxy Project [16] offers a website (https://usegalaxy.org) as well as the ability to install the Galaxy software on a local computer cluster. The "blue collar bioinformatics" pipeline available at https://github.com/chapmanb/bcbio-nextgen also offers the ability to run on a local cluster or on the cloud, but installing and running the pipeline still require considerable expertise in computer programming. In this chapter, we will not detail the use of these pipeline packages, but instead, describe a protocol leading the user through the separate steps that might be executed within them. In this way, the reader can expect to gain a better understanding of the software being used and the steps being carried out.

2 Materials

This protocol outlines the steps required to analyze sequence reads, either single- or paired-end, from Illumina GAII or HiSeq sequence analyzers, in order to create files in VCF format containing genotypes of SNVs and small insertions and deletion variants. The required materials to follow this protocol are:

1. A high-performance computer or computer cluster running the Linux operating system (see below for hardware requirements).
2. Sequence reads in FASTQ format for samples of interest. Alternatively, Subheading 3.6.1 gives instructions for downloading a sample dataset containing sequence from the National Center for Biotechnology Information (NCBI).
3. Installed software for sequence analysis:
   - SRA Toolkit—http://www.ncbi.nlm.nih.gov/Traces/sra/sra.cgi?view=software (optional, for use in obtaining sample data from the NCBI Sequence Read Archive)
   - Samtools—http://www.htslib.org
   - bamUtil—http://genome.sph.umich.edu/wiki/BamUtil (only required for PCR duplicate removal if the installed version of samtools is 1.0 or later)
   - Novoalign—http://www.novocraft.com/products/novoalign/

- Variant and genotype calling software (one of the following):
  - Freebayes—https://github.com/ekg/freebayes
  - Bam2mpg—http://research.nhgri.nih.gov/software/bam2mpg/
- tabix—http://sourceforge.net/projects/samtools/files/tabix/
- ANNOVAR—http://www.openbioinformatics.org/annovar/
- VarSifter—https://github.com/teerjk/VarSifter

Familiarity with running programs within a Linux environment and experience with a scripting language such as Perl, Python, or Ruby are helpful, although not strictly necessary. While it is possible to perform these analyses on a single computer with at least eight gigabytes of physical memory available for processing, samples are processed faster using a high-performance computer cluster equipped with job scheduling software, so that jobs can be parallelized to decrease throughput time for each sample. In addition, considerable amounts of disk space are required. For extended analyses, 50 or more gigabytes per mammalian sample for whole exomes, and on the order of a terabyte per sample for mammalian whole genomes will provide a generous amount of room, especially when running these programs for the first time, but eventually, deletion and compression of unnecessary files used early in the process can reduce these disk requirements considerably.

In addition, example processing scripts and updates regarding software usage changes are maintained in the github repository available at http://github.com/nhansen/MMBVariantCalling.

3 Methods

Figure 1 shows the steps to be followed in this single-sample variant calling pipeline. First, sequence reads are aligned to a set of reference sequences which define the loci that will be interrogated for variants. The alignment software (novoalign) produces alignment files in "BAM" format [17], detailing each read's mapped location within the reference, as well as the base-by-base alignment of reads against the reference sequence itself. Pairs of library inserts with common aligned reference endpoints are assumed to come from multiple PCR copies of the same original library fragment ("PCR duplicates"), so in order to obtain a more accurate sampling of the chromosomes present in the DNA sample, reads representing the extra copies are removed from the dataset. Next, a variant caller is run on the alignment file to predict reference loci where at least one chromosome contains sequence different from the reference

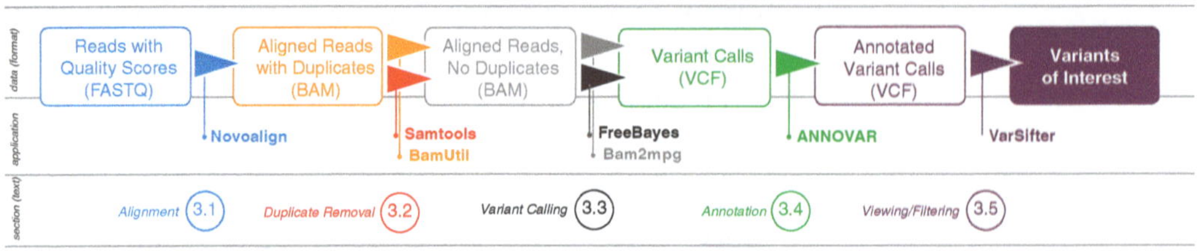

Fig. 1 Diagram of the variant calling workflow. *Circled numbers* beneath each step refer to the section of the chapter in which they are described

Table 1
Approximate memory and CPU requirements for programs run in this protocol

	Whole exome		Whole genome	
Program	**Memory**	**CPU time**[a]	**Memory**	**CPU time**[a]
Novoalign[b]	8 Gb	200 h	8 Gb	2000 h
Duplicate removal	< 1 Gb	40 min	< 1 Gb	6 h
Freebayes	< 100 Mb	5 h	<200 Mb	3 h
Bam2mpg[c]	8 Gb	18 h	8 Gb	300 h
ANNOVAR	<1 Gb	5 min	< 1 Gb	3 h

Whole exome statistics were measured using the paired-end dataset with accession SRR1611184 available from the NCBI Sequence Read Archive

[a] All reported CPU times are approximate

[b] Novoalign processing times are total time. Licensed versions of novoalign, when run on *n* CPUs, have total elapsed times *n* times faster than reported CPU time

[c] On whole genome sequences, bam2mpg should be run on separate regions of up to ten million bases in size and results recombined to conserve memory

sequence. These differences are generally SNVs and DIVs, but can also be complex substitutions which change multiple adjacent bases in a way that is difficult to ascribe to simple mutation events. All discovered variant alleles are typically reported, along with the genotypes of the sample being analyzed, in VCF format [18]. VCF files are then annotated to classify variants according to their locations within genes (5′ or 3′ untranslated region, coding sequence, or intronic sequence) and their effect on the translated protein (synonymous, nonsynonymous, stop-gain, stop-loss, frameshift, or splicing) for protein-coding genes [19]. Finally, annotated variants in VCF format can be viewed and filtered using specialized viewing software [20].

When executing each of the steps detailed in the sections that follow, attention should be paid to the estimated required computer resources (both physical memory and CPU time) provided in Table 1, and longer running jobs should be divided into multiple shorter jobs in the ways described within some sections of the text.

3.1 Alignment of Reads

3.1.1 FASTQ Files

Generally, unaligned next generation sequencing data are stored in FASTQ format. FASTQ-formatted files have four lines per sequence read, containing

1. The read name preceded by the character "@".
2. The sequence of the read, directed from 5′ to 3′.
3. A line beginning with the character "+".
4. A line containing ASCII-encoded quality scores for each base.

If a sample's library or libraries were sequenced in a paired-end run, there should be two FASTQ files for each sample lane, containing entries for read 1 or read 2 of each pair, and each file's read entries should appear in the same order as the other's.

Information on obtaining FASTQ-formatted sequence data from the NCBI's Sequence Read Archive (SRA) can be found in Subheading 3.6.1.

3.1.2 Running Novoalign

Before running novoalign on sequences from a new species, an index of the reference genome for that species must first be created using the "novoindex" command. Steps for obtaining a reference file and preparing a genome index for novoalign are described in Subheadings 3.6.2 through 3.6.4. Once these steps have been completed, the resulting index file, which has the file extension ".ndx", can be used to align any set of FASTQ-formatted reads against that particular genome reference sequence.

Novoalign can be run as licensed or unlicensed software. The program will search for its license file in all directories in the user's Linux path, as well as the directory in which the executable itself resides. Obtaining a paid license for novoalign allows the user to read FASTQ files that have been compressed with the "gzip" utility, and also allows novoalign to be run on multiple threads of a single computer, which can dramatically decrease the storage requirements and elapsed run time to process whole exome and whole genome sequences. Alternatively, FASTQ files can be split into multiple smaller files, and run separately on different processors, after which the resulting BAM files can be merged.

For example, if the FASTQ files are named "SRR1611184_1.fastq.gz" and "SRR1611184_2.fastq.gz", novoalign is run with the command:

```
# Run novoalign, piping output to samtools to make a sorted BAM file:
novoalign -F STDFQ -o SAM -d hg38.ndx -f SRR1611184_1.fastq.gz
  SRR1611184_2.fastq.gz | samtools view -uS - |
  samtools sort - NA12878
```

which will produce a BAM-formatted file with the name "NA12878.bam", containing reads aligned to the hg38 reference.

When using an unlicensed copy of novoalign, or when analyzing a large dataset such as the one obtained from whole genome sequencing, it is usually better to split large FASTQ files into multiple smaller ones and then run novoalign on each smaller FASTQ file separately. A script "split_fastq.pl" is provided for this purpose in the scripts subdirectory of the github repository at http://github.com/nhansen/MMBVariantCalling.

```
# Split fastq file obtained from NCBI into 40 separate smaller files
# --nogzip option will create uncompressed files and should only
# be used when running novoalign without the paid license
split_fastq.pl SRR1611184_1.fastq.gz 40 --nogzip --dir split_40/
split_fastq.pl SRR1611184_2.fastq.gz 40 --nogzip --dir split_40/
```

Once the read sequences for a sample are in multiple FASTQ files, novoalign can be run on each pair of FASTQ files as described above, and the resulting BAM files can be merged using samtools:

```
samtools merge NA12878.merge.bam NA12878.1.bam NA12878.2.bam [...]
```

Note that in the samtools merge command, the output file precedes the BAM files you are merging in the list of arguments.

3.2 Removal of PCR Duplicates

Since variant calling software will generally assume read pairs represent an unbiased random sampling of DNA from cells, it is important to remove extra copies of PCR-amplified DNA molecules, or "PCR duplicates," from the BAM file before calling variants. Prior to the release of samtools version 1.0, the most common way to remove duplicate reads was to use samtools's "rmdup" command, but samtools versions 1.0 and later no longer have a working rmdup command, so "option 2" below describes the use of the bamUtil command "dedup" as an alternative.

3.2.1 Duplicate Removal Option 1

If you are running a version of samtools prior to version 1.0 (e.g., version 0.1.19), run the available rmdup command:

```
# remove PCR duplicates using samtools
samtools rmdup NA12878.merge.bam NA12878.final
```

This will create a new BAM file with duplicates removed called "NA12878.final.bam". If libraries were sequenced with only a single end (not paired), one should include the "-s" option to samtools rmdup, and note that duplicate removal will generally remove more reads than would be the case for paired ends.

3.2.2 Duplicate Removal Option 2

If earlier versions of samtools are not available, one can use the bamUtil package, available at http://genome.sph.umich.edu/wiki/BamUtil, to remove PCR duplicates. Once bamUtil has been installed, users can run:

```
# remove PCR duplicates using bamUtil
bam dedup --in NA12878.merge.bam --out NA12878.final.bam --rmDups
```

After either Option 1 or Option 2 has been completed, the final BAM file "NA12878.final.bam" should be indexed for fast access to genomic regions of interest.

```
# Index BAM file for fast retrieval of genomic regions
samtools index NA12878.final.bam
```

3.3 Variant and Genotype Calling

Both Freebayes and bam2mpg can be used to obtain accurate variant calls from duplicate-free BAM files. Table 1 shows that Freebayes requires far less CPU time and physical memory than bam2mpg, but Freebayes also provides less informative genotype quality scores for sites where the program predicts no variation. This latter point is critical when the goal is to determine whether a sample is truly free of variation in a region without a variant call, or whether more sequence coverage is needed to "test" a sample for a particular variant. Instructions are given below for creating BED files of confidently covered regions with bam2mpg.

3.3.1 Variant Calling with Freebayes

Freebayes [11] is a haplotype-based caller for producing VCF-formatted single nucleotide and small deletion/insertion variant calls from BAM-formatted sequences. Run Freebayes on your final BAM file with the command:

```
freebayes --genotype-qualities -f hg38.mfa NA12878.final.bam >
  NA12878.vcf
```

Freebayes variants represent local haplotypes, so two side-by-side SNVs will be represented in the VCF file by a phased "multi-nucleotide polymorphism," or MNP. Phased MNPs can be annotated more accurately than side-by-side SNVs.

3.3.2 Variant Calling with bam2mpg

Another accurate, yet easy to use, variant caller is bam2mpg [12]. In addition to calling variants in VCF format, bam2mpg also assigns a genotype quality score to genomic sites that are called homozygous reference. In this way, it is possible to create BED-formatted files delineating regions that have adequate sequence coverage to determine with high probability that no variant is present at a locus. This can be valuable in calculating accurate population variant allele frequencies for rare variants.

To generate VCF files NA12878.snv.vcf and NA12878.div.vcf with bam2mpg, run:

```
bam2mpg --bam_filter '-q31' --qual_filter 20 --only_nonref
  --snv_vcf NA12878.snv.vcf --div_vcf NA12878.div.vcf hg19.mfa
  NA12878.final.bam
```

To obtain a BED file of regions determined to have adequate depth of coverage for calling diploid variants, one must not use the

"–only_nonref" option when running bam2mpg. This will create larger files than the "variant only" VCF files (on the order of 5 gigabytes for a whole exome sample, or 15 gigabytes for a whole genome sample), but the resulting "MPG" formatted output file can be used as input to the script "mpg2bed.pl" to create a BED-formatted file of adequately covered regions:

```
bam2mpg --bam_filter '-q31' --qual_filter 20 --mpg NA12878.mpg.out.gz
  hg19.mfa NA12878.final.bam
mpg2bed.pl --minscore 10 sample.mpg.out.gz > NA12878.mpg.bed
```

3.3.3 Compression and Random Access with VCF Files

Especially if they contain invariant loci, VCF files can be large, so to save disk space, they should be compressed with the block compression tool "bgzip," which is available as part of the "tabix" package. Since bgzip retains block structure in its compression of VCF files, it is also possible to index files of variants, and quickly access entries in arbitrary parts of the genome with a handful of simple commands:

```
bgzip NA12878.vcf # create the compressed file sample.vcf.gz
tabix -p vcf NA12878.vcf.gz # index the VCF file
tabix NA12878.vcf.gz chr13:32889617-32973809 # retrieve BRCA2 variants
```

3.4 Annotation of VCF Files

The genomic position and exact sequence change represented by a variant is rarely informative all by itself. To evaluate variants and gain insights into their impact on a gene's function, it is helpful to view these variants in the context of various gene annotations. The program ANNOVAR [19] can be used to compare variant locations to annotation sets downloaded from the ANNOVAR website and the University of California at Santa Cruz's (UCSC) Genome Browser [21]. Specifically, lists of "gene-based" (describing variants' impact on gene models from NCBI's RefSeq, UCSC's known gene, or the ENSEMBL gene set) and "filter-based" (annotating specific variants with a wide variety of allele frequency, pathogenicity, and conservation-related measures) can be downloaded and used to add annotations to VCF files for viewing and filtering.

For example, to add gene annotation from the National Center of Biotechnology's "RefGene" database to a variant VCF file, first download the "refGene" database from the ANNOVAR website:

```
# Download RefSeq gene annotations
annotate_variation.pl --buildver hg38 -downdb -webfrom annovar refGene
  humandb/
```

Annotating variants with allele frequencies from different populations can also be helpful in determining whether a particular variant might be pathogenic. So also download the Exome Aggregation Consortium's "ExAC" database:

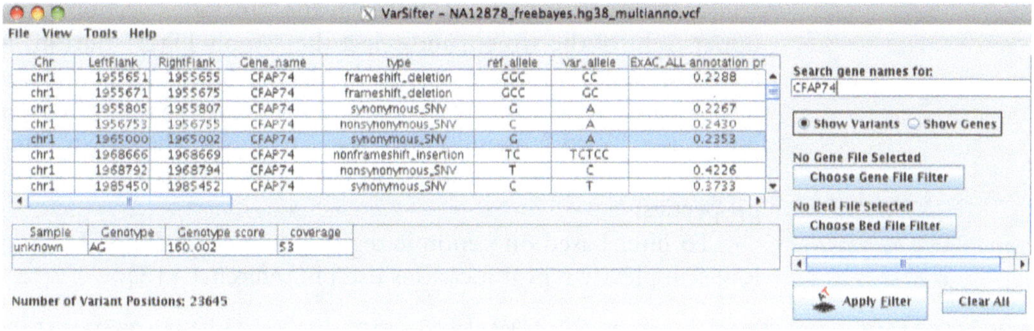

Fig. 2 VarSifter display of NA12878's CFAP74 gene variants

```
# Download ExAC variant annotations
annotate_variation.pl --buildver hg38 -downdb -webfrom annovar exac03
  humandb/
```

Finally, run ANNOVAR's table_annovar.pl script using these databases to add INFO fields to the variant VCF file:

```
# Add RefGene and ExAC annotations to the VCF file
table_annovar.pl --vcfinput NA12878.rmdup.freebayes.vcf.gz humandb/
  -buildver hg38 -remove -protocol refGene,exac03
  -out NA12878_freebayes -operation g,f humandb/
```

This will create a VCF file called NA12878_freebayes.hg38_multianno.vcf, in which all of the Freebayes variants are annotated with their RefSeq locations (on a "gene" basis, as indicated by the -operation "g" value) and with population frequencies from the Exome Aggregation Consortium's ExAC database on a "filter" basis, suitable for viewing and filtering with VarSifter.

3.5 Viewing and Filtering Variants

Figure 2 shows the graphical interface of the viewing tool VarSifter [20]. Using checkboxes and custom queries, VarSifter makes it possible to filter annotated VCF files based on variants' impact on protein sequence, population allele frequency (e.g., from large sequencing projects such as ExomeGO), consistency with known Mendelian inheritance schemes, or potential to be pathogenic from databases like HGMD [22].

To open a VCF file in VarSifter, type:

```
java -Xmx1G -jar VarSifter<version>.jar
```

where "version" is the version number of VarSifter installed, and the memory requirement specified after "-Xmx" may need to be increased for larger VCF files. Open the VCF file to be viewed, and select the name of the INFO fields containing the mutation type and gene name (for RefSeq annotations added by ANNOVAR, these fields have identifiers "ExonicFunc.RefGene" and "Gene.RefGene," respectively). VarSifter then allows the user to select which INFO annotations in the VCF file to display.

To filter variants, check boxes in the "Include" section at the upper right of the screen, and select the "Apply Filter" button. In addition, regular expressions for particular genes can be entered into "Search gene names for" box, and the "Custom query" option under the "View" menu allows users to create their own filters based on genotypes, variant scores, and numerous other variant properties.

To filter based on genomic region, download the BED file of low-complexity repeat locations used in reference [15]:

```
wget https://github.com/lh3/varcmp/blob/master/scripts/LCR-hs38.bed.gz?raw=true
```

An option to filter based on an uncompressed BED file's genomic locations using the "Choose BED File Filter" button allows the user to view only the variants which lie within the specified regions. The file of low copy repeats, above, can be used to find variants which are likely false positives.

3.6 Supplementary Protocols

3.6.1 Obtaining Sequence Data from NCBI

The NCBI SRA acts as a repository for sequence data. After a user locates a sequence dataset of interest, NCBI recommends the use of its "SRA Toolkit" to download data. With the SRA Toolkit and Aspera Connect/ascp installed, a sample dataset for this method can be downloaded using the "prefetch" command, and converted to FASTQ format using fastq-dump:

```
# Download the .sra file and necessary NCBI reference sequences into
# your home directory's "ncbi" subdirectory
prefetch SRR1611184
# Convert .sra file into fastq format, separating read 1 and read 2
# into two files, and gzip'ing to save space
fastq-dump -I --gzip --split-3 SRR1611184
```

If a novoalign license has not been purchased, the "–gzip" option should be omitted from the fastq-dump call, and larger, uncompressed FASTQ files will be generated.

3.6.2 Downloading a Genome's Reference Sequence

The most convenient way to obtain a reference sequence for an organism is often to download FASTA files from the University of California, Santa Cruz's Genome Browser. Since these files can be moderately large (e.g., 1 gigabyte for a gzipped file of the sequences from NCBI's Build GRChr38 of the human genome), it is best to use the rsync command:

```
# Find reference sequences for your genome build of interest at
# ftp://hgdownload.cse.ucsc.edu/goldenPath
# rsync places gzipped fasta file in current directory
rsync -a -P
  rsync://hgdownload.cse.ucsc.edu/goldenPath/hg38/bigZips/hg38.fa.gz ./
```

3.6.3 Removing Alternate Haplotypes and Masking Pseudo-Autosomal Regions in the Reference

Newer builds of the human reference sequence (e.g., hg19, hg38) contain "alt" sequences containing alternate haplotypes of regions of the human genome that are highly polymorphic in the population. Although some aligners, like bwa-mem, will consider these alternate haplotype sequences differently in their alignment algorithms, variant and genotype callers typically don't know how to handle these alignments yet, so the value in aligning to alternate haplotypes is still unclear, and entries for alternate haplotypes should usually be removed from the reference FASTA file before formatting an alignment index. If a user is particularly interested in one of the regions for which common alternate haplotypes exist, he or she should explore possibilities for custom, region-specific analyses, which are not covered in this protocol.

In addition to removing alternate haplotypes from the reference sequence, it is also helpful to mask the pseudo-autosomal regions, or "PARs," of the organism's sex chromosomes. These PARs are identical for the X and the Y chromosomes, and like the alternate haplotypes above, they can confuse alignment software because they appear to be exact repeats. For human reference builds, the coordinates of these regions are available from the UCSC Genome Browser.

A simple Perl script for removing alternate haplotype sequences from a reference assembly and masking the PARs on chromosome Y for human is included in the "scripts" subdirectory of the github repository at http://github.com/nhansen/MMBVariantCalling, and can be run by typing:

```
# Specify output fasta as an "mfa", or "multi-FASTA" file:
prepare_reference_fasta.pl --build hg38 --input hg38.fa.gz
  -output hg38.mfa
```

3.6.4 Formatting a Novoalign Database

Before running novoalign, it is necessary to first have a FASTA-formatted database of reference (i.e., chromosome) sequences, against which the program will align the reads. Once a particular build of the organism's genome has been selected (e.g., hg19, also known as "GRCh37"), a FASTA file of chromosome sequences can be downloaded from a resource like the UCSC Genome Browser. If a reference database has not been created ahead of time, it can be created using the "novoindex" command of novoalign:

```
# Format an hg38 index for novoalign (requires 8Gb of memory
# for a human genome)
novoindex -n hg38 hg38.ndx hg38.mfa
```

4 Notes

- The program novoalign is available with a free license. However, with a paid license, it can be run using multiple threads, which can considerably speed up processing time, especially for whole genome sequences. Without a paid license, it is also necessary to use uncompressed FASTQ files as input to novoalign, which necessitates a greater use of disk space.

- When running novoalign on paired-end reads, the reads in the FASTQ file for read number one must be in the same order as their pairs in the FASTQ file for read number two. Most paired FASTQ files are correctly ordered, but a mismatch in ordering will result in an error of the form "Error: Read headers do not match at Record #...".
- If reads are greater than 100 bases in length, novoalign can be run with the option "-t 400" to decrease run time without significant loss of alignment accuracy.
- The program Freebayes should be installed using "git clone" with the recursive option, which will download necessary external repositories that are not included in the distributed tarball.

Acknowledgments

Artwork for Fig. 1 was provided by DXYN Studios. This work was supported by the Intramural Research Program of the National Human Genome Research Institute, National Institutes of Health. The content is solely the responsibility of the author and does not necessarily represent the official views of the National Human Genome Research Institute or the National Institutes of Health.

References

1. International Human Genome Sequencing Consortium (2001) Initial sequencing and analysis of the human genome. Nature 409 (6822):860. http://dx.doi.org/10.1038/35057062
2. Bentley DR, Balasubramanian S, Swerdlow HP, Smith GP, Milton J, Brown CG, Hall KP, Evers DJ, Barnes CL, Bignell HR, Boutell JM, Bryant J, Carter RJ, Keira Cheetham R, Cox AJ, Ellis DJ, Flatbush MR, Gormley NA, Humphray SJ, Irving LJ, Karbelashvili MS, Kirk SM, Li H, Liu X, Maisinger KS, Murray LJ, Obradovic B, Ost T, Parkinson ML, Pratt MR, Rasolonjatovo IMJ, Reed MT, Rigatti R, Rodighiero C, Ross MT, Sabot A, Sankar SV, Scally A, Schroth GP, Smith ME, Smith VP, Spiridou A, Torrance PE, Tzonev SS, Vermaas EH, Walter K, Wu X, Zhang L, Alam MD, Anastasi C, Aniebo IC, Bailey DMD, Bancarz IR, Banerjee S, Barbour SG, Baybayan PA, Benoit VA, Benson KF, Bevis C, Black PJ, Boodhun A, Brennan JS, Bridgham JA, Brown RC, Brown AA, Buermann DH, Bundu AA, Burrows JC, Carter NP, Castillo N, Chiara M, Catenazzi E, Chang S, Neil Cooley R, Crake NR, Dada OO, Diakoumakos KD, Dominguez-Fernandez B, Earnshaw DJ, Egbujor UC, Elmore DW, Etchin SS, Ewan MR, Fedurco M, Fraser LJ, Fuentes Fajardo KV, Scott Furey W, George D, Gietzen KJ, Goddard CP, Golda GS, Granieri PA, Green DE, Gustafson DL, Hansen NF, Harnish K, Haudenschild CD, Heyer NI, Hims MM, Ho JT, Horgan AM, Hoschler K, Hurwitz S, Ivanov DV, Johnson MQ, JamesT, Huw Jones TA, Kang GD, Kerelska TH, Kersey AD, Khrebtukova I, Kindwall AP, Kingsbury Z, Kokko-Gonzales PI, Kumar A, Laurent MA, Lawley CT, Lee SE, Lee X, Liao AK, Loch JA, Lok M, Luo S, Mammen RM, Martin JW, McCauley PG, McNitt P, Mehta P, Moon KW, Mullens JW, Newington T, Ning Z, Ling Ng B, Novo SM, O'Neill MJ, Osborne MA, Osnowski A, Ostadan O, Paraschos LL, Pickering L, Pike AC, Pike AC, Chris Pinkard D, Pliskin DP, Podhasky J, Quijano VJ, Raczy

C, Rae VH, Rawlings SR, Chiva Rodriguez A, Roe PM, Rogers J, Rogert Bacigalupo MC, Romanov N, Romieu A, Roth RK, Rourke NJ, Ruediger ST, Rusman E, Sanches-Kuiper RM, Schenker MR, Seoane JM, Shaw RJ, Shiver MK, Short SW, Sizto NL, Sluis JP, Smith MA, Ernest Sohna Sohna J, Spence EJ, Stevens K, Sutton N, Szajkowski L, Tregidgo CL, Turcatti G, Vandevondele S, Verhovsky Y, Virk SM, Wakelin S, Walcott GC, Wang J, Worsley GJ, Yan J, Yau L, Zuerlein M, Rogers J, Mullikin JC, Hurles ME, McCooke NJ, West JS, Oaks FL, Lundberg PL, Klenerman D, Durbin R, Smith AJ (2008) Accurate whole human genome sequencing using reversible terminator chemistry. Nature 456(7218):53. doi:10. 1038/nature07517

3. The 1000 Genomes Project Consortium (2012) An integrated map of genetic variation from 1,092 human genomes. Nature 491 (7422):56. http://dx.doi.org/10.1038/nature11632
4. Hoadley KA, Yau C, Wolf DM, Cherniack AD, Tamborero D, Ng S, Leiserson MD, Niu B, McLellan MD, Uzunangelov V, Zhang J, Kandoth C, Akbani R, Shen H, Omberg L, Chu A, Margolin AA, van't Veer LJ, N. Lopez-Bigas, Laird PW, Raphael BJ, Ding L, Robertson AG, Byers LA, Mills GB, Weinstein JN, Waes CV, Chen Z, Collisson EA, Benz CC, Perou CM, Stuart JM (2014) Multiplatform analysis of 12 cancer types reveals molecular classification within and across tissues of origin. Cell 158(4):929. doi:http://dx.doi.org/10.1016/j.cell.2014.06.049. http://www.sciencedirect.com/science/article/pii/S0092867414008769
5. Nielsen R, Paul JS, Albrechtsen A, Song YS (2011) Genotype and SNP calling from next-generation sequencing data. Nat Rev Genet 12 (6):443. doi:10.1038/nrg2986
6. Li H, Durbin R (2010) Fast and accurate long-read alignment with Burrows-Wheeler transform. Bioinformatics 26(5):589. doi:10.1093/bioinformatics/btp698
7. Langmead B, Trapnell C, Pop M, Salzberg SL (2009) Ultrafast and memory-efficient alignment of short DNA sequences to the human genome. Genome Biol 10(3):R25. doi:10.1186/gb-2009-10-3-r25
8. Li R, Yu C, Li Y, Lam TW, Yiu SM, Kristiansen K, Wang J (2009) SOAPZ: an improved ultrafast tool for shot real alignment. Bioinformatics 25(15):1966. doi:10.1093/bioinformatics/btp336
9. Smith TF, Waterman MS (1981) Identification of common molecular subsequences. J Mol Biol 147(1):195
10. Needleman SB, Wunsch CD (1970) A general method applicable to the search for similarities in the amino acid sequence of two proteins. J Mol Biol 48(3):443
11. Garrison E, Marth G (2012) Haplotype-based variant detection from short-read sequencing. arXiv:1207.3907V2 [q-bio.GN]. http://arxiv.org/abs/1207.3907
12. Teer JK, Bonnycastle LL, Chines PS, Hansen NF, Aoyama N, Swift AJ, Abaan HO, Albert TJ, Margulies EH, Green ED, Collins FS, Mullikin JC, Biesecker LG (2010) Systematic comparison of three genomic enrichment methods for massively parallel DNA sequencing. Genome Res 20(10):1420. doi:10.1101/gr.106716.110
13. Ewing B, Green P (1998) Base-calling of automated sequencer traces using phred. II. Error probabilities. Genome Res 8(3):186
14. Van der Auwera GA, Carneiro MO, Hartl C, Poplin R, Del Angel G, A. Levy-Moonshine, Jordan T, Shakir K, Roazen D, Thibault J, Banks E, Garimella KV, Altshuler D, Gabriel S, DePristo MA (2013) From FastQ data to high confidence variant calls: the Genome Analysis Toolkit best practices pipeline. Curr Protoc Bioinformatics 11(1110):11.10.1. doi:10.1002/0471250953. bi1110s43
15. Li H (2014) Toward better understanding of artifacts in variant calling from high-coverage samples. Bioinformatics 30(20):2843. doi:10.1093/bioinformatics/ btu356
16. Goecks J, Nekrutenko A, Taylor J (2010) Galaxy: a comprehensive approach for supporting accessible, reproducible, and transparent computational research in the life sciences. Genome Biol 11(8):R86. doi:10.1186/gb-2010-11-8-r86
17. Li H, Handsaker B, Wysoker A, Fennell T, Ruan J, Homer N, Marth G, Abecasis G, Durbin R (2009) The Sequence Alignment/map format and SAM tools. Bioinformatics 25 (16):2078. doi:10.1093/ bioinformatics/btp352
18. Danecek P, Auton A, Abecasis G, Albers CA, Banks E, DePristo MA, Handsaker RE, Lunter G, Marth GT, Sherry ST, McVean G, Durbin R (2011) The variant call format and VCF tools. Bioinformatics 27(15):2156. doi:10.1093/bioinformatics/btr330
19. Wang K, Li M, Hakonarson H (2010) ANNOVAR: functional annotation of genetic variants from high-through put sequencing data. Nucleic Acids Res 38(16):e164. doi:10.1093/nar/gkq603
20. Teer JK, Green ED, Mullikin JC, Biesecker LG (2012) Var Sifter: visualizing and analyzing exome-scale sequence variation data on a

desktop computer. Bioinformatics 28(4):599. doi:10.1093/bioinformatics/btr711

21. Karolchik D, Hinrichs AS, Furey TS, Roskin KM, Sugnet CW, Haussler D, Kent WJ (2004) The USSC Table Browser data retrieval tool. Nucleic Acids Res 32(Database issue): D493. doi:10. 1093/nar/gkh103

22. Stenson PD, Mort M, Ball EV, Shaw K, Phillips A, Cooper DN (2014) The Human Gene Mutation Database: building a comprehensive mutation repository for clinical and molecular genetics, diagnostic testing and personalized genomic medicine. Hum Genet 133 (1):1. doi:10.1007/s00439-013-1358-4

Chapter 12

Genome-Scale Analysis of Cell-Specific Regulatory Codes Using Nuclear Enzymes

Songjoon Baek and Myong-Hee Sung

Abstract

High-throughput sequencing technologies have made it possible for biologists to generate genome-wide profiles of chromatin features at the nucleotide resolution. Enzymes such as nucleases or transposes have been instrumental as a chromatin-probing agent due to their ability to target accessible chromatin for cleavage or insertion. On the scale of a few hundred base pairs, preferential action of the nuclear enzymes on accessible chromatin allows mapping of cell state-specific accessibility in vivo. Such accessible regions contain functionally important regulatory sites, including promoters and enhancers, which undergo active remodeling for cells adapting in a dynamic environment. DNase-seq and the more recent ATAC-seq are two assays that are gaining popularity. Deep sequencing of DNA libraries from these assays, termed genomic footprinting, has been proposed to enable the comprehensive construction of protein occupancy profiles over the genome at the nucleotide level. Recent studies have discovered limitations of genomic footprinting which reduce the scope of detectable proteins. In addition, the identification of putative factors that bind to the observed footprints remains challenging. Despite these caveats, the methodology still presents significant advantages over alternative techniques such as ChIP-seq or FAIRE-seq. Here we describe computational approaches and tools for analysis of chromatin accessibility and genomic footprinting. Proper experimental design and assay-specific data analysis ensure the detection sensitivity and maximize retrievable information. The enzyme-based chromatin profiling approaches represent a powerful and evolving methodology which facilitates our understanding of how the genome is regulated.

Key words Chromatin remodeling, DNase-seq, ATAC-seq, High-throughput sequencing, Computational genomics, Genomic footprinting

1 Introduction

Chromatin exerts significant regulation of the genome through tight packaging of DNA in the nucleus of a eukaryotic cell, preventing access of transcription factors and other proteins to their cognate sites [1, 2]. Accessibility at promoters, enhancers, or silencers is actively maintained and dynamically altered in a cell- and condition-specific manner [3–7]. Chromatin accessibility can be measured by the susceptibility of DNA either to cleavage by nucleases such as DNase I [8] or to transposition [9]. For example,

Ewy Mathé and Sean Davis (eds.), *Statistical Genomics: Methods and Protocols*, Methods in Molecular Biology, vol. 1418, DOI 10.1007/978-1-4939-3578-9_12, © Springer Science+Business Media New York 2016

DNase I hypersensitive sites (DHSs) are defined as the regions particularly prone to cutting by DNase I, and they represent regions with an "open chromatin" structure. DNase I hypersensitivity coupled with high-throughput sequencing (DNase-seq) has been used to provide genome-wide identification of functional regulatory elements [8, 10]. More recently, the assay for transposase-accessible chromatin using sequencing (ATAC-seq) was developed as a simpler method that can be performed on a small number of cells. Each assay generates a continuous high-resolution profile of chromatin accessibility along the genome in a given cell state [9]. DNase-seq and ATAC-seq have been shown to produce very similar signal profiles, in contrast to the poor concordance between DNase-seq and FAIRE (formaldehyde assisted isolation of regulatory elements)-seq [11]. FAIRE may not permit sensitive detection of regulatory regions due to high background signals. Here we focus on the enzyme-based chromatin assays DNase-seq and ATAC-seq and discuss computational analysis methods that extract epigenetic information from the data generated.

If a DNase-seq library is sequenced deeply to yield a large number (>300 million) of reads, the genomic loci which are highly occupied by transcription factors may be identified as narrow regions of protection against DNase I cleavage, termed "footprints" [12–14]. Although the cost of sequencing becomes an issue in practice, sufficient tag coverage allows pinpointing of specific binding sites at the nucleotide resolution. However, the detection of protein footprints and inferring the identity of factors are technically and computationally more challenging in comparison to the detection of accessible regions.

This chapter provides a description of the procedures that we have been employing to analyze DNase-seq and ATAC-seq data. Surveys of existing methods mostly cover analysis tools for ChIP-seq or RNA-seq [15], with fewer studies comparing different analysis methods for DNase-seq [3, 16–18]. The chapter is divided into two parts based on the resolution of analysis: First on analyses pertaining to chromatin accessibility on the scale of 100 bp to 1 kb, and the other on analyses of transcription factor footprints on the bp scale. Within each part, the algorithms are roughly categorized into different types of analyses: (1) generation of browser tracks for visual exploration; (2) detection of significant regions (hotspots or footprints) based on a background probability model and calculation of statistical measures; (3) artifact adjustment and filtering; (4) annotation of the identified regions with respect to other genomic features or related data; (5) downstream analyses and useful plotting strategies for delineating meaningful patterns from the combined set of regions across multiple conditions or time points.

2 Analysis of Chromatin Accessibility

2.1 Assay Protocols, Biases, and Data Reproducibility

It is worthwhile to note that distinct protocols exist under the same term "DNase-seq" (Table 1). Depending on the DNase-seq protocol, there are different data features and biases that one needs to take into consideration for the analysis and interpretation of the data. To distinguish between the protocols in this chapter, we denote the size selection-based methods as "DNase-seq I" and "DNase-seq II," according to the sequencing type. We designate the biotin end-labeling method as "DNase-seq III." With DNase-seq I and II, the reads (aka tags) come from the ends of the DNA fragments within accessible chromatin which are cleaved and released. The size selection for 100–500 bp range enriches for fragments that are doubly cut by DNase I. With DNase-seq III, individual DNA ends are labeled with biotin and captured for single-end sequencing. Interestingly, the sample processing protocols DNase-seq I/II and DNase-seq III produce different DNA sequence bias patterns [19]. Adjusting for the sequence-dependent cleavage bias becomes important for analysis of cut counts and TF footprint detection (Subheading 3.3).

ATAC-seq utilizes a completely different approach by inserting sequencing adaptors directly to accessible chromatin using a transposase. In contrast to DNase I whose DNA cleavage activity is used to mark open chromatin, this assay relies on transposition as the primary molecular reaction for targeting and sampling open chromatin. Therefore, reaction kinetics and targeting preferences are likely to be distinct from DNase-based methods. Despite the difference, the correlation, at least at the level of chromatin accessibility, between ATAC-seq and DNase-seq I was reported to be as high

Table 1
Enzyme-based chromatin assays

Assay	Protocol feature	Sequencing	Notes	References
DNase-seq I (UW)	Size selected fragments released by two genomic hits	Single-end	Specific for doubly cut chromatin	[8]
DNase-seq II (DNase-FLASH)	Size selected fragments released by two genomic hits	Paired-end	Length analysis reveals nucleosome occupancy/positioning information	[33]
DNase-seq III (Duke)	End capture with biotin	Single-end	Background signal from single-strand nicks	[10]
ATAC-seq	Fragments with transposed sequencing adaptors at both ends	Paired-end	Length analysis reveals nucleosome occupancy/positioning information; Mitochondrial DNA contamination	[9]

as that between DNase-seq I and III [9]. The correlation between ATAC-seq and DNase-seq III was slightly lower.

Current high throughput sequencing of a single or multiplexed sample routinely produces hundreds of millions of sequence reads of 35–100 bp in length from a lane. Quality-filtered sequence reads are then aligned to the reference genome. The regions densely populated with reads are putative DHSs or accessible chromatin regions. Even though accessibility data generated from proper experimental design are reproducible and visually convincing, there are systematic biases that should be corrected. For example, a proportion of the reads may not align to the reference genome simply because the genome of the cells used for the experiment is structurally different from the reference genome, containing aberrations such as polyploidy, translocations, or other mutations. Amplified regions would contribute more to the DNA sample and deleted regions would not produce any sequence reads.

Another source of sequencing data bias arises from the fact that the genomic locations of the sequenced fragments are inferred from finding the "best match" in the genome sequence. However, the accuracy of aligning a read back to the genome varies greatly depending on the sequence and read length. Hence it is necessary to consider the read "mappability" (Fig. 1). A given n-mer sequence read may occur at a unique location or at multiple genomic positions under a preset mismatch tolerance. Although reads with multiple genomic matches can, in principle, be probabilistically mapped, a common alignment approach allows only one genomic coordinate for each read and discards reads that cannot be uniquely mapped. The procedure creates "dark spots" across the genome and directly affects the background probability of observing reads at any given position in the genome (Fig. 1).

Identification of the genomic regions where reads are significantly enriched over the background must take into account these and other sources of bias and artifacts in the sequencing data. The objective of an algorithm for detecting accessible regions is to find all of the truly read-enriched sites while minimizing the false positive rate (Subheading 2.3).

2.2 Building a Profile for Data Visualization in a Genome Browser

Visualization of the data is important for assessing data quality and for confirming results from a global analysis. It is useful to note that there are a few different approaches even for this apparently simple practice. First, there are multiple ways of generating the data tracks depending on how the read distribution is summarized. A density profile or a coverage map can be generated by calculating (i) the number of reads overlapping each genomic bin of fixed size (ranging from 1 bp to 20 bp, for example), (ii) the number of reads whose starting nucleotides are in each bin, (iii) the number of reads whose fragments (extended from the starting nucleotide into the genomic sequence by a fixed length) overlap each genomic bin, or

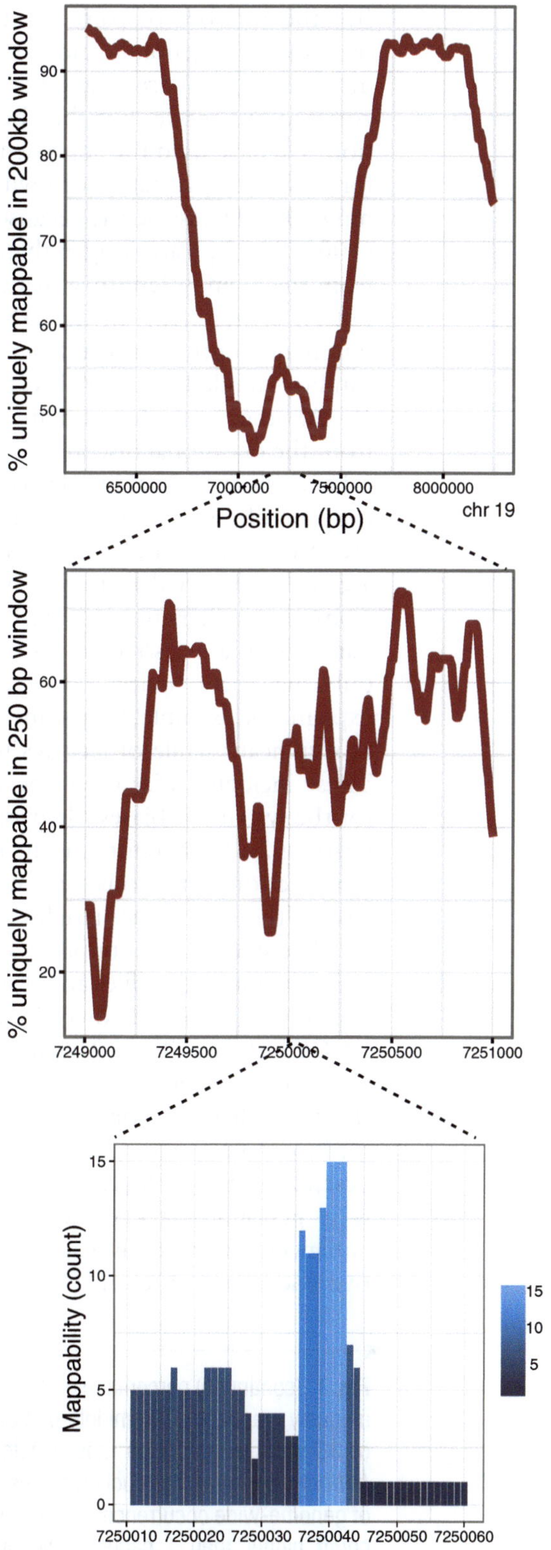

Fig. 1 Mappability of sequence reads is far from uniform, across the scales from Mb to bps. The *top plot* shows the fluctuation of mappability along mm9 as the

(iv) the number of paired-end reads whose spanned fragments overlap each genomic bin, etc. The differences resulting from methods (i)–(iv) are negligible when the tracks are browsed in a zoomed-out mode, but they become noticeable at a high resolution. We have used the approach (iii) for DNase-seq I data and (iv) for DNase-seq II data, both with a nucleotide resolution (no binning). For DNase-seq I, because the size-selected DNA fragments (100–1000 bp) are longer than the sequence reads (35–100 bp) from the 5′ ends, length adjustment is made to estimate the distribution of fragments in the DNase-treated sample. If the average fragment length is known from the sample QC, each sequence tag can be extended in the 3′ direction up to that length.

Another consideration for generating data track files stems from the occasional disconcordance between results from a statistical analysis and visual impressions from the browser tracks. For the purpose of assessing data quality, minimally processed data tracks are often used to display the "raw data" as well as the anomalies that are to be excluded from any systematic analysis, such as artifacts from repeat elements and PCR-duplicated reads. Such unadjusted data tracks are also used to convey the final analysis results in published data figures. However, the unadjusted data may not explain, for example, why some weakly accessible sites are detected as significant while other sites with similar read densities are not. These incidents arise often, because a detection algorithm adjusts for the systematic biases when assessing statistical significance (Subheading 2.3). Therefore, using adjusted data tracks might produce visualization more consistent with the results from statistical analyses, although this approach is not widely used.

There are several publicly accessible browser tools that accept users' genomic data files and display them in the context of annotation tracks such as known genes, ncRNAs, repeat elements, and ENCODE data (Table 2). The University of California Santa Cruz Genome Browser has been popular and their website also provides the Table Browser from which one can download public data tracks for incorporation into further correlative analyses (http://genome.ucsc.edu/cgi-bin/hgTables). Integrated Genome Browser (IGB) is a genomic data browser which has undergone significant enhancements recently, supporting many file formats. The Integrative Genomics Viewer (IGV) and the Washington University

Fig. 1 (continued) percentage of 35-mers in a 200 kb moving window which are uniquely mappable. The middle plot displays the mappability as the percentage of 35-mers in a 250 bp moving window which are uniquely mappable. The *bottom plot* shows the nucleotide-resolution mappability itself, i.e., the number of genome-wide occurrences of each 35-mer. The positions with the mappability count higher than 1 cannot have any reads mapped from commonly used parameter settings of an alignment tool

Table 2
Genome browsers for data visualization

Browser	Data types	File format for upload	Features	URL
UC Santa Cruz Genome Browser	ChIP-seq, RNA-seq, DNase-seq, 4C	bigwig, wig, bed, bigbed, bedgraph, gff, gtf, bam	Preloaded annotation tracks including ENCODE data, comparative genomics, etc.; allows mirror installation	http://genome.ucsc.edu
Integrated Genome Browser (IGB)	ChIP-seq, RNA-seq, whole-genome seq, 4C, microarray	bam, sam, sgr, bigwig, wig, bed, bedgraph, bgr, chp, fasta, gff	Originally developed by Affymetrix for tiling array data; released as open-source with similar capabilities as UCSC browser	http://bioviz.org/igb/
Integrative Genomics Viewer (IGV)	ChIP-seq, RNA-seq, whole-genome seq, SNP, variants	bam, sam, bed, bedgraph, bigwig, fasta, gff, gtf	Similar capabilities as UCSC browser; 1000 human genomes available	http://www.broadinstitute.org/igv/
WashU Epigenome Browser	ChIP-seq, DNA methylation (bisulfite seq), 5C, Hi-C, ChIA-PET	bam, bigbed, bigwig, tabix	Preloaded ENCODE data; juxtaposition of distant genomic regions, diagonal heatmaps and circlet plots for long-range interaction data; open-source	http://epigenomegateway.wustl.edu/

Epigenome Browser have assay-specific capabilities for certain data types that other browsers do not provide. Hence, investigators who generate such data may benefit from the customized data exploration tools from these browsers.

2.3 Region Detection Algorithm

There are only a few algorithms specifically developed to identify accessible chromatin regions from DNase-seq (protocols I, II, and III) data, while numerous software packages exist now for calling peaks or enriched sites from ChIP-seq data. We have developed and described a one-pass algorithm for detecting "hotspots" in detail elsewhere, and refer the reader to [16] and the accompanying source code "DNase2Hotspots" and documentation at http://sourceforge.net/projects/dnase2hotspots.

Here we briefly outline the core components of the algorithm. DNase2Hotspots finds hotspots, or local enrichment of reads in a 250 bp target window relative to a local background (surrounding 200 kb window), based on the binomial distribution. The usage of a local background, rather than a genome-wide uniform background, adjusts for the local fluctuations in read distributions

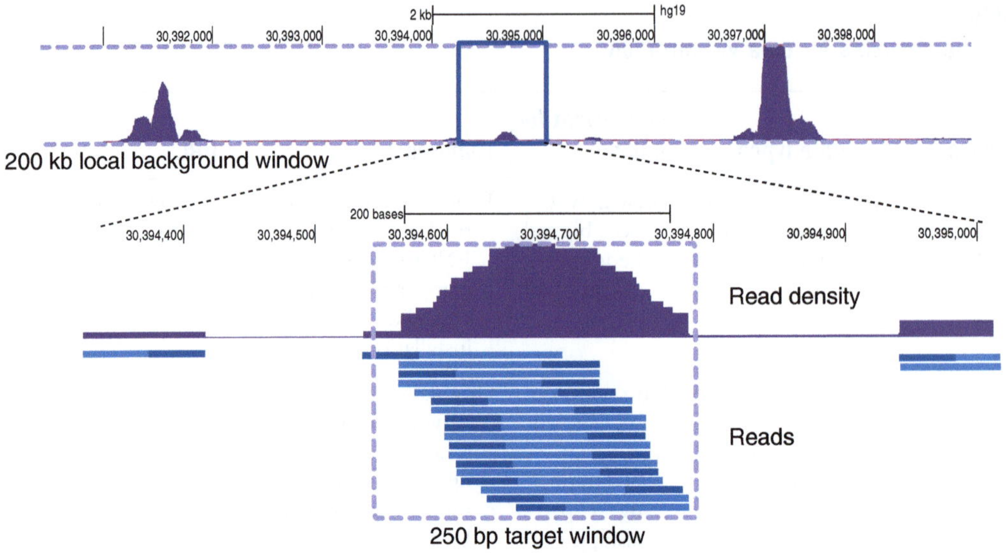

Fig. 2 DNase2Hotspots assesses the enrichment of extended reads within a target window by computing the binomial z score with respect to the local background window. The *top track* shows a part of the larger 200 kb background window. The *bottom track* shows the distribution of individual reads (*dark blue*) and the estimated fragments (*light blue*) which are extended from the single end reads of DNase-seq I or III. For DNase-seq II and ATAC-seq, the ends of the fragments are defined by the paired end reads. The maximum read density or the average read density can be associated with each hotspot as a quantitative measure of chromatin accessibility

reflecting large-scale differences in chromatin accessibility or copy number variations. Significance of read enrichment within the target window over the local background is assessed using a binomial z score from counting expected read occurrences only at uniquely mappable genomic coordinates (Figs. 1 and 2). Hence the mappability is incorporated directly into the z score. An unthresholded hotspot is defined as a contiguous cluster of 250 bp windows whose z scores are nominally significant, i.e., greater than 2. The final z score threshold is imposed based on an empirically calculated false discovery rate (FDR). When the analysis calls for stringent region calling, hotspots are selected with 0 % FDR, i.e., the z score threshold is set by the minimum absolute value that does not allow any hotspots called from the randomized data. If it is desirable to include a larger number of accessible sites with a higher sensitivity of detection, then hotspots can be called with a higher FDR, such as 1 % or 5 %.

The ENCODE group at the University of Washington had developed the original hotspot detection algorithm which uses a two-pass procedure to capture weakly accessible sites that can be masked by nearby big DHSs. The ENCODE program is currently available at http://www.uwencode.org/proj/hotspot/.

F-seq was developed by the authors of DNase-seq III [20, 21] and an updated version is available at https://github.com/aboyle/F-seq. It is not unusual for the same data to produce significantly different sets of hotspots or peaks depending on the detection algorithm. To reconcile the different sets without relying on a single detection method, sometimes an ad hoc combination of the different sets is used to obtain the final set of accessible regions from the data [3].

2.4 Region Annotation and Integrative Analyses

Much of the biologically meaningful data analyses are performed during this stage of the analysis. When there are several accessibility profiles from different experimental conditions or cell states, it is very useful to have a "master set" of hotspots derived from reconciling the boundaries of overlapping hotspots. Essentially the same site may show up from multiple biological samples as hotspots with slightly different start and end coordinates. There can be different ways of defining the boundaries of hotspots to construct the master set for subsequent analyses. For example, each hotspot in the master set which represents overlapping hotspots detected in individual samples can be defined as their union, intersection, or union of the top three accessible sites, etc. The determination of the master hotspots is necessary for a comparative analysis which reveals chromatin accessibility changes across the samples or during a time course, based on a single convenient measure per hotspot. We have been using the maximum read density or the average read density as such a measure which reflects the extent of accessibility at each hotspot. Cluster analyses or supervised classification methods can then be applied to discern distinct patterns of chromatin behavior.

Once the hotspots are obtained, it is often desirable to annotate the sites with genomic information such as the closest genes, the distance to TSSs, or whether they overlap with regions found from other related data [16]. For instance, one can examine the proportions of accessible sites located at promoters, introns, or intergenic regions, or the extent of overlap with regions exhibiting other enhancer marks or repressive chromatin marks.

2.5 Motif Analysis on Hotspots

Motif analysis allows a higher resolution examination of the underlying genomic regions than any methods purely based on hotspots which can range up to a few kilobases. The presence of a TF binding motif element indicates a potential protein binding event within the accessible sites (*see* Subheading 3 for further discussions). There are two common types of sequence motif analyses that can be performed on the set of DNA sequences from a specific subset of the identified hotspots. One method is scanning the sequences for the presence of motifs for known TFs [22] (FIMO is available at http://meme.nbcr.net/meme/doc/fimo.html). It requires prior knowledge of TF binding motifs but the computation is straightforward.

Another analysis aims at discovery of novel motifs enriched in the target DNA sequences, which is computationally very intensive due to the large number of accessible sites that are often used as search input. A strategy to handle the computational demand is to reduce the total DNA content of the input set by narrowing down to the strongest signal regions, i.e., peaks or local summits. Deciding which regions to focus on critically affects the output motifs that are found to be enriched from the regions. One common caveat is selecting the top DHSs ranked by read density. Often the sites that produce the highest DNase-seq signal are constitutively open AT-rich regions whose accessibility may be governed more by their sequence characteristics than by dynamic chromatin regulation. By choosing the top 200 DHSs, for example, the investigator may only find simple repeat sequences that tend to avoid nucleosomes. Cell type-dependent and TF-specifically regulated sites are likely to reside in hotspots of modest read density. For this reason, we remove the top DHSs and include as many hotspots as possible for each genomic set of interest by limiting the searchable DNA sequences onto the narrow peaks within the hotspots [16]. For preparing input, the UCSC Table Browser can be used to extract the DNA sequences of specified genomic regions in the FASTA format.

The widely used de novo motif discovery tool MEME [23] uses an expectation maximization algorithm (http://meme.nbcr.net/meme/tools/meme). DREME is another discovery tool developed by the authors of MEME [24]. HOMER is a different tool that has gained popularity for ease of use [25]. These discovery tools seem to have different sensitivity for finding certain types of motifs. Therefore, it is recommended that users should try more than one method to discover a wider class of motifs. The enriched motifs found from the discovery step can be batch-queried against the known TF binding motifs available in motif databases such as JASPAR or UniPROBE, using the motif comparison tool TomTom (available at the same site for the MEME suite) [26].

3 Analysis of TF Footprints

TF footprinting aims to detect sites bound by all protein factors from the same biological sample with a nucleotide precision [12–14]. To find TF footprints, one looks for narrow regions (8–30 bp) on which cleavage (or transposition in the case of ATAC-seq) is significantly reduced in comparison to the immediately surrounding regions (Fig. 3). The analysis requires ultra-deep sequencing to achieve reasonable coverage of cleavage events for all the hotspots in the genome.

3.1 Data Requirement

The same experimental protocols for DNase-seq or ATAC-seq are used to generate data for the purpose of genomic TF footprinting. However, additional data standards are imposed to determine the suitability of the data for higher resolution analyses. First, the depth

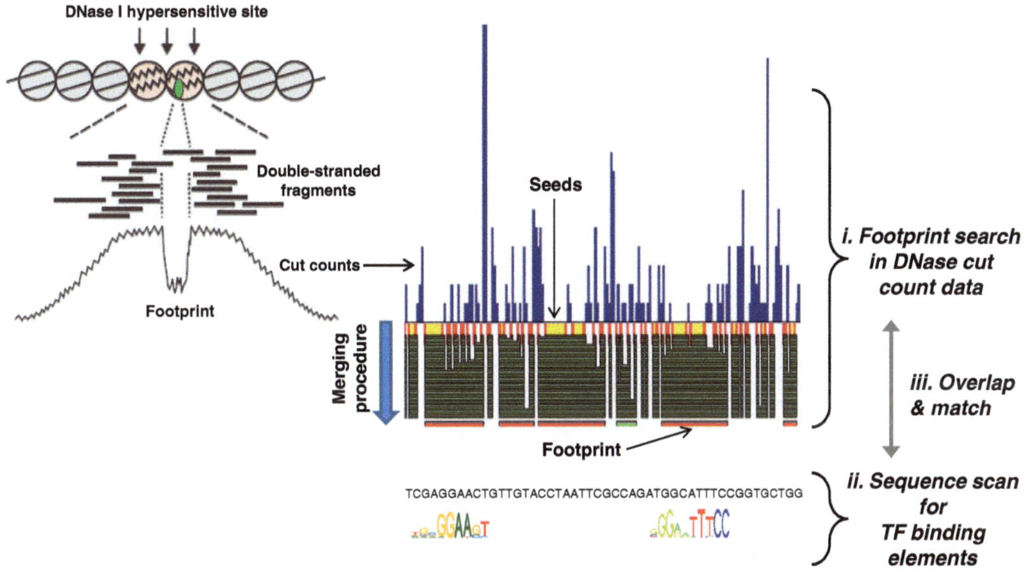

Fig. 3 Illustration of TF footprints which can be detected as protected regions from DNase cleavage and the corresponding narrow "valleys" in the cut count profile. DNase2TF begins with data-derived seeds and merges neighboring candidate regions until the significance of depletion no longer improves. Putative footprints are overlaid with known TF binding elements in the genome and assigned to best candidate TFs

of sequencing should be sufficient to provide at least 300 million uniquely mapped reads for a mammalian genome. Depending on the complexity of the sequencing library and the level of contaminating mitochondrial or other irrelevant DNA, the actual number of sequence reads needed may be much higher than the final target value. It is worth noting that, despite the decreasing cost and improved throughput of sequencing, currently feasible sequencing depths do not generally allow robust and reproducible detection of *individual* footprints for mammalian genomes.

Second, the data quality, as measured by the enrichment of nuclease activity within accessible chromatin, is useful to estimate the "signal-to-noise" ratio. We have used a quality score, similar to SPOT of the ENCODE team at the University of Washington, which is defined as proportion (ranging from 0 to 1) of reads overlapping FDR-unthresholded nominally defined hotspots. Datasets with low quality scores due to high background may be excluded or at least flagged for cautious data interpretation. Datasets with the quality score higher than 0.5 are generally considered to meet the suitability for TF footprinting analysis (Fig. 4).

3.2 Cut Count Profiles

Although the cut count profile is generally thought to convey the raw data, there are data features which result in a few variant definitions that can potentially affect the visual representation of putative footprints. First, since the exact location of a DNase

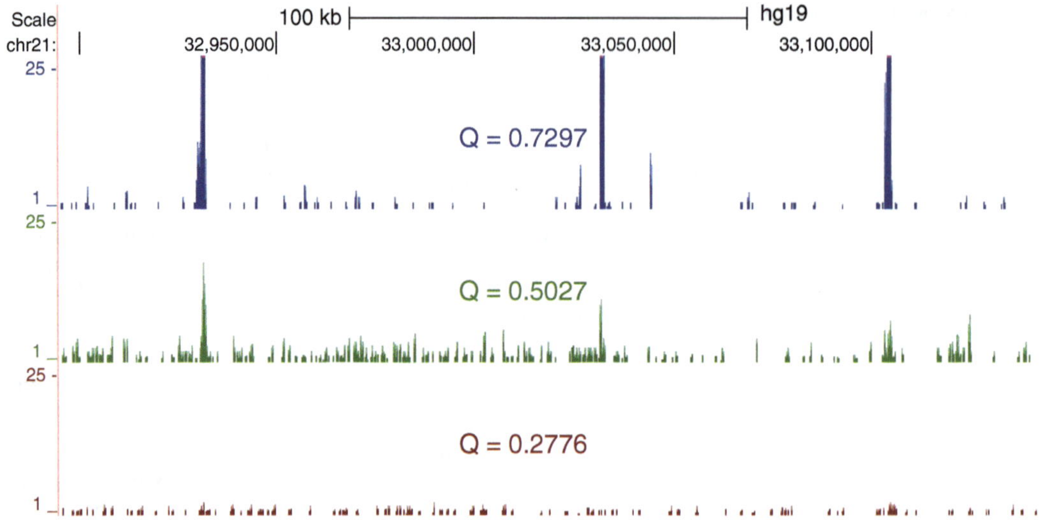

Fig. 4 Quality score is a genome-wide measure of read enrichment within the relevant regions versus the nonspecific background. Defined as the proportion of reads falling within nominally called hotspots, the Q score ranges from 0 to 1 and can be useful for deciding whether to advance to ultra-deep sequencing and footprinting analysis. Shown are read density tracks of example DNase-seq I samples with a range of quality scores

I cleavage event is *between* nucleotides, not *at* a nucleotide, cut counts are truly assigned to mid-points between bp coordinates. However, if the cut count data is to be uploaded to a browser, the obligatory assignment to integer coordinates necessitates a convention for the 1 bp-offset to be introduced to either the forward or the reverse direction reads [18]. While the choice is arbitrary, a consistent convention should be used throughout subsequent analyses.

ATAC-seq has an additional correction step to account for the distance between the sites of the sequencing adaptor insertion and the transposase binding [9]. The plus strand reads are shifted by +4 bps and the minus strand reads by −5 bps.

Analogous to the issue of raw versus adjusted data which was discussed in Subheading 2.2, the cut count profile may be generated to display the enzyme bias-corrected profile (*see* also Subheading 3.3). The choice depends on whether the resulting plot is intended to show technical features from the particular nuclease used to generate the data.

3.3 Artifacts from the Enzyme Bias on Sequence Patterns at Cut Sites

We and others have independently demonstrated that the sequence bias of DNase cleavage is quite pronounced [18, 27, 28], despite the previous assumption that DNase I cuts DNA in a sequence nonspecific manner. The cleavage bias generates distinct cut signatures when the cut count is averaged over TF binding motif elements. The cut signatures arise purely from the DNA sequence bias of DNase I, encoded in the tetramers or the hexamers surrounding the cleavage site, and are observed in deproteinized DNA [18, 27]. Analogously, distinct sequence biases have been observed for the transposase used in the ATAC-seq assay [9]. These findings raise doubts about the original interpretation of the cut signatures as reflecting the exact nucleotides bound and protected by sequence-specific proteins [14, 29, 30].

3.4 Cut Count Analysis When Matching ChIP-seq Data Are Available for TFs of Interest

To assess the true cut profiles at binding sites for a given TF, a reference dataset needs to be compiled: DNase-seq, ChIP-seq of the TF from a matching biological sample, and well-characterized PWM(s) for the TF. Then the TF binding motif elements called by FIMO can be separated based on whether they overlap ChIP peaks. The average cut count profiles can be computed over the two sets of motif elements (bound versus unbound) to delineate the effect of TF binding. The use of the bias-adjusted cut count profiles (Subheadings 3.2 and 3.3) may suppress the enzyme-specific artifacts and facilitate such comparisons.

3.5 Detection of Putative Footprints and Limitations in Inference of TF Occupancy In Vivo

The first footprint detection program, developed by Stamatoyannopoulos and coworkers, was used to identify footprints in DNase-seq data from *S. cerevisiae* [13]. The software does not scale well with large mammalian genomes. More recent detection programs have been developed based on completely different algorithms. The Wellington algorithm was designed to increase specificity of footprint calls by analyzing the plus and minus strands separately [31]. CENTIPEDE takes a different approach by making use of the a priori TF binding motifs, sequence conservation, and epigenetic marks [32]. However, the additional information available for making predictions about binding does not seem to result in higher accuracy [18]. We have developed an efficient computational algorithm that adjusts for the enzyme bias and read mappability [18]. The software package implementing the footprint detection algorithm is available as "DNase2TF" (http://sourceforge.net/projects/dnase2tfr) (Fig. 5).

Despite the progress, it remains difficult to detect individual TF footprints with an acceptable accuracy and reproducibility. The high quality of the data necessary for TF footprinting and ultra-deep sequencing remain as nontrivial technical bottlenecks. One should also acknowledge the inherent limitation of TF footprinting arising from lack of footprint depths for TFs with short DNA binding residence times [18].

3.6 Sequence Motif Analysis

Even though a comprehensive TF discovery analysis is generally not possible with current tools, some novel TFs may still be found from significantly enriched footprints. For example, detected footprints which do not overlap any matches from known TF binding motifs can be analyzed separately for enrichment of de novo motifs.

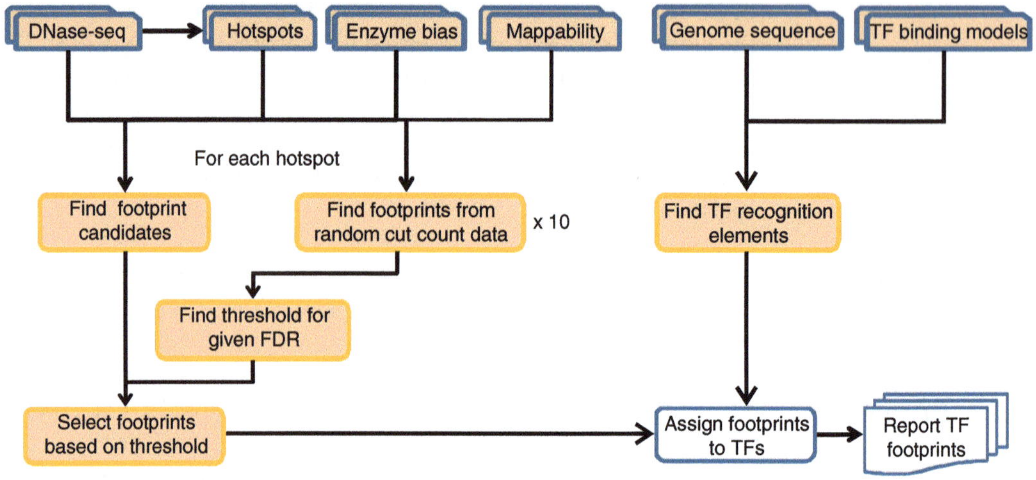

Fig. 5 Flow chart for DNase2TF and TF motif matching. Hotspots are precomputed as the set of regions within which the algorithm searches for footprints. For each hotspot, the reads are randomized to estimate the local FDR ten times. The FDR-thresholded footprints are called and matched with known TF binding motif elements

Since the genomic regions called as putative footprints are much narrower than accessible regions called as hotspots, the search for de novo motifs becomes more focused and specific.

References

1. Voss TC, Hager GL (2014) Dynamic regulation of transcriptional states by chromatin and transcription factors. Nat Rev Genet 15:69–81
2. Felsenfeld G, Groudine M (2003) Controlling the double helix. Nature 421:448–453
3. Hsiung CC, Morrissey CS, Udugama M, Frank CL, Keller CA, Baek S, Giardine B, Crawford GE, Sung MH, Hardison RC, Blobel GA (2015) Genome accessibility is widely preserved and locally modulated during mitosis. Genome Res 25:213–225
4. Morris SA, Baek S, Sung MH, John S, Wiench M, Johnson TA, Schiltz RL, Hager GL (2014) Overlapping chromatin-remodeling systems collaborate genome wide at dynamic chromatin transitions. Nat Struct Mol Biol 21:73–81
5. Thurman RE, Rynes E, Humbert R, Vierstra J, Maurano MT, Haugen E, Sheffield NC, Stergachis AB, Wang H, Vernot B, Garg K, John S, Sandstrom R, Bates D, Boatman L, Canfield TK, Diegel M, Dunn D, Ebersol AK, Frum T, Giste E, Johnson AK, Johnson EM, Kutyavin T, Lajoie B, Lee BK, Lee K, London D, Lotakis D, Neph S, Neri F, Nguyen ED, Qu H, Reynolds AP, Roach V, Safi A, Sanchez ME, Sanyal A, Shafer A, Simon JM, Song L, Vong S, Weaver M, Yan Y, Zhang Z, Zhang Z, Lenhard B, Tewari M, Dorschner MO, Hansen RS, Navas PA, Stamatoyannopoulos G, Iyer VR, Lieb JD, Sunyaev SR, Akey JM, Sabo PJ, Kaul R, Furey TS, Dekker J, Crawford GE, Stamatoyannopoulos JA (2012) The accessible chromatin landscape of the human genome. Nature 489:75–82
6. Siersbaek R, Nielsen R, John S, Sung MH, Baek S, Loft A, Hager GL, Mandrup S (2011) Extensive chromatin remodelling and establishment of transcription factor 'hotspots' during early adipogenesis. EMBO J 30:1459–1472
7. John S, Sabo PJ, Thurman RE, Sung MH, Biddie SC, Johnson TA, Hager GL, Stamatoyannopoulos JA (2011) Chromatin accessibility pre-determines glucocorticoid receptor binding patterns. Nat Genet 43:264–268
8. John S, Sabo PJ, Canfield TK, Lee K, Vong S, Weaver M, Wang H, Vierstra J, Reynolds AP, Thurman RE, Stamatoyannopoulos JA (2013) Genome-scale mapping of DNase I hypersensitivity. Curr Protoc Mol Biol Chapter 27:Unit 21.27
9. Buenrostro JD, Giresi PG, Zaba LC, Chang HY, Greenleaf WJ (2013) Transposition of native chromatin for fast and sensitive epigenomic profiling of open chromatin, DNA-binding proteins and nucleosome position. Nat Methods 10:1213–1218

10. Boyle AP, Davis S, Shulha HP, Meltzer P, Margulies EH, Weng Z, Furey TS, Crawford GE (2008) High-resolution mapping and characterization of open chromatin across the genome. Cell 132:311–322
11. Song L, Zhang Z, Grasfeder LL, Boyle AP, Giresi PG, Lee BK, Sheffield NC, Graf S, Huss M, Keefe D, Liu Z, London D, McDaniell RM, Shibata Y, Showers KA, Simon JM, Vales T, Wang T, Winter D, Zhang Z, Clarke ND, Birney E, Iyer VR, Crawford GE, Lieb JD, Furey TS (2011) Open chromatin defined by DNaseI and FAIRE identifies regulatory elements that shape cell-type identity. Genome Res 21:1757–1767
12. Boyle AP, Song L, Lee BK, London D, Keefe D, Birney E, Iyer VR, Crawford GE, Furey TS (2011) High-resolution genome-wide in vivo footprinting of diverse transcription factors in human cells. Genome Res 21:456–464
13. Hesselberth JR, Chen X, Zhang Z, Sabo PJ, Sandstrom R, Reynolds AP, Thurman RE, Neph S, Kuehn MS, Noble WS, Fields S, Stamatoyannopoulos JA (2009) Global mapping of protein-DNA interactions in vivo by digital genomic footprinting. Nat Methods 6:283–289
14. Neph S, Vierstra J, Stergachis AB, Reynolds AP, Haugen E, Vernot B, Thurman RE, John S, Sandstrom R, Johnson AK, Maurano MT, Humbert R, Rynes E, Wang H, Vong S, Lee K, Bates D, Diegel M, Roach V, Dunn D, Neri J, Schafer A, Hansen RS, Kutyavin T, Giste E, Weaver M, Canfield T, Sabo P, Zhang M, Balasundaram G, Byron R, MacCoss MJ, Akey JM, Bender MA, Groudine M, Kaul R, Stamatoyannopoulos JA (2012) An expansive human regulatory lexicon encoded in transcription factor footprints. Nature 489:83–90
15. Pepke S, Wold B, Mortazavi A (2009) Computation for ChIP-seq and RNA-seq studies. Nat Methods 6:S22–S32
16. Baek S, Sung MH, Hager GL (2012) Quantitative analysis of genome-wide chromatin remodeling. Methods Mol Biol 833:433–441
17. Tsompana M, Buck MJ (2014) Chromatin accessibility: a window into the genome. Epigenetics Chromatin 7:33
18. Sung MH, Guertin MJ, Baek S, Hager GL (2014) DNase footprint signatures are dictated by factor dynamics and DNA sequence. Mol Cell 56:275–285
19. Koohy H, Down TA, Hubbard TJ (2013) Chromatin accessibility data sets show bias due to sequence specificity of the DNase I enzyme. PLoS One 8, e69853
20. Boyle AP, Guinney J, Crawford GE, Furey TS (2008) F-Seq: a feature density estimator for high-throughput sequence tags. Bioinformatics 24:2537–2538
21. Song L, Crawford GE (2010) DNase-seq: a high-resolution technique for mapping active gene regulatory elements across the genome from mammalian cells. Cold Spring Harb Protoc 2010:pdb.prot5384
22. Grant CE, Bailey TL, Noble WS (2011) FIMO: scanning for occurrences of a given motif. Bioinformatics 27:1017–1018
23. Bailey TL, Williams N, Misleh C, Li WW (2006) MEME: discovering and analyzing DNA and protein sequence motifs. Nucleic Acids Res 34:W369–W373
24. Bailey TL (2011) DREME: motif discovery in transcription factor ChIP-seq data. Bioinformatics 27:1653–1659
25. Heinz S, Benner C, Spann N, Bertolino E, Lin YC, Laslo P, Cheng JX, Murre C, Singh H, Glass CK (2010) Simple combinations of lineage-determining transcription factors prime cis-regulatory elements required for macrophage and B cell identities. Mol Cell 38:576–589
26. Gupta S, Stamatoyannopoulos JA, Bailey TL, Noble WS (2007) Quantifying similarity between motifs. Genome Biol 8:R24
27. He HH, Meyer CA, Hu SS, Chen MW, Zang C, Liu Y, Rao PK, Fei T, Xu H, Long H, Liu XS, Brown M (2014) Refined DNase-seq protocol and data analysis reveals intrinsic bias in transcription factor footprint identification. Nat Methods 11:73–78
28. Lazarovici A, Zhou T, Shafer A, Dantas Machado AC, Riley TR, Sandstrom R, Sabo PJ, Lu Y, Rohs R, Stamatoyannopoulos JA, Bussemaker HJ (2013) Probing DNA shape and methylation state on a genomic scale with DNase I. Proc Natl Acad Sci U S A 110:6376–6381
29. Neph S, Stergachis AB, Reynolds A, Sandstrom R, Borenstein E, Stamatoyannopoulos JA (2012) Circuitry and dynamics of human transcription factor regulatory networks. Cell 150:1274–1286
30. Stergachis AB, Neph S, Sandstrom R, Haugen E, Reynolds AP, Zhang M, Byron R, Canfield T, Stelhing-Sun S, Lee K, Thurman RE, Vong S, Bates D, Neri F, Diegel M, Giste E, Dunn D, Vierstra J, Hansen RS, Johnson AK, Sabo PJ, Wilken MS, Reh TA, Treuting PM, Kaul R, Groudine M, Bender MA, Borenstein E, Stamatoyannopoulos JA (2014) Conservation of

trans-acting circuitry during mammalian regulatory evolution. Nature 515:365–370

31. Piper J, Elze MC, Cauchy P, Cockerill PN, Bonifer C, Ott S (2013) Wellington: a novel method for the accurate identification of digital genomic footprints from DNase-seq data. Nucleic Acids Res 41, e201

32. Pique-Regi R, Degner JF, Pai AA, Gaffney DJ, Gilad Y, Pritchard JK (2011) Accurate inference of transcription factor binding from DNA sequence and chromatin accessibility data. Genome Res 21:447–455

33. Vierstra J, Wang H, John S, Sandstrom R, Stamatoyannopoulos JA (2014) Coupling transcription factor occupancy to nucleosome architecture with DNase-FLASH. Nat Methods 11:66–72

Part IV

Tools

Chapter 13

NGS-QC Generator: A Quality Control System for ChIP-Seq and Related Deep Sequencing-Generated Datasets

Marco Antonio Mendoza-Parra, Mohamed-Ashick M. Saleem, Matthias Blum, Pierre-Etienne Cholley, and Hinrich Gronemeyer

Abstract

The combination of massive parallel sequencing with a variety of modern DNA/RNA enrichment technologies provides means for interrogating functional protein–genome interactions (ChIP-seq), genome-wide transcriptional activity (RNA-seq; GRO-seq), chromatin accessibility (DNase-seq, FAIRE-seq, MNase-seq), and more recently the three-dimensional organization of chromatin (Hi-C, ChIA-PET). In systems biology-based approaches several of these readouts are generally cumulated with the aim of describing living systems through a reconstitution of the genome-regulatory functions. However, an issue that is often underestimated is that conclusions drawn from such multidimensional analyses of NGS-derived datasets critically depend on the quality of the compared datasets. To address this problem, we have developed the NGS-QC Generator, a quality control system that infers quality descriptors for any kind of ChIP-sequencing and related datasets. In this chapter we provide a detailed protocol for (1) assessing quality descriptors with the NGS-QC Generator; (2) to interpret the generated reports; and (3) to explore the database of QC indicators (www.ngs-qc.org) for >21,000 publicly available datasets.

Key words Next-generation sequencing, Massive parallel sequencing, Quality control, ChIP-sequencing, Galaxy, Database

1 Introduction

The rapid development of next-generation sequencing (NGS) technologies poses multiple challenges to the bioinformatics-based analyses of the enormous amounts of data, which still increase exponentially. While in the past years computational efforts focused mostly on the description of local enrichment confidence in NGS-generated profiles, a number of other critical issues that limit the conclusions drawn from multi-profile comparisons have been neglected or only incompletely addressed. NGS technology can be

Ewy Mathé and Sean Davis (eds.), *Statistical Genomics: Methods and Protocols*, Methods in Molecular Biology, vol. 1418, DOI 10.1007/978-1-4939-3578-9_13,

used to address a variety of molecular events. This includes the epigenetic histone modifications and transcription factor–chromatin association (ChIP-seq), DNA methylation profiling (MeDIP-seq and related technologies), analysis of transcriptional activity either by RNA-Polymerase II profiling or by evaluating the nascent transcriptomes (GRO-seq and similar technologies), or studying the association of RNAs with (particular) ribosomes by ribosome capture technologies followed by RNA-seq. It is important to emphasize that the integrative analysis of several different datasets requires particular attention to profile pattern variability, which is highly affected by the choice of antibodies, sequencing depth, cross-linking procedures, immunoprecipitation (IP) efficiency, and many other parameters.

For this reason, we have developed the NGS-QC Generator [1], a bioinformatics-based quality control (QC) system that uses the raw NGS data sets to (1) infer a set of global QC indicators that reveal the comparability of different NGS data sets; (2) provide local QC indicators to judge the robustness of cumulative read counts ("peaks or islands") in a particular region; (3) provide guidelines for the choice of the optimal sequencing depth for a given target; and finally (4) to have quantitative means of comparing different antibodies and antibody batches for ChIP-seq and related antibody-driven studies.

Briefly, this computational approach is based on the assumption that under optimal conditions NGS-generated profiles might conserve the relative enrichment patterns when a subset of the total sequenced/mapped reads are used for their reconstruction. In this context, it generates read count intensity (RCI) profiles from randomly selected subsets of the total originally mapped reads (TMRs) associated to the NGS-profile under study and evaluates the divergence from the theoretically expected read count intensities recovery after sampling relative to the original profile (described as dRCI or RCI dispersion). While most if not all NGS-generated datasets diverge systematically from the hypothesized "ideal behavior," the extent of such divergence represents a quantifiable descriptor of the quality of any NGS-generated profile.

In this chapter, we provide a step-by-step procedure for evaluating the quality of any ChIP-seq and related dataset by taking advantage of the multiple random samplings procedure embedded into the NGS-QC Generator. Furthermore, we provide to the reader a guide to interpret the quality reports generated by this tool. Finally we provide a global survey of the NGS-QC Generator database currently hosting quality grades for more than 21,000 publicly available datasets, such that the reader is able to compare publicly available datasets in the context of their quality grades, and to use this information to reveal the performance of "ChIP-seq grade" antibodies.

2 Materials

This chapter is dedicated to the use of the NGS-QC Generator tool and its associated database, both available through the Web access (www.ngs-qc.org). Users do not need to install any software; datasets to be processed are uploaded to a dedicated FTP server. For the purpose of this protocol, several datasets are available in the "Shared Data/Data Libraries".

2.1 Data Description: Preliminary Datasets Quality Evaluation (FastQC)

For the purpose of this protocol, we have selected the ChIP-seq datasets GSM733737 and GSM733720 (Table 1) from the publicly available ENCODE [2] histone modifications collection in the Gene Expression Omnibus (GEO) repository [3]. Both datasets were generated using antibodies targeting the active histone modification mark H3K4me3 on HepG2 and NHEK cell-lines respectively. A typical H3K4me3 profile is expected to display sharp and highly enriched occupancy of reads around Transcription Start Sites (TSS) of active genes while the rest of the genome including inactive genes could contain background noises. Each of the selected datasets has multiple replicates but for demonstration we have selected "Replicate3" and "Replicate1" from NHEK and HepG2, respectively (highlighted in Table 1).

While the NGS-QC Generator takes aligned datasets as input, we consider that the evaluation of the quality of the sequencing itself is also an essential part of the analysis. Thus, to illustrate the procedure to follow, we have downloaded raw sequence data SRR227526 and SRR227563 from the Short Read Archive (SRA) database [4] corresponding to the datasets GSM733737 and GSM733720 respectively. Throughout this chapter, use of terms NHEK and HepG2 data refers to the replicate 3 of NHEK and replicate 1 of HepG2 data respectively.

Table 1
Summary of the ChIP-seq datasets used in the illustrations

GEO ID	SRA ID	Replicate	Histone mark	Cell line	Raw reads	% Aligned reads	% Unique reads
GSM733720	*SRR227525*	Rep1	H3K4me3	NHEK	20855214	58.939908	93.03
GSM733720	*SRR227527*	Rep2	H3K4me3	NHEK	7121361	75.289583	95.92
GSM733720	*SRR227526*	*Rep3*	*H3K4me3*	*NHEK*	*6304701*	*69.269772*	*85.32*
GSM733737	*SRR227563*	*Rep1*	*H3K4me3*	*HepG2*	*10228152*	*87.698785*	*91.15*
GSM733737	*SRR227564*	Rep2	H3K4me3	HepG2	11600046	83.196291	95.92

2.2 Data Description: Datasets Format for NGS-QC Generator Analysis

NGS-QC Generator accepts the most commonly used formats of alignment, BAM and BED. In fact, most of the aligners reports the alignment output in a huge plain text SAM file format, which after compression and sorting generates BAM files. BED is a minimal format to represent the alignment information where only six columns (chromosome, start, end, score, id, strand) are used to represent each alignment. As this minimal information is enough for NGS-QC Generator, it is highly advised to convert other formats to BED format. Furthermore, providing sorted and compressed BED files to the NGS-QC Generator might be suitable since it saves time during the ftp upload and speed the analysis as well.

There are different tools and methods available to convert alignment file to an acceptable format. Samtools [5] can be used to convert SAM to sorted BAM as following:

```
samtools view -bS input.sam | samtools sort - output.srt
```

Use –q option to exclude alignments with low mapping quality in conversion, and with versions >=0.1.19 multithreading is available with -@ option. Use the input file name in the place of input.sam and output.srt and the output will be generated with automatic suffix .bam.

In a similar manner, Bedtools [6] can be used to convert BAM to BED:

```
bedtools bamtobed -i input.bam | sort -k 1,1 -k 2,2n -k 3,3n
-k 6,6 | gzip -f >output.srt.bed.gz
```

Finally, a simple awk script can be used to convert from bowtie default format to bed format:

```
awk -F "\t" 'BEGIN{OFS="\t";}{len=$4+length($5); print
$3,$4-1,len,$1,"0",$2}' input.bow bam | sort -k 1,1 -k
2,2n -k 3,3n -k 6,6 | gzip -f >output.srt.bed.gz
```

This script can be used to convert from any other alignment text files by changing the appropriate columns in the script. These are the corresponding fields from bowtie output:

\$3 = Chromosome; \$4 = Start position; \$5 = Aligned sequence; \$1 = Read ID; \$2 = Strand.

The identity of the different columns can be retrieved by examining the first lines of the file to convert by using the "head" command and change the script and variable numbers accordingly (*see* **Note 1**).

3 Methods

3.1 Preliminary Sequencing Quality Assessment

Before stepping into the analysis of experiment quality based on enrichment with the NGS-QC Generator, it is important to check the quality of the DNA sequencing itself, as this factor can directly influence the results.

We currently use FastQC [7] to evaluate the DNA sequencing quality and potential contaminations. FastQC aims to provide a simple way to do some quality control checks on raw sequence data coming from high throughput sequencing pipelines prior to alignment. It can be used through simple GUI interface or through command-line for easy batch processing and the tool can handle different formats like FASTQ, SAM, and BAM file. FastQC generates a HTML output containing a variety of graphical displays illustrating the quality of the data under different aspects. Here we highlight the important modules that require close investigation from FastQC results to assess better the quality of the data.

3.1.1 Sequence Quality

"Per base sequence quality" and "Per sequence quality scores" are the two important plots that has to be paid attention to understand the sequencing quality of the data. The first plot is a boxplot illustrating the average, median, and distribution of quality per base from all the reads in the data (Fig. 1a). As Illumina's sequencing chemistry is sequencing per cycle approach, at each cycle one base of all reads are sequenced, thus the 36 cycle kit yields all reads with 36 bp length. While more recent versions of Illumina sequencing kits generate much longer reads (50 to even 150 bp length); a FastQC analysis will still be helpful to identify potential abnormalities during the sequencing process.

Hence it is expected to have dip or hike in quality at or from a particular base on in most of the sequences due to technical issues. This pattern can be easily observed in "Per base sequence quality" box plot with average and distribution of quality per base from all reads. It is important to remember that due to the reagents "burning" out over the period, quality of bases towards the end of the reads result in the gradual decrease of average quality in the plot.

While an average view of the read sequences provides a first way to evaluate sequence quality, the presence of reads with poor quality levels will affect the alignment accuracy to the reference genome (false or multiple alignment outcomes). One can address this problem by using the "Per sequence quality scores" plot which shows the distribution of each read average quality to get an idea about altogether bad reads.

Unlike the "per base distribution" approach, the analysis in a "per sequence quality" context allows to estimate the fraction of reads with low quality. It is important to note from Fig. 1a, that the NHEK dataset presents an overall poor sequencing quality, which is also reflected by the lower percentage of mapped reads in comparison to that observed for HepG2 datasets (Table 1).

Note that tools like FASTX-toolkit [http://hannonlab.cshl.edu/fastx_toolkit/], NGSQC-toolkit [8], or SeqQC [http://genotypic.co.in/Products/7/Seq-QC.aspx] are dedicated to trim and filter reads presenting low quality scores (*see* **Note 2**).

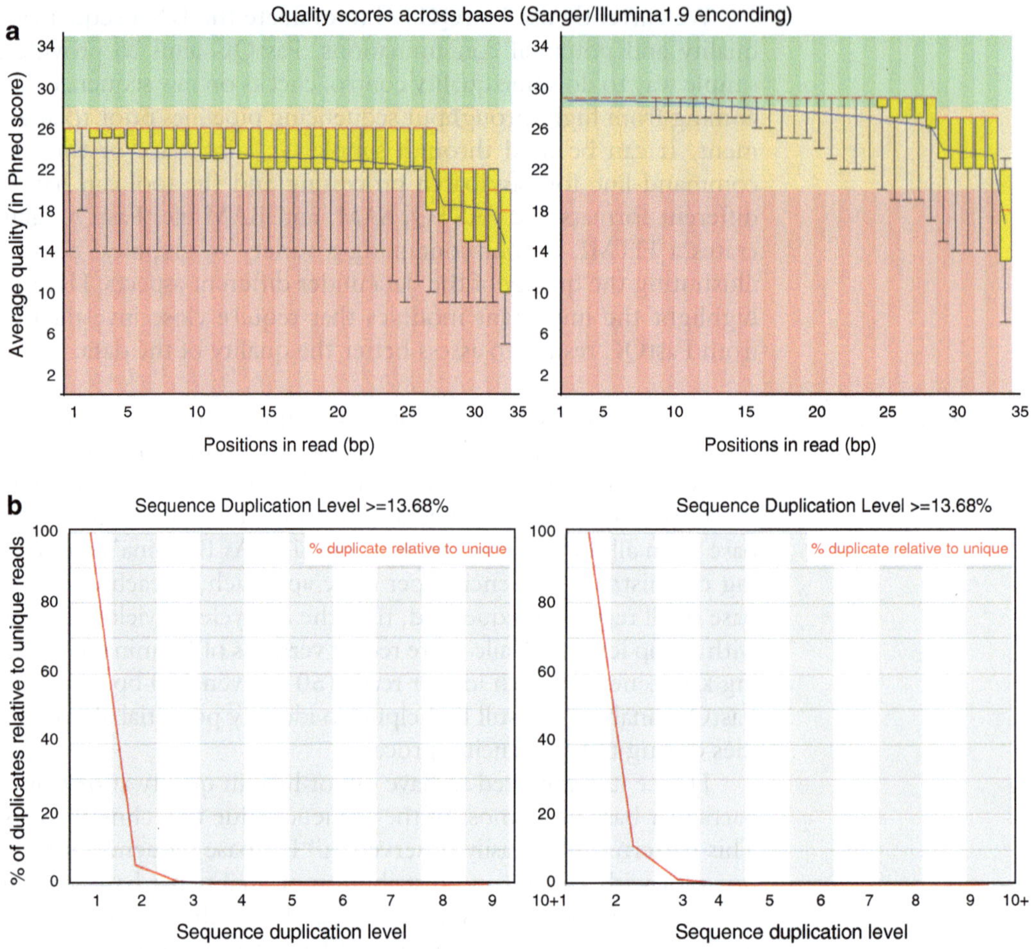

Fig. 1 (**a**) FastQC generated "Per base sequence quality" plot for GSM733720 NHEK (*left panel*) and GSM733737 HepG2 datasets (*right panel*). (**b**) FastQC generated "Sequence Duplication Level" for both datasets as in (**a**). Both the data show similar harmless level of clonal reads

3.1.2 Clonal Reads

"Sequence duplication level" is important aspect to understand the quality of data rather than the sequencing itself. When there is low amount of starting material to start with or by over-amplifying the fragments during library preparation could lead to high levels of sequence duplication. As clonal reads bias the analysis by over-representing the same fragment, they can generate false-positive enrichments. Hence, it is highly recommended to remove duplicate reads and keep only one copy.

Also it is important to point out that the currently existing methods to detect clonal reads from single-end sequencing assays may overestimate their presence. Clonal reads are defined as reads aligned with the same start and end position; indeed an analysis in which the two reads generated by paired-end sequencing from each DNA were treated separately as single reads demonstrated that reads considered as clonal in a single-end sequencing context can have different alignment coordinates in the corresponding pair-read analysis (data not shown).

It is highly recommended to remove clonal reads after alignment as sequencing errors could bias the similarity estimation among reads when it is carried out prior alignment. Hence, the FastQC sequence duplication level can be used for a rough assumption only, as it is done with raw data and only the first million reads are considered for the calculation. This deviation can be observed between FastQC sequences duplication level from Fig. 1b with the actual unique reads count from Table 1 (*see* **Note 3**).

3.1.3 Adapter Contamination

When fragmented DNA molecules are shorter than the applied sequencing length, DNA sequencing will continue over the adapter sequence at the end of the process. This is rare as most of the chromatin immunoprecipitated DNA is in average larger than several hundreds of nucleotides but when it happens (e.g., over-sonication; sequencing library primer-dimer contamination) it can be detected by the counts of "Overrepresented sequences." If there are bulk of reads with adapter contamination, tools like Cutadapt [9], FASTX-toolkit, NGSQC-toolkit, or SeqQC can be used to trim the adapter sequence but retain the actual DNA sequence.

3.2 Exploring the NGS-QC Generator Portal

NGS-QC can be accessed through the portal www.ngs-qc.org, which gives access to different elements of NGS-QC method. Among the various components available in the main page we can cite the direct access to the customized Galaxy instance hosting the NGS-QC Generator tool, the access to the NGS-QC database as well as to a detailed tutorial describing the principles in use for assessing quality scores and their interpretation in the context of the enrichment behavior (Fig. 2).

3.3 Accessing to the NGS-QC Generator Tool via the Dedicated Galaxy Instance

3.3.1 An Overview of the Galaxy Interface

Galaxy [10–12] is an open, Web-based platform for biological data analysis. It enables to run bioinformatics tools in a user-friendly point-and-click environment. The NGS-QC Generator is available through a local Galaxy server reachable at http://galaxy.ngs-qc.org/. It is worth to mention that any user requires to create a login account such that a dedicated space is allocated to his jobs in the NGS-QC Generator associated server. Once logged in, users can access to the main Galaxy interface composed of three sections (as illustrated in Fig. 3):

1. A panel containing the list of available tools, grouped by field of analysis (left side).
2. A central panel where the interaction with the selected tools is enable as well as the display of the generated results.
3. A panel containing the history of data, requested jobs and results (right side). It is important to mention that this last panel

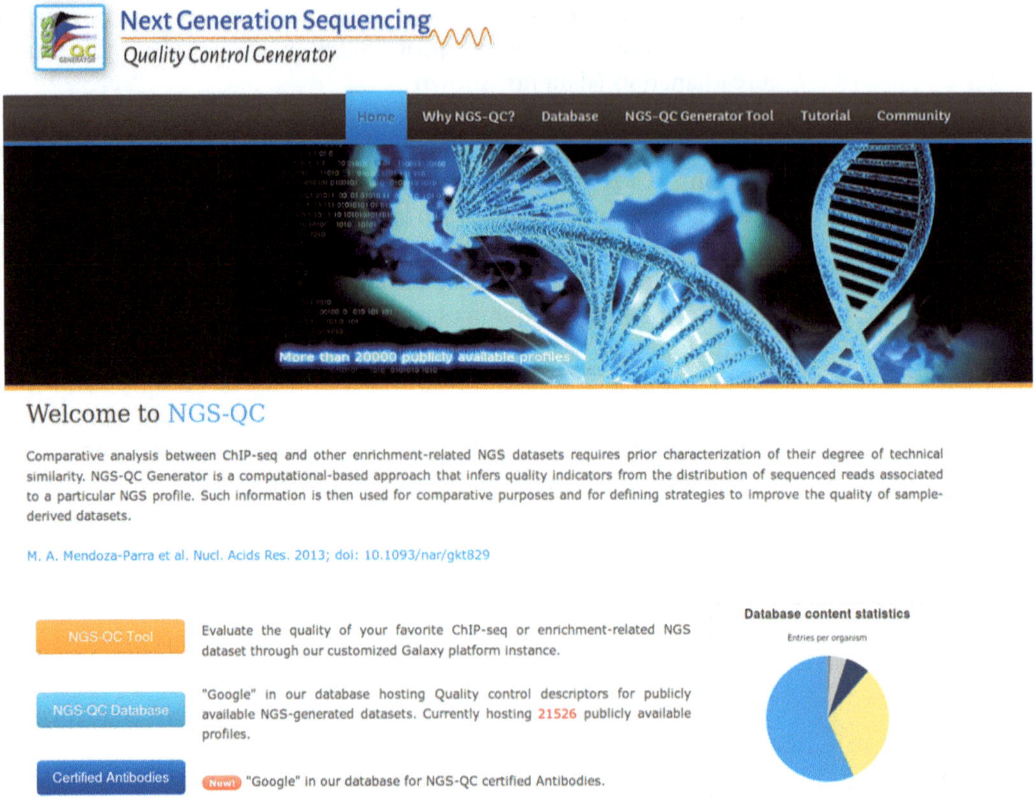

Fig. 2 Home page of ngs-qc.org website. Notice that as part of the main display, access to the NGS-QC Generator tool and to the dedicated database is available. Furthermore, an access to the database content statistics is illustrated such that visitors could have a glance over the variety of the publicly available datasets currently hosted by this website

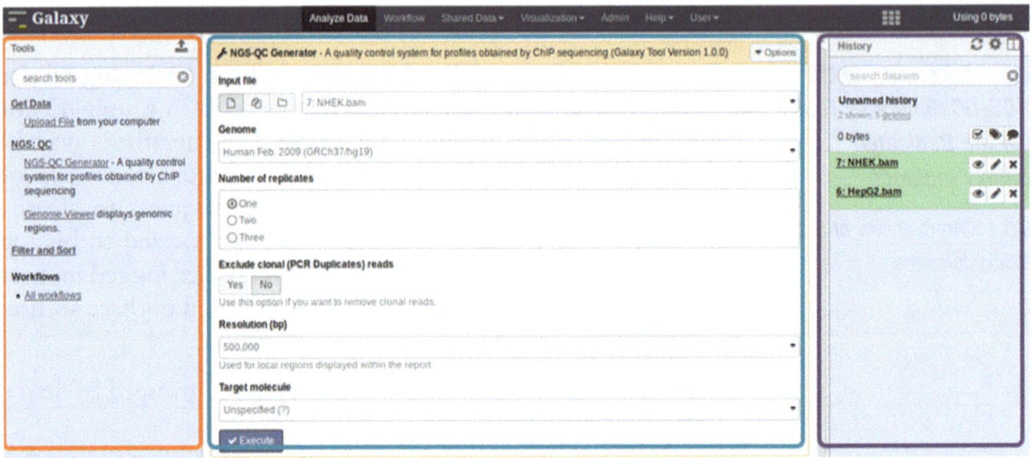

Fig. 3 Illustration of a classical Galaxy interface composed by the "tools" panel (*left side*, demarcated in *red*), the central panel where tool parameters and results are displayed, and the history panel (*right side in purple*) containing the datasets available. In this example, the current history panel contains two datasets imported from the "Shared Libraries"

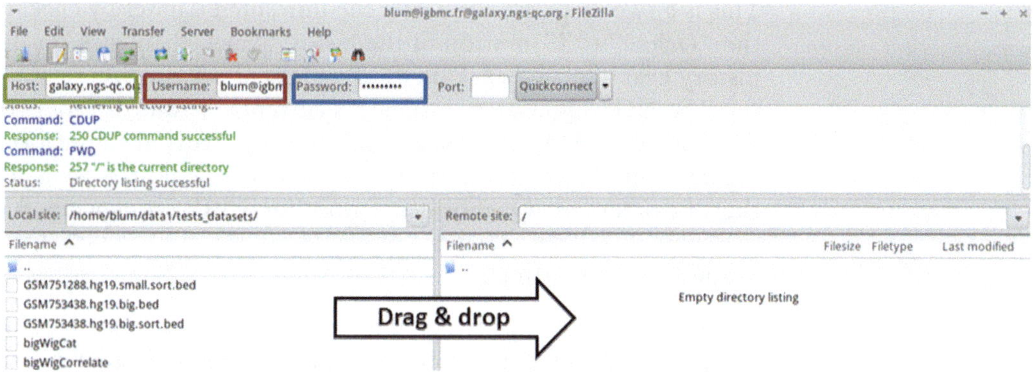

Fig. 4 Screenshot of the FileZilla interface. An FTP connection can be established by filling the "Host" field with "galaxy.ngs-qc.org" and the Galaxy user/password in the "Username"/"Password" fields, respectively, then clicking the "Quickconnect" button. Dragging files from the *left-side panel* and dropping them onto the *right-side panel* will start the upload

provides the history of every dataset/job associated to the user, thus a continuous cleaning of its content is a good practice to avoid overloading the allocated space in our servers (see below).

Running a tool on Galaxy requires importing the dataset of interest into their history. Locally stored files can be uploaded either through the user's Web browser, or by using our FTP server; nevertheless we might strongly recommend using the second option in the case of large size datasets (larger than 2Gbs). In the case of first option, locally stored files can be uploaded to Galaxy by clicking on "Get Data" in the tool panel, and select "Upload File". Either one can drag and drop files from their desktop file manager to the Galaxy interface, or select files from the local disk by clicking the "Choose local file" button. Once selected, click the "Start" button to start the upload.

To upload files by FTP, a connection to our FTP server with an FTP client is required (e.g., *FileZilla* [https://filezilla-project.org/]; an open source and cross-platform tool). In the FTP client, the host has to be given as "galaxy.ngs-qc.org" and users should use their Galaxy credentials for login (Fig. 4). Once the files are uploaded, they can be imported into the history by clicking on "Get Data", "Upload File", and "Choose FTP file" on the Galaxy left panel. Datasets successfully imported will be shown in green color box in the history panel (right side).

Though our Galaxy server is equipped with high computing capacity and storage space for handling multiple users, we cannot keep each and every dataset uploaded. For this reason, files, which are older than a month will be automatically removed from Galaxy. In case users are running out of space, they can reduce their disk usage by permanently deleting datasets they do not need anymore.

This is a two steps action: delete the unwanted dataset by clicking the "Delete" ("X") button in the history panel, then click on the history gear icon and click on "Purge deleted datasets".

The NGS-QC Galaxy instance is currently hosting a variety of tools generated by the Galaxy community (Text manipulation, Operate on Genomic Intervals) but also two dedicated tools for the assessment of Quality descriptors for ChIP-sequencing and enrichment related datasets, available under the "NGS: QC" group in the tool panel:

1. The *NGS-QC Generator* computes global and local quality descriptors by performing multiple random sub-samplings of the mapped reads retrieved in the provided dataset, to then use them for reconstructing and comparing their enrichment patterns with the original profile.
2. The *Local QCs viewer* allows users to visualize the read count intensity levels retrieved in their datasets in the context of the computed local quality descriptors. Regions to display can be chosen by the user either by (a) gene names, (b) genomic coordinates or (c) let the tool to select these regions based on local QC score ranking.

3.3.2 Performing a Global Quality Control Analysis

As mentioned earlier, we are going to run NGS-QC Generator on NHEK and HepG2 datasets, available in the "Shared libraries". To use these datasets from shared libraries, users have to import these datasets into their history panel as following:

1. Click "Shared Data" at the top of the page;
2. Navigate to "Data libraries", "Examples";
3. Select both "NHEK" and "HepG2" datasets;
4. Finally select "Import to current history" in the "For selected datasets" menu; click the "Go" button, and go back to the main page by clicking on "Analyze Data".

Now the datasets are into the history, we can run the NGS-QC Generator:

1. Click on "NGS-QC Generator" under "NGS: QC" in the tool panel
2. In the central panel, select one or more (batch mode) datasets, and the genome assembly; in this case hg19. The following options are also available:
   (a) Generate up to three technical replicates, to evaluate the results variability.
   (b) Remove the clonal reads (potential PCR duplicates; by default it is not removed).

(c) Exclude background noise (users can switch-off this option if interested in assessing its influence in QC scoring; but by default it is active).

(d) Select the resolution for the read-counts intensity representation of displayed genomic regions.

(e) Select the sample's antibody target. If the user provides the target identity, the output report will contain a scatter-plot showing a comparison of the evaluated dataset's quality with those retrieved in the NGS-QC collection.

3. Click the "Execute" button to launch the job.

For each run the NGS-QC Generator produces two outputs which are made available in the history panel. The first one is an HTML page linking to the PDF report that summarizes the quality descriptors, and a ZIP archive containing several supplementary files.

A NGS-QC Generator report (Fig. 5) is composed by three major elements:

1. *Dataset information* where characteristics of the dataset under analysis like the total mapped reads, the fraction of unique mapped reads (i.e., without the clonal sequences), reads' length size, the genome assembly, and the target molecule (when available) are indicated.
2. *QC parameters* where a detailed overview of the various parameters used for generating the corresponding QC report are listed such that it could be reproduced if required.
3. *QC results* where a compilation of the global quality indicators assessed in the context of the considered mapped reads are displayed (either with or without clonal reads). The result panel is complemented with a scatter-plot displaying the read counts per genomic region (500 bp bin; *x*-axis) in comparison to the recovered counts after multiple reads' random subsampling (90 %, 70 %, and 50 % subsampled reads; *y*-axis). Furthermore a global QC certification score (from "AAA" to "DDD" for designating from high and low quality datasets) is stamped over the main page of the QC report such that the quality of the analyzed dataset is expressed in a rather intuitive manner without the need of getting deep into the assessed quality scores.

Each QC report is further complemented with series of genomic regions displays illustrating the enrichment patterns associated to the dataset under analysis complemented by the demarcation of genomic areas presenting high enrichment robustness (local QC indicators; heatmap display). In addition, when the target molecule identity is provided, a scatter-plot displaying the total mapped reads

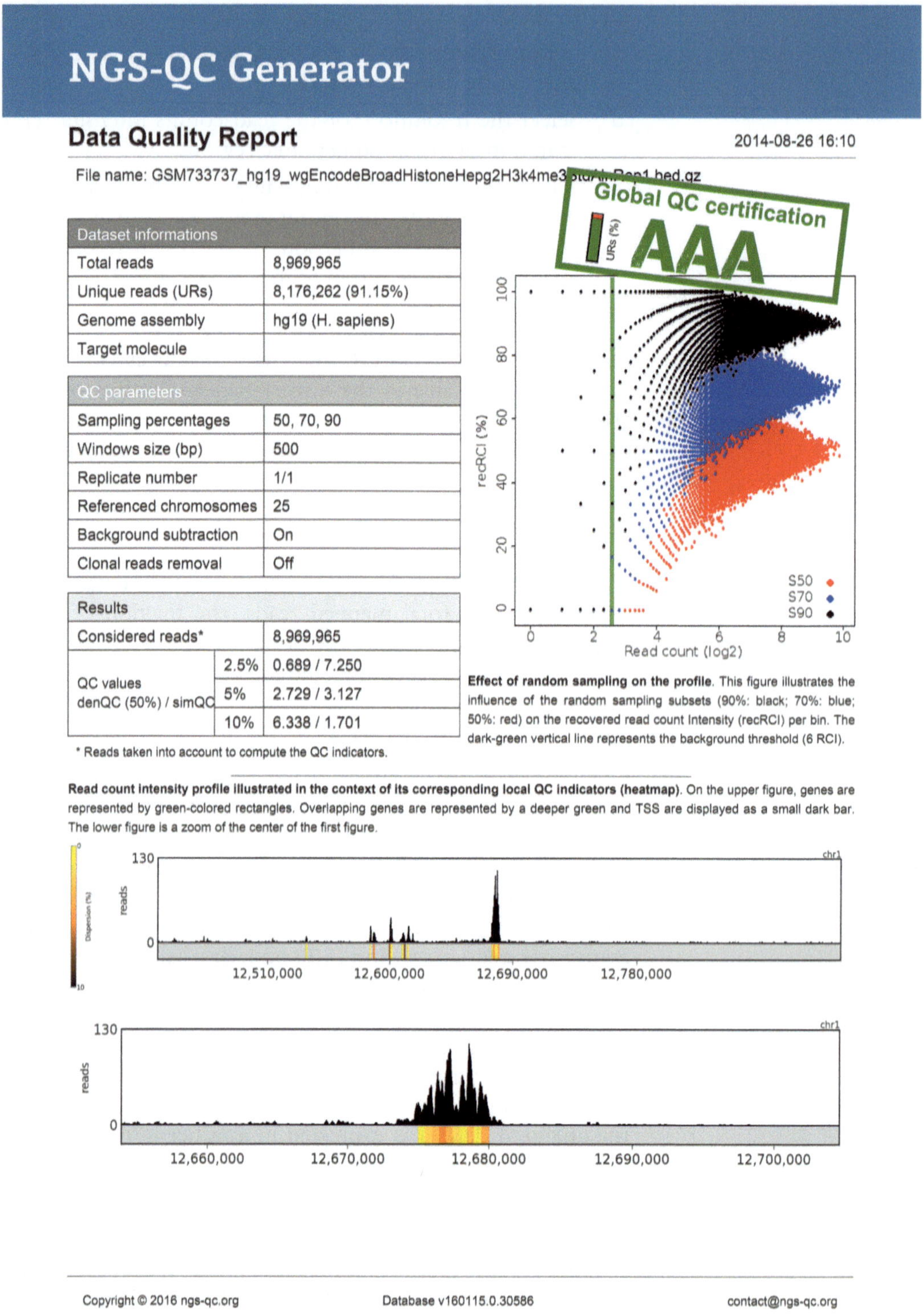

NGS-QC Generator

Data Quality Report

2014-08-26 16:10

File name: GSM733737_hg19_wgEncodeBroadHistoneHepg2H3k4me3StdAlnRep1.bed.gz

Dataset informations	
Total reads	8,969,965
Unique reads (URs)	8,176,262 (91.15%)
Genome assembly	hg19 (H. sapiens)
Target molecule	

QC parameters	
Sampling percentages	50, 70, 90
Windows size (bp)	500
Replicate number	1/1
Referenced chromosomes	25
Background subtraction	On
Clonal reads removal	Off

Results		
Considered reads*		8,969,965
QC values denQC (50%) / simQC	2.5%	0.689 / 7.250
	5%	2.729 / 3.127
	10%	6.338 / 1.701

* Reads taken into account to compute the QC indicators.

Effect of random sampling on the profile. This figure illustrates the influence of the random sampling subsets (90%: black; 70%: blue; 50%: red) on the recovered read count Intensity (recRCI) per bin. The dark-green vertical line represents the background threshold (6 RCI).

Read count intensity profile illustrated in the context of its corresponding local QC indicators (heatmap). On the upper figure, genes are represented by green-colored rectangles. Overlapping genes are represented by a deeper green and TSS are displayed as a small dark bar. The lower figure is a zoom of the center of the first figure.

Copyright © 2016 ngs-qc.org Database v160115.0.30586 contact@ngs-qc.org

Fig. 5 First page of the NGS-QC report for the HepG2 dataset (GSM733737; replicate 1). On the *upper right side* of the page, the global quality stamp is displayed along with a colored bar representing the proportion of unique reads(URs) in the dataset

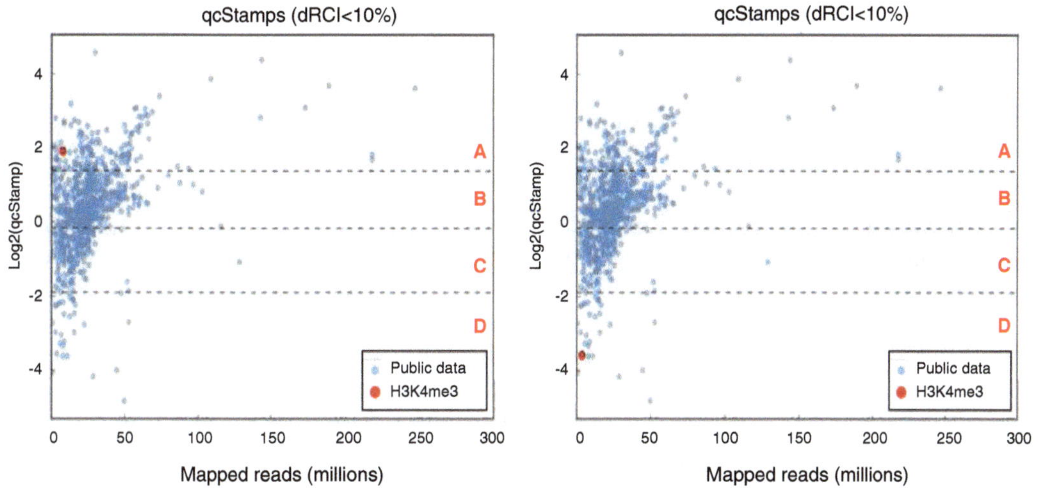

Fig. 6 Quality scores (*y*-axis) relative to their total mapped reads (*x*-axis) assessed for several public datasets (*blue*) in comparison to the dataset under evaluation (*red*). In the *left panel* this analysis has been performed for the HepG2 dataset; while in the *right side* it is illustrated for the NHEK dataset. This type of displays are included int he last page of the QC-report when the target molecule identity is provided

versus the quality descriptors assessed for entries available in the NGS-QC database in comparison with evaluated dataset is included at the end of the QC report (Fig. 6).

In addition to the QC report a set of supplementary files are generated (ZIP archive format). This supplementary folder contains further genomic regions display and local QC regions in either wiggle or BED format files such that they can be uploaded into a genome browsers like IGB [13] or on the UCSC Genome Browser [14]. Furthermore, a text file, referred here as "local Qci file", containing all genomic regions presenting enriched regions with high robustness (dRCI < 10 %) is also included in this supplementary file. This last file is also created in the current history, since it is required to generate multiple genomic regions displays (*see* Subheading 3.3.3).

3.3.3 Visualize Local Enrichment Patterns in the Context of Their Quality

Considering the interest of users to visualize genomic regions as a way to confirm the potential low or high quality of the dataset under study, we have developed a convenient way to display them together with a demarcation enriched regions based on their robustness or local QC indicators score. For it, our new tool called "*local QCs viewer*", takes alignment file (BED or BAM format) and the previously generated local QCi file as input to produce up to 25 genomic regions displays (PDF report format), with three selection strategies. By default, regions to be displayed are selected based on the local QC scores, the read count intensities and the presence of genes near a highly enriched region. Users have also the possibility to provide a list of genes or genomic positions as a way to customize these displays. As performed for the QC reports, genomic regions display enrichment patterns associated to the dataset under analysis complemented by the demarcation of genomic areas presenting high enrichment robustness (local QC indicators; heatmap display) (Fig. 7).

3.4 Exploring the NGS-QC Generator Database

3.4.1 A General Overview

In addition to the access to NGS-QC Generator tool, users can retrieve a large collection of quality indicators computed for a variety of publicly available datasets in the dedicated website. To access it, users have to select either "Database" on the top navigation bar or "NGS-QC Database" on the main page of the NGS-QC portal. There are two ways to browse our results; either by using the proposed query panel or through the interactive boxplots table located below. The *query panel* (Fig. 8a) allows specifying the request by multiple options like the model organism, target molecule, quality grades and also a public identifier from GEO database (GSM or GSE) or from ENCODE consortium (wgEncode). Importantly, each of these query options can be used in combinations. For example users can query for all the qualified datasets corresponding to mouse (*Mus musculus*) and human (*Homo sapiens*) model systems targeting the histone modification mark H3K4me3 and presenting quality grades between "A" and "B" (Fig. 8).

The *boxplot table* displays quality scores (QC-Stamp; dRCI < 10 %) distribution assessed over the whole database content (currently more than 21,000 datasets) as well as the discretized QC-STAMP intervals (from A to D). Furthermore, the quality score distribution per target molecule are displayed such that users might have a global overview of their associated quality scores. It is worth to mention that in addition to display each of these distributions under a boxplot format, a violin plot is also displayed such that a more detailed view of the distribution of population is available. Each target molecule in boxplot is complemented by a legend indicating its identity as well as the number of entries available (target molecules represented by less than 10 datasets are categorized as "Others"). By clicking on any of the target molecule boxplots, users are redirected to the results panel associated to that target molecule such that further refinement queries could be performed if required.

For instance by clicking on the boxplot associated to histone modification mark H3K4me3, users are redirected to the results page where the QC scores for the datasets associated to this target are displayed.

The results page (Fig. 9) is composed of five elements:

1. *Query panel* displaying the current request made.
2. *Boxplot table* displaying the QC scores distribution for each of the targets included in the request.

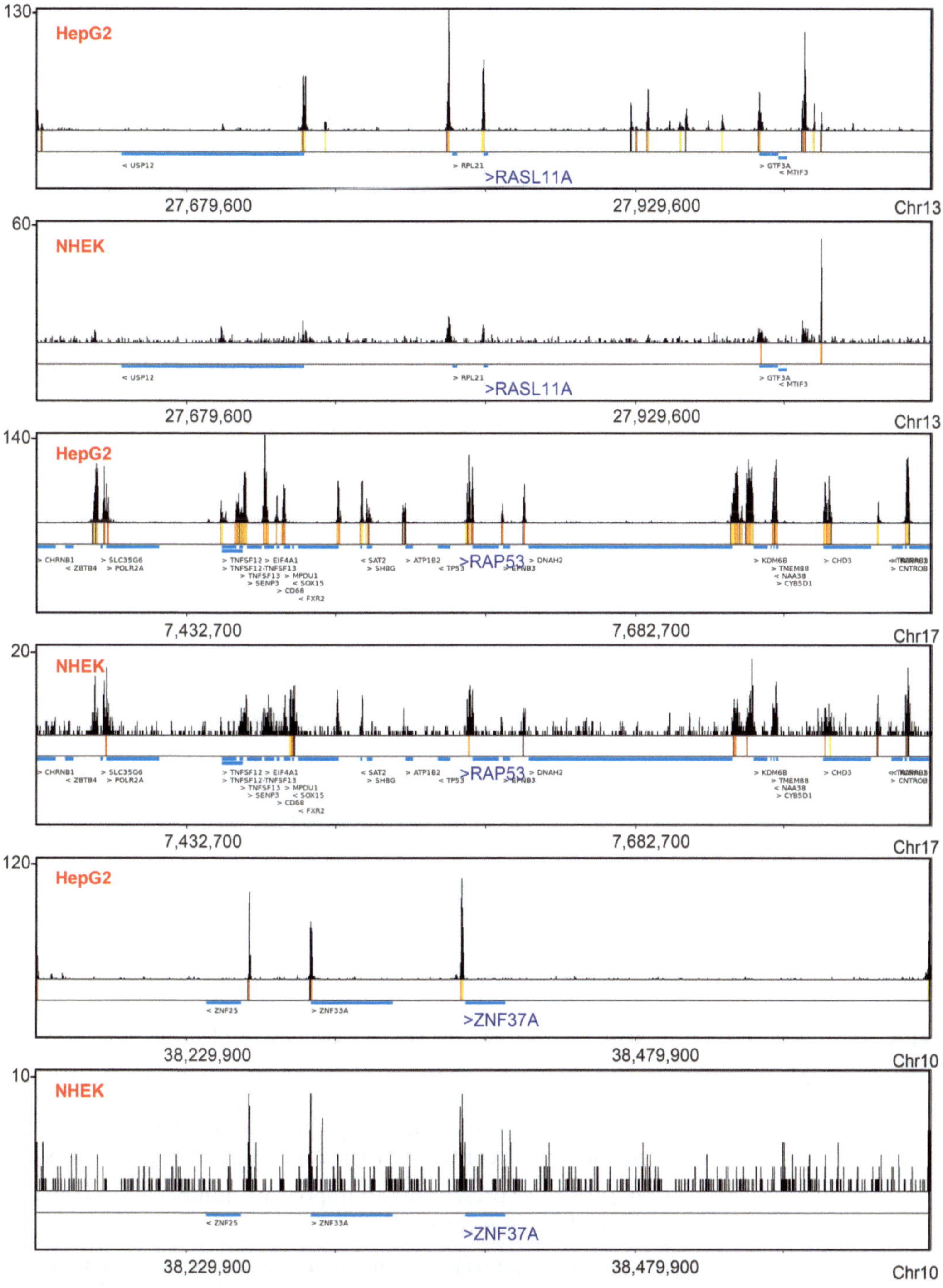

Fig. 7 Comparison of three genomic regions from HepG2 (*top*) and NHEK (*bottom*). Regions were selected using the RASL11A, WRAP53, and ZNF37A genes. Each plots contains three parts: (1) *top*, the representation of the read-counts intensity, (2) *center*, the position of the local regions with a dispersion between 0 % (*yellow*) and 10 % (*black*) and the genes' positions. The HepG2 plots present conserved local qc regions with highly peak intensity. On the contrary, NFEK plots are poor in enriched regions

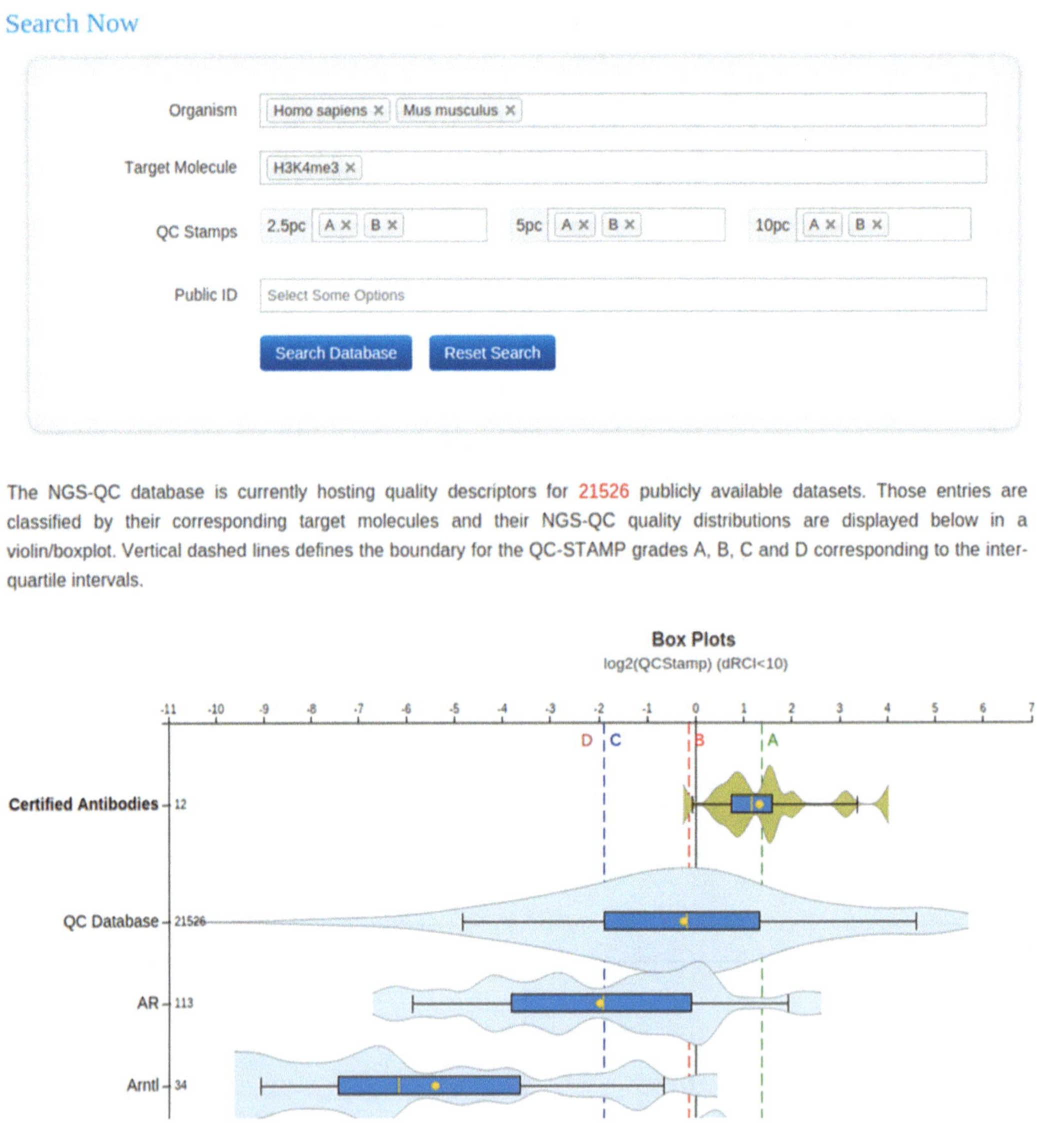

Fig. 8 Display illustrating the database page showing the search panel (*top*) and boxplots/violin plots table (*bottom*)

3. a *Scatter-plot* displaying the quality scores (QC-stamp) for each dataset in the context of their total mapped reads (TMRs).
4. *Results table* presenting an important number of information for each dataset retrieved.
5. *Refinement panel* providing further query options to be applied over the initial request to refine the results to specific interest.

All the described elements are meant to provide complementary information to the user, as well as means to narrow down the mining to users' specific interest. For instance, when multiple

Fig. 9 Display illustrating the results obtained after performing a query in the NGS-QC database. (**a**) scatter-plot displaying the QC-indicators relative to the total mapped reads. (**b**) Boxplot/vioplots displaying the different target molecule retrieved on the query. (**c**) Results table including several additional information for each dataset (*right panel*) and the refinement panel (*left panel*)

targets are part of the initial query, the boxplot panel provides the possibility to select one of them (Fig. 9b). Furthermore, the *scatter-plot* display provides a powerful way to visualize the quality scores associated to each dataset in the context of their related TMRs. For instance, in the context of the query targeting the histone modification marks H3K27me3, the user can pass the pointer over the scatter-plot to have a pop-up displaying element like the dataset identifier, its associated TMRs, the target identity, the model organism and its computed quality grade (Fig. 10). Considering that the quality scores are computed at three different read count intensity dispersion conditions (dRCI), each of them are

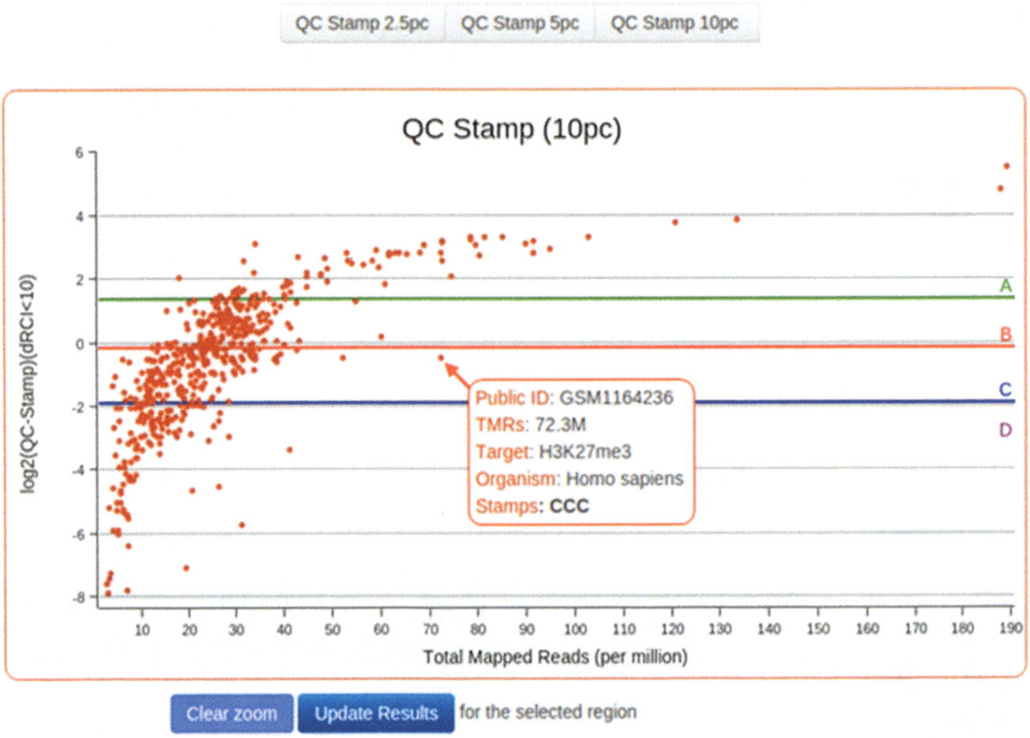

Fig. 10 Scatter-plot illustrating the quality scores computed for several H3K27me3 generated datasets in the context of their total mapped reads. The identity of one of these datasets presenting more than 70 million TMRs are highlighted as done in the NGS-QC Generator website

summarized in one of the letters corresponding to the QC-STAMP grade (e.g., A, A and A—represented in short AAA—for dRCI < 2.5 %, <5 % and <10 % respectively). In this context, users can switch the displays among all three corresponding scatter-plots using the buttons located on the top of the panel. Furthermore, users can zoom on each scatter-plot to perform for further refinement of the results for a desired group of datasets. To do so, users might select the "Zoom" icon located on the right top corner of the plot and then press and hold mouse button to select the region. Alternatively, zooming on the scatter-plot is possible by scrolling the wheel mouse button.

The *results table* (Fig. 9c), located under the scatter-plot provides further information for each of the datasets retrieved like public identifier (ID); study ID, associated publication ID (PMID), model organism, data type, target molecule, TMRs complemented by the fraction of unique mapped reads (i.e., by excluding the clonal reads), global quality grades (QC-Stamps) and a Global QC indicators report available in a pdf format. Further supplementary information is available under a ZIP file format available for downloading. Taken in consideration that users might systematically wish to associate a qualified dataset to an article describing it, users can access to the

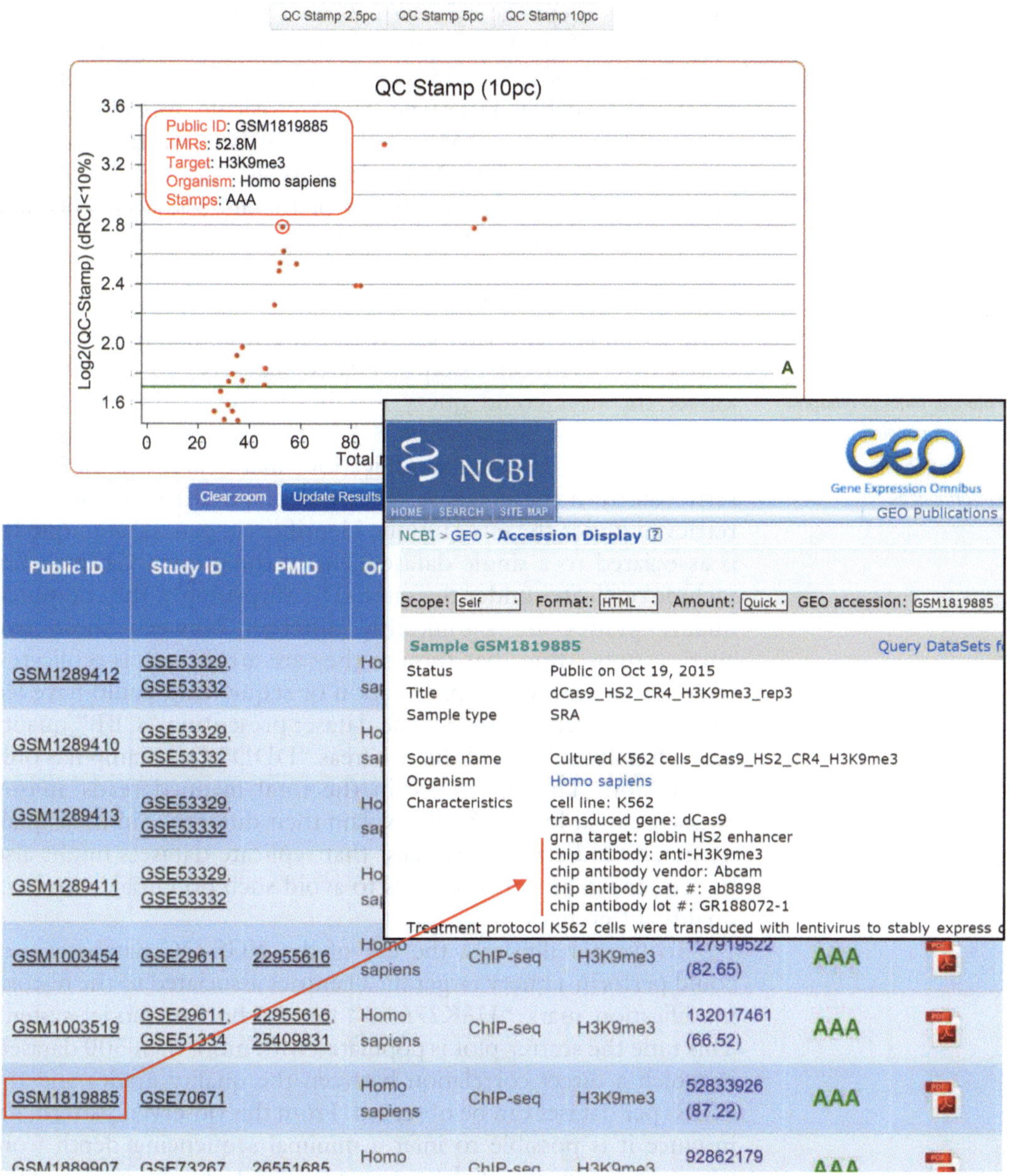

Fig. 11 Example illustrating the procedure by which the NGS-QC database can be used for retrieving the antibody in use for high quality grades datasets. A refined query has been performed for retrieving high quality datasets associated to the histone mark H3K9me3; followed by the identification of the best dataset ("AAA" QC grade with the least TMRs). Finally the corresponding GSM ID link has been used to explore over the original information retrieved in GEO

abstract of the related publication (when available) by placing the mouse pointer over the corresponding PMID.

Finally, some times the initial query might include too many outputs, such that users might wish to focus their view on a defined subset; thus a separate *refinement panel* is provided at the left of the results table (Fig. 9c). Importantly the refinement panel is composed by a list of 13 elements aiming at providing a large flexibility for this task such as Data type, target molecule, model organism, sequencing platform, TMRs, submission date, Abstract content, Author and Public ID. Other elements like the Cell line/tissue, the potential cell/tissue treatment or the antibody reference and batch available under a nonacademic access (*see* below).

3.4.2 The NGS-QC Database as a Source of Information for Providing Potential Hints to Interpret ChIP-Sequencing Assays with Low Quality Performances

An important application of the database is to provide potential hints to understand the reasons for the low quality grade associated to certain ChIP-seq or related datasets. To illustrate this aspect the user could query for the dataset "GSM733754" as described in the Subheading 3.4.1. In result, two entries associated to the histone modification mark "H3K27me3" are retrieved from the database, which correspond to two replicates retrieved under the same unique identifier (in general a unique ID is associated to a single dataset; but in some cases GEO entries include replicates under the same ID). Surprisingly, the computed quality grades are significantly different between these two entries, suggesting that though they are attributed as replicates, technical differences in preparation or sequencing could have led to these differences. In fact the dataset presenting "CBB" quality grade has 26 million TMRs, whereas "DDD" QC-stamp has only 9.2 million. This difference in the total mapped reads among datasets can at least partially explain their difference in their quality grades and strongly suggest that replicate datasets might also be sequenced at similar depths to avoid such potential sources of quality differences.

To further illustrate the use of the NGS-QC database, user could perform a query targeting all entries associated to the histone modification mark "H3K27me3" for the human model system. This time the scatter-plot is populated with more than 500 datasets in which a direct correlation between the quality grades and the TMRs per dataset can be observed. From this observed pattern, for instance it is possible to infer a minimal sequencing depth from which the majority of the retrieved datasets might present high quality grades (e.g., for TMRs higher than 30 million, most of the datasets are "CCC" or higher). It is important to mention that even for datasets generated with high TMRs levels, significant differences in quality scores can be observed. For instance, as illustrated in Fig. 10, datasets generated with more than 70 million reads could still present "CCC" quality grades.

3.4.3 Identifying ChIP-seq Grade Antibodies from the NGS-QC Database Content

While this in silico approach could provide guidelines concerning the sequencing-depth in use for a given factor, it is not guaranteed that all higher sequencing-depth experiment will be successful. In fact several other parameters like the antibody in use, the cell line/

tissue under analysis, as well as the performance of the experimenter are crucial factors in ChIP-seq pipeline that could have influence in the quality. To illustrate this important aspect, users might query for datasets generated with the histone modification mark H3K9me3 in the context of human studies (Fig. 11). Then, by taken advantage of the zoom option available in the results scatter-plot we can refine the query to datasets presenting quality grades of "A". Finally, the refined output can be sorted based on TMRs using the results table such that high quality datasets (AAA) generated with the least number of TMRs might appear on the top of the result table panel. At this point, users might be interested on retrieving the Antibody source that gave rise to such optimal results, this is indeed possible by accessing the corresponding public repository page in GEO database by using the link embedded in the public ID displayed in the results table. Additionally, we are currently annotating the antibody source per QC-certified datasets in the NGS-QC database, such that users might directly access to this information through the previously described refinement panel.

4 Conclusions and Perspectives

As illustrated in the previous sections, the NGS-QC Generator and the QC database represent a powerful solution to evaluate the quality of any ChIP-seq and enrichment-related datasets. In april 2015, the qualities of more than 21,000 datasets available in the public domain have been assessed and the scientific community has full access to the corresponding quality reports (www.ngs-qc.org). This database is continuosly expanded. Indeed at the time of the final version of this document, the NGS-QC database bypassed the 30,000 processed datasets; covering more than 80 % of the publicly available ChIP-seq datasets in GEO (January 2016). We are currently developing additional modules, such as variable background detection and correction in cancer cells due to local DNA amplification or reduction, and expand its quality evaluation system to other types of datasets, like RNA-seq and all its related variants, as well as to chromatin conformation capture assays (e.g., Hi-C; ChIAPET).

While our methodology can currently detect poor quality datasets without necessarily providing an explanation for it, we are investing efforts in the evaluation of the antibodies in use. On one hand we have used text mining of all analyzed GEO entries to identify the cell line/tissue and antibody source and batch associated to each dataset (when available). We can thus provide a predictive analytics view of antibody performances based on several thousands of entries retrieved from the public domain. Considering that public datasets necessarily include variations originating from different experimenter skills and/or sample treatment conditions, we have recently set up an standardized automated pipeline for certifying antibodies dedicated to ChIP-seq applications. This last point is of great interest as there is currently no quantitative system apart from the NGS-QC Generator for certifying ChIP-seq quality grade antibodies. Indeed, the various commercial products labeled as "ChIP-seq grade" by their manufacturers receive their labels from a nonquantitative, subjectively chosen screenshot of a ChIP-seq profile, which has only very limited value to predict antibody performance along the entire genome.

We are convinced that a quantitative evaluation of the quality of enrichment-base datasets needs to be integrated in any dataset deposited in the public domain in the future. Similarly commercially available antibodies should have a quantified and independently certified quality performance associated to each antibody batch. Only this way the huge amount of information present in the public domain can be adequately extracted and waste of effort, time, and money can be saved when only high quality datasets are used for comparative or integrative analysis. Obviously, there is resistance towards the use of rigorous quality assessment tools by those performing these assays but at the end the science and the scientific community will benefit enormously from integrating such tools as a routine process as it has been done in the past for microarray analyses.

5 Notes

1. With coreutils (version>=8.6), −parallel option is available to speed up the sort command.
2. Preprocessing of raw data prior to alignment (i.e., trimming/filtering low quality or adapter-contaminated reads) is necessary if one is going to use end-to-end alignment approaches where there is a possibility of reads failing to align with long chunks of wrongly called bases (i.e., bad quality) or adapter contaminations at the end. Recent versions of alignment tools have "local alignment" approaches and low quality ends will be automatically clipped if not matching with the reference genome.
3. In experiments where enzymatic digestion is used for fragmenting DNA, the frequency of clonal reads might be enhanced due to a potential nuclease DNA-sequence specificity. In this scenario the removal of clonal reads is not advised.

Acknowledgements

This work was supported by funds from SATT/Conectus, the Fondation pour la Recherche Médicale (FRM), the Alliance Nationale pour les Sciences de la Vie et de la Santé–Institut Thématique Multi-organismes Cancer–Institut National du Cancer (INCa) grant "Epigenomics of breast cancer" and "EpiPCa," the Ligue National Contre le Cancer (to H.G.; Equipe Labellisée).

References

1. Mendoza-Parra MA, Van Gool W, Saleem MAM, Ceschin DG, Gronemeyer H (2013) A quality control system for profiles obtained by ChIP sequencing. Nucleic Acids Res 41, e196
2. Bernstein BE, Birney E, Dunham I, Green ED, Gunter C, Snyder M (2012) An integrated encyclopedia of DNA elements in the human genome. Nature 489:57–74. doi:10.1038/nature11247
3. Barrett T, Wilhite SE, Ledoux P, Evangelista C, Kim IF, Tomashevsky M et al (2013) NCBI GEO: archive for functional genomics data sets—update. Nucleic Acids Res 41: D991–D995. doi:10.1093/nar/gks1193
4. Kodama Y, Shumway M, Leinonen R (2012) The Sequence Read Archive: explosive growth of sequencing data. Nucleic Acids Res 40: D54–D56. doi:10.1093/nar/gkr854
5. Li H, Handsaker B, Wysoker A, Fennell T, Ruan J, Homer N et al (2009) The Sequence Alignment/Map format and SAMtools. Bioinformatics 25:2078–2079. doi:10.1093/bioinformatics/btp352
6. Quinlan AR, Hall IM (2010) BEDTools: a flexible suite of utilities for comparing genomic features. Bioinformatics 26:841–842. doi:10.1093/bioinformatics/btq033
7. Andrews S. FastQC: a quality control tool for high throughput sequence data [Internet]. http://www.bioinformatics.babraham.ac.uk/projects/fastqc/. citeulike-article-id:11583827
8. Patel RK, Jain M (2012) NGS QC toolkit: a toolkit for quality control of next generation sequencing data. PLoS One 7, e30619. doi:10.1371/journal.pone.0030619
9. Martin M (2011) Cutadapt removes adapter sequences from high-throughput sequencing reads. EMBnet J 17(1). Next Gener Seq Data Anal. http://journal.embnet.org/index.php/embnetjournal/article/view/200/479
10. Goecks J, Nekrutenko A, Taylor J (2010) Galaxy: a comprehensive approach for supporting accessible, reproducible, and transparent computational research in the life sciences. Genome Biol 11:R86. doi:10.1186/gb-2010-11-8-r86
11. Giardine B, Riemer C, Hardison RC, Burhans R, Elnitski L, Shah P et al (2005) Galaxy: a platform for interactive large-scale genome analysis. Genome Res 15:1451–1455. doi:10.1101/gr.4086505
12. Blankenberg D, Von Kuster G, Coraor N, Ananda G, Lazarus R, Mangan M et al (2001) Galaxy: a web-based genome analysis tool for experimentalists. Curr Protoc Mol Biol. doi:10.1002/0471142727.mb1910s89
13. Helt GA, Nicol JW, Erwin E, Blossom E, Blanchard SG, Chervitz SA et al (2009) Genoviz Software Development Kit: Java tool kit for building genomics visualization applications. BMC Bioinformatics 10:266. doi:10.1186/1471-2105-10-266
14. Kent WJ, Sugnet CW, Furey TS, Roskin KM, Pringle TH, Zahler AM et al (2002) The Human Genome Browser at UCSC. Genome Res 12:996–1006. doi:10.1101/gr.229102

Chapter 14

Operating on Genomic Ranges Using BEDOPS

Shane Neph, Alex P. Reynolds, M. Scott Kuehn, and John A. Stamatoyannopoulos

Abstract

The bulk of modern genomics research includes, in part, analyses of large data sets, such as those derived from high resolution, high-throughput experiments, that make computations challenging. The BEDOPS toolkit offers a broad spectrum of fundamental analysis capabilities to query, operate on, and compare quantitatively genomic data sets of any size and number. The toolkit facilitates the construction of complex analysis pipelines that remain efficient in both memory and time by chaining together combinations of its complementary components. The principal utilities accept raw or compressed data in a flexible format, and they provide built-in features to expedite parallel computations.

Key words Bioinformatics, Sequencing, Algorithm, Overlap, Genomics, Compression, Parallelization

1 Introduction

Investigators seek to answer a range of genomics questions by analyzing experimental data. With modern, highly multiplexed sequencing technologies, the volume of data produced through a single experiment is unprecedented and continues to grow quickly [1]. Massive data sets generated from sequencers are commonplace, with publicly available, whole-genome experiments in human alone numbering in the thousands [2]. Analyzing these and similarly large data sets can stress computational resources to the point of failure. By imposing structure on input data, the BEDOPS analysis toolkit addresses this concern directly for many common analysis tasks [3].

One fundamental analysis is drawing associations between feature sets. As a simple example, a researcher might work with high-throughput sequencing reads following a ChIP-seq experiment for the CTCF transcription factor (TF). Significantly populated regions make up an important set of features for the experiment, as it shows where CTCF is mostly likely bound in that cell or tissue type under the experimental conditions. Annotating overlapping or

Ewy Mathé and Sean Davis (eds.), *Statistical Genomics: Methods and Protocols*, Methods in Molecular Biology, vol. 1418, DOI 10.1007/978-1-4939-3578-9_14, © Springer Science+Business Media New York 2016

nearby genomic features, such as genes or the binding sites for a different TF, can provide insight into CTCF activity.

Among other capabilities, BEDOPS offers a variety of methods to draw associations between data sets. The *bedops* tool applies overlap and difference operations to include or exclude genomic elements from one set, based on their full or partial overlaps with regions in another set. Other associations can be drawn with *closest-features*, which finds those features from a secondary source that fall nearest to the regions of interest. Statistical summaries with *bedmap* can involve various calculations on associated features, such as the mean signal value from a second data source over the regions of interest.

The toolkit includes utilities to convert efficiently common genomic data formats, such as BAM, GFF, and WIG, to the BED format required by BEDOPS. These tools make it easy to integrate data from a variety of sources into one framework for analysis. A data management utility archives BED in a deeply compressed form that offers descriptive metadata and random access to data by chromosome, which reduces disk space overhead and makes it simple to parallelize analysis pipelines.

2 Methods

2.1 Genomic Data Formats

The BEDOPS toolkit works with data formatted as Browser Extensible Data (BED), which is a text-based, tab-delimited format used for storing position and annotation information along the genome. The accepted format relaxes some of the constraints found in the original BED specification, which was developed for use with the UCSC Genome Browser [4]. In contrast to the original UCSC BED, the extended format accepts general measurements and not only integer values, allows 0- or 1-based coordinate indexing, and places no constraint on the type or number of columnar data that may exist beyond a small default minimum of 3–5 columns. These modest extensions to BED offer broad versatility for analyses and data format conversions.

A bewildering array of open and commercial representations of genomic data exist, yet data found in many other popular formats can be recast as extended BED without loss of information. In order to enable BEDOPS to work with data stored in other formats, we include the *convert2bed* tool, along with a convenience wrapper script per input format, to transform the data into extended BED.

The binary BAM format and its text-based analogue, SAM, are often used to store low-level sequence alignment information [5]. The GTF and GFF file types store gene annotations. The open Variant Call Format (VCF) handles variant, deletion, and insertion data with associated genotype information. The PSL format from

UCSC stores positional alignment results. The UCSC WIG file format, which itself includes more than one variation, contains genomic positions with continuous signal data. The information in these and other data formats can be converted in their entirety into extended BED for use with BEDOPS.

Various data formats call for different interpretations of genomic coordinates, which often leads to confusion when integrating data from a variety of sources. Some formats use 0-based indexing while others use 1-based coordinate indexing, where the 5′-most base (forward strand) of a chromosome is labeled, respectively, as position 0 or 1. Positions along a chromosome are labeled with increasing integer values starting from the label of that first base. Even closely related formats, such as BAM (0-based) and SAM (1-based), often differ in their coordinate indexing specifications. The UCSC Genome Browser tool requires 1-based BED inputs for proper data visualization, while the same suite explicitly defines BED as a 0-based coordinate format. Adding further confusion, popular data conversion utilities that are not part of the BEDOPS toolkit inconsistently handle indexing shifts between various formats.

As a small extension to the original BED format definition, BEDOPS does not check or specify coordinate indexing. In short, BEDOPS fully supports 0-based and 1-based coordinate indexing simply through non-interpretation, as all underlying algorithms are unaffected by that choice. If inputs have a 0-based index, for example, then any output results will also have a 0-based index. However, coordinate indexing can be changed explicitly with the *bedops* utility through its *--range* option. The next program call example shows how to use *bedops* to produce a 0-based output from a 1-based input. The subsequent call shows how to achieve a coordinate shift in the opposite direction. In each case, the two signed integers, separated by a colon and part of the *--range* argument, are added to the start and end coordinate positions, respectively, of each row of input.

```
$ bedops --everything --range -1:-1 one-based-input.bed \
  > zero-based-output.bed
$ bedops --everything --range 1:1 zero-based-input.bed \
  > one-based-output.bed
```

There is yet another interpretation related to genomic coordinates that also differs between various genomic formats. The location of a genomic feature typically includes chromosome information, as well as the start and end coordinate positions of that feature on the chromosome. Are the start and end positions inclusive to this interval? For example, given a genomic feature on chromosome 2 with start position 101 and end position 120, should this be interpreted as an interval that includes base positions 101 and 120, or perhaps only the base positions in between these two end points? The UCSC BED format specifies a half-open interval interpretation, where the start coordinate is part of the interval and the end coordinate is 1 nucleotide beyond the interval range, concisely denoted as *[start, end)*. In this example, the BED specification dictates that base positions 101 through 119 are part of the interval, inclusively, while base position 120 is not. The BEDOPS toolkit honors this interpretation. This meaning simplifies interval length calculations, which can be derived by subtracting the half-open coordinates. In the example, the number of base positions in the interval is 19, which is also the difference of 120 and 101.

Different genomic formats use different indexing schemes and interval interpretations. One must ensure that all data sources used together as inputs to BEDOPS are created using consistent coordinate interpretations. To assist, the *convert2bed* tool transforms a number of incongruent genomic formats to BED using a 0-based index with an appropriate half-open range. Also, the *bedops* utility can help by moving data between 0-based and 1-based indexing as needed.

2.2 Pipelining with Streams

BEDOPS adopts common Unix design principles [6] in segregating the functionality of individual tools and using standard Unix data streams to handle input and output between tools. A single process works on input data and prints output and error messages to data streams (Fig. 1). Multiple processes can be glued or chained together with the Unix "pipe" operator to construct larger and more complex analyses (Fig. 2). In a Unix pipeline, chunks of input data are operated upon in one process, and generated results are progressively fed to the next process, and so on to any number of processes, until there is no more input data to consume. In this fashion, Unix pipe operators allow a person to create an expressive and self-documenting unit of code.

The BEDOPS utilities work with and produce results in sorted BED format. Results can be directed to a file or sent directly to the next application through a piped data stream. Since pipes do not involve disk accesses, streaming data between utilities offers performance benefits over constructing and maintaining temporary, intermediate files. Further, on modern systems with multiple

Fig. 1 A single, generic Unix process with three standard data streams: input ("*stdin*"), output ("*stdout*") and error ("*stderr*"). The process receives data on *stdin*, printing its normal output to *stdout* and any error and log messages to *stderr*

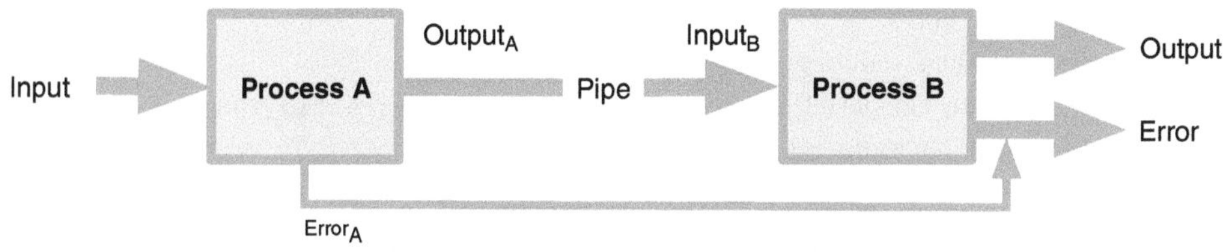

Fig. 2 Two generic Unix processes connected with a Unix pipe. *Process A* receives data on *stdin*, passes its output data stream to the input stream of *Process B*, which in turn prints normal output to the final *stdout* stream. Error and log messages from *Processes A* and *B* are both printed to *stderr*

Fig. 3 Example of a single Unix process, passing forward-stranded transcription start sites (TSSs) to the standard input stream (*stdin*) of *bedops*. In turn, *bedops* makes 1 kb upstream windows of each TSS and prints them to its standard output stream (*stdout*). Any error messages are printed to the standard error stream (*stderr*)

processors, tools linked via pipes can run simultaneously to improve overall throughput.

As an example, the *bedops* program can produce a filtered subset of some genomic input data which is then "piped" downstream to any of a number of BEDOPS or Unix tools. The refined subset might be piped into *bedmap* to investigate statistical relationships with another data set, or to *closest-features* to find the nearest gene annotations. Those extended results could further be piped on to the *starch* utility to be compressed and archived for later retrieval, or to the Unix *tee* utility before finding the genomic complement with yet another call to *bedops*.

To give a more concrete example of how one might progressively build an analysis set, we start by creating a 1 kilobase window upstream of a subset of forward-stranded transcription start sites (TSSs) found in *TSSs.bed*, writing this to a file called *1kb_windows_upstream_of_forward_stranded_TSSs.bed* (Fig. 3).

```
$ awk '$6 == "+"' TSSs.bed \
  | bedops --range 1000:0 --everything - \
  >1kb_windows_upstream_of_forward_stranded_TSSs.bed
```

The *awk* statement, a standard tool found on popular Unix-like systems, shows that strand information for each row is expected to be in the 6th column for this example (though BEDOPS has no such restriction). The pipe operator ('|') sets up a Unix pipe in between the utilities. A hyphen ('-') used in a *bedops* call tells the program to check its standard input stream for incoming data. Reverse strand TSS upstream padding is similar, and all stranded results can be combined together by calling *bedops* once more.

```
$ awk '$6 == "-"' TSSs.bed \
  | bedops --range 0:1000 --everything - \
  | bedops --everything 1kb_windows_upstream_of_forward_stranded_TSSs.bed - \
  >1kb_upstream_padding_TSSs.bed
```

We are likely less interested in the windowed regions themselves than in their biologically relevant attributes. Perhaps the selected genes found in *TSSs.bed* harbor an enrichment of predicted motif binding sites in their upstream regulatory regions, which may highlight the importance of a particular transcription factor in regulating many of the genes.

Rather than writing all of the windowed regions to disk as shown, we can pipe them directly into yet another operation to map on motif predictions found in a BED-formatted text file called *motif_predictions.bed*, and save that extended result to file. The *motif_predictions.bed* file has at least these four columns of information: chromosome, start coordinate, end coordinate, and the predicted motif model's name.

```
$ awk '$6 == "-"' TSSs.bed \
  | bedops --range 0:1000 --everything - \
  | bedops --everything 1kb_windows_upstream_of_forward_stranded_TSSs.bed - \
  | bedmap --echo --echo-map-id-uniq - motif_predictions.bed \
  > motif_IDs_1kb_upstream_padding_TSSs.bed
```

The use of the hyphen character with *bedmap* tells the program that data are available from its standard input (*stdin*) stream. Here, those data come from the standard output (*stdout*) stream of the upstream *bedops* process (Fig. 4). The *--echo* option for *bedmap* causes the program to repeat the information found in its first input (the *Reference*) to its output. The *--echo-map-id-uniq* option instructs *bedmap* to add another output column that lists all unique motif model identifiers, separated by semicolons, that overlap a windowed region of the *Reference* input.

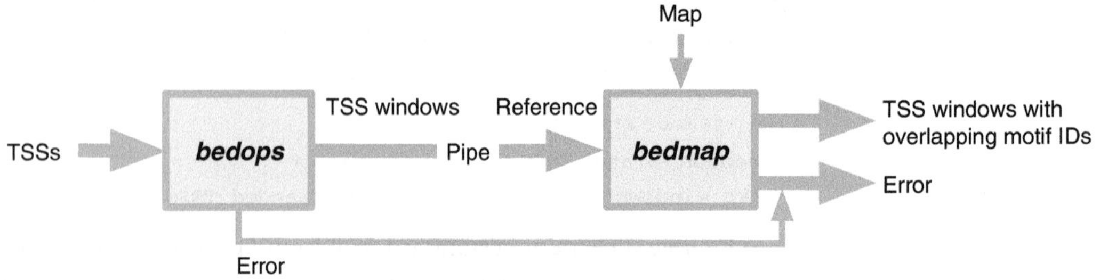

Fig. 4 Two Unix processes running *bedops* and *bedmap*, connected with a Unix pipe. The *bedops* process operates on TSS data received on *stdin* and passes padded TSS windows over its output data stream. These data are received on the input stream of the *bedmap* process, which is configured to treat them as the source of *Reference* set elements. Selected mapping operations are applied and the end result—padded windows with unique motif model names—is written to the final *stdout* stream. Error and log messages from *bedops* and *bedmap* are both printed to the *stderr* stream

Representing standard input with the hyphen character is also the convention in many core Unix utilities, and we can just as easily pipe the standard output from BEDOPS applications to these utilities. For example, we can add calls to the *tr*, *sort*, and *uniq* utilities to build a unique list of motif identifiers that overlap at least one buffered region in our entire TSS data set.

```
$ awk '$6 == "-"' TSSs.bed \
  | bedops --range 0:1000 --everything - \
  | bedops --everything 1kb_windows_upstream_of_forward_stranded_TSSs.bed - \
  | bedmap --echo-map-id-uniq - motif_predictions.bed \
  | tr ';' '\n' \
  | sort - \
  | uniq - \
  > overlapping_motif_IDs_unique_windowed_TSSs.txt
```

Piping data in between BEDOPS and Unix applications is a common and powerful technique to construct complex, yet readable analysis pipelines. With the exception of *unstarch*, all BEDOPS applications can read from a standard input stream, and all write to a standard output stream.

The Unix *bash* shell supports a process substitution feature which makes it syntactically simple to read data streams from more than one upstream process. For instance, if we have BAM-formatted sequencing reads which we want to relate to VCF-formatted variants, say to retrieve SNP names that overlap our sequencing reads, we can use a pair of process substitutions in one BEDOPS command.

```
$ bedmap --echo --echo-map-id-uniq <(bam2bed < reads.bam) <(vcf2bed < variants.vcf) \
  > reads_with_unique_IDs_of_overlapping_SNPs.bed
```

Process substitution in the *bash* shell allows passing the standard output streams of a pair of format conversion tasks directly to *bedmap*, without the wasteful generation and subsequent cleanup of intermediate BED files.

In conjunction with GNU *make*, *rake*, *snakemake*, Luigi, or other pipeline construction toolkits, we can easily abstract an arrangement of piped operations and process substitutions to automate larger and longer-running data analysis tasks (Fig. 5). Operational units ("targets" in GNU *make* terminology) can process data in streams to improve overall performance. They can help with the readability, development, and maintenance of pipelines. One might design long-running tasks to be restartable from an intermediate run point upon interruption, and modifications to downstream pipeline parameters can be tested without regenerating upstream input data.

```
all: all_snps_in_genes.bed

all_snps.bed: snps_1.bed snps_2.bed snps_3.bed
    bedops -m $^ > $@

all_snps_in_genes.bed: all_snps.bed genes.bed
    bedops -e 1 $^ > $@

...
```

Fig. 5 Example snippet of a GNU *make* makefile. Running *make all* attempts to build the file dependency *all_snps_in_genes.bed* from dependencies *all_snps.bed* and *genes.bed*, but the *all_snps.bed* file does not yet exist. The build process therefore looks at the *all_snps.bed* target and creates it from merging files *snps_1.bed*, *snps_2.bed*, and *snps_3.bed* with *bedops*, before creating *all_snps_in_genes.bed*

3 Core Functionality

Relating and operating on genomic intervals is the principal business of BEDOPS. Overlap and proximity relationships among data sets are determined, allowing for general annotations, statistical evaluations, and set-based operations.

3.1 Set-Based Operations

The *bedops* tool offers a host of efficient overlap and set-based operations that can be applied across any number of sorted BED inputs in a single program call. Operations to filter data based upon overlap relationships, generate regions between existing data points, merge intervals across data sets, and many others are all readily available.

Determining the overlaps between genomic data sets is useful for answering research questions, some specific real-world examples of which include:

- Generating SNP variants located within the open chromatin regions as measured with DNaseI hypersensitivity [7].
- Building a list of exemplar regions to represent partially overlapping regions in two or more input data sets [8].
- Filtering out elements that overlap unmappable regions across the genome [9].
- Determining the degree of overlap in peak signal regions between heterogeneous high-throughput experiments [10].
- Building a list of TFs that overlap a set of proximal promoters [11].

With *bedops*, we can define explicitly what it means for two regions to overlap, either through a number of nucleotides specification or as a fraction of a feature's length. For example, in order to use higher-confidence regions in downstream analyses, we can pull out the subset of ChIP-seq peaks in a data set that overlap a replicate experiment's peak regions by 50 % or more.

In addition to determining overlaps, which can be viewed as an element membership operation, *bedops* offers a host of set-like operations for genomic regions. These include the multiset union, element non-membership, complement, symmetric difference, and intersection, among others. All *bedops* operations efficiently handle any number of data inputs at once. For example, one can determine the set of genomic intervals specific to any one of 100 input files using:

```
$ bedops --symmdiff file1.bed file2.bed ... file100.bed \
> intervals_only_from_any_single_file.bed
```

Range adjustments can be made in conjunction with any set-like operations of *bedops*. These coordinate modifications can be symmetric or asymmetric to shrink or expand input regions before applying an operation. The *bedops* utility offers flexibility and very low memory overhead by working with and producing sorted data. More information about sorting is provided in the section on *sort-bed*.

3.2 Proximity

The *closest-features* utility identifies the nearest upstream and downstream elements between two sorted data sources. It can report the signed distances, choose the single nearest element, or exclude overlapping elements in its report. This can help to:

- Build a histogram of the distances between a set of SNPs of interest and their closest DNase-seq peaks in the K562 cell line.
- Identify the nearest predicted motif binding site for a set of NFE2 ChIP-seq peaks in HepG2 cell line, where the distances between features can help to differentiate between direct and indirect binding of the TF.
- Locate the nearest origin of replication to a set of OMIM genes in a disease study.
- Determine the minimum distance between features of any two data sets, optionally ignoring interactions that directly overlap.

As discussed in the previous section on pipelines, the *closest-features* tool can read and write standard input and output. We can pipe the results of calling this tool back into the tool itself to find progressively further elements with a little help from *bedops*. To demonstrate, we show the process for searching for the name of the nearest gene (found in *genes.bed*) to each element in a set of regions of interest (*roi.bed*).

```
$ closest-features --closest --no-ref roi.bed genes.bed \
  | cut -f4 - \
  > closest_gene_name_to_roi.txt
```

We can extend this procedure to find only the next closest gene.

```
$ closest-features --closest --no-ref roi.bed genes.bed \
  | sort-bed - \
  | bedops --not-element-of genes.bed - \
  | closest-features --closest --no-ref roi.bed - \
  | cut -f4 - \
  > second_closest_gene_name_to_roi.txt
```

This process can be repeated until the *n*th nearest element is located, and columns from the resulting text files can be joined into a matrix of regions-of-interest and their nearest *n* genes.

3.3 Statistics and Annotations

The *bedmap* tool associates an element from a *Reference* data set with any number of elements from a second set, *Map*, where associations are determined using a rich set of overlap options. This program is the most feature-rich and general tool for analysis in BEDOPS. Like the *bedops* program, *bedmap* supports overlap specifications using either a number of nucleotides or the proportion of a feature's size, though the latter includes several powerful variants available only with *bedmap*. The program includes options to compute statistics from the overlapping elements and to group the annotations of those elements.

The *bedmap* program can process data for further downstream signal analysis and answer questions about related overlapping features, such as:

- Count how many disease-associated SNPs are found within the open chromatin regions of a cancer tissue sample for comparison against a similar count in a matched sample.
- Smooth experimental sequencing results into binned windows for further signal analysis on a broader scale [12].
- Group TF names that have predicted binding sites at intron-exon boundaries.
- Calculate statistical features of ChIP-seq signals over methylated sites.

Any score-based *bedmap* operation is applied on the measurement or score components of mapped BED elements. For example, one might want to calculate the median or standard deviation of values of all elements from the *Map* input that overlap an element from the *Reference* set. The *bedmap* tool offers these and other common score-based operations, such as trimmed means, variance, coefficient of variation, mean, median absolute deviation, the *k*th-order statistic, sum, and others.

Other *bedmap* operations report the mapped BED elements themselves or selected aspects of the mapped elements, such as a list of their measurement values or identifiers. Yet other operations count how many elements map onto each *Reference* element, determine the genomic range of all overlapping elements, or calculate the fraction of a *Reference* element's bases covered by *Map* element bases.

There exists a core distinction between what the *bedops --element-of* operator produces and what *bedmap* reports. The *bedops --element-of* operator is a filter that gives back a subset of the elements from a *Reference* data set, specifically the subset that overlaps entries from other inputs. In contrast, *bedmap* reports, per *Reference* element, information about the overlapping elements from the *Map* file.

Like the other utilities, *bedmap* works efficiently with sorted inputs of any size. A person can choose any number of options to compute in a single call to the program, which reduces input/output (I/O) overhead and helps to simplify pipeline scripts. Each specified calculation leads to another appended output column, separated by a default '|' symbol. The next program call calculates a number of phyloP [13] conservation score statistics (*phylop_conservation.bed*) over selected intronic regions (*introns.bed*) to produce a six-column output table. The phyloP conservation file has at least five columns, and the conservation score value is in the fifth column.

```
$ bedmap --mean --stdev --mad --kth 0.25 --median --kth 0.75 \
  introns.bed \
  phylop_conservation.bed \
  > phylop_sequence_conservation_statistics_of_introns.txt
```

4 Working Efficiently with Big Data

There are strategies to working more effectively on whole-genome and similarly large-scale data sets, colloquially described as "Big Data", and BEDOPS enables several practical methodologies. For instance, partitioning input data sets into smaller disjoint subsets often can make it possible to solve focused subproblems in parallel, reducing the overall run-time. If the inputs are ordered, then many algorithms can capitalize on that structured information and avoid reading in all data at once. Finally, if inputs are compressed and indexed, then disk access and network I/O overhead can be reduced. Employing some or all of these strategies can make finding solutions to large-scale bioinformatics problems more tractable.

4.1 Parallelization

In combination with popular open-source distributed computing tools like GNU Parallel, Open Grid Scheduler and Hadoop, BEDOPS applications can quickly partition large, whole-genome analyses into smaller, per-chromosome problems that can be distributed and computed separately, by simply specifying the *--chrom* option along with a chromosome name. With this option, a BEDOPS application immediately jumps to and works with data only from the requested chromosome.

Once all of the smaller calculations are complete, a suitable utility can collate results. Performing tasks in parallel can reduce overall computation time to nearly that required to complete the largest subproblem. Often, this is the time taken to process the largest chromosome, which typically contains the greatest number of input elements in practice.

A generic *map-reduce* approach employs three steps to parallelize work:

1. Build a list of distinct chromosome names in the input file(s).
2. Loop through the list, launching a BEDOPS task on a computational unit or node with the *--chrom* operand, which focuses work on a specified chromosome.
3. Collate all smaller results into a final answer.

BEDOPS includes scripts that show this technique in speeding up the conversion of an indexed BAM file to a BED or Starch (compressed and indexed BED) file, as well as quickly generating a Starch-formatted archive from a BED input.

More concretely, when converting indexed BAM to BED with the parallelized version of *bam2bed*, the procedure is as follows:

1. Use *samtools* [5] on the BAM input to generate a list of chromosome names.
2. Loop through the list of names, farming out per-chromosome BAM-to-SAM extraction tasks to a computational node. On each node, convert the extracted per-chromosome SAM data to BED with *sam2bed.*
3. Concatenate the per-chromosome BED files into one final result with the Unix core utility, *cat.*

4.2 Sorting

The principal utilities of BEDOPS require ordered BED inputs. In exchange, operations are efficient in both time and memory, and they usually produce sorted results that are immediately available for further BEDOPS operations. As an additional benefit to working with sorted data, the toolkit works with any species' genome without modification.

The BEDOPS toolkit provides *sort-bed*, a tool to put data in the requisite order. In practice, the tool is considerably faster than the equivalent operation using GNU *sort* or similar tools, and it includes an option to limit the use of RAM so that even the largest data sets can be sorted safely. The program accepts any practical number of input data sets and sends a single, final result to its output data stream. BEDOPS defines proper order first by a lexicographical sort on chromosome names, followed by a numerical sort on the start coordinates, with a further numerical sort on the end coordinates when needed.

```
$ sort-bed unsorted_dataset.bed > sorted_dataset.bed
```

The tool also works with data coming from standard input, where a hyphen instructs the program to read from its standard input data stream.

```
$ cat unsorted_dataset.bed \
  | sort-bed - \
  > sorted_dataset.bed
```

While sorting can be a relatively expensive operation in extreme cases, it is almost always a one-time, upfront cost applied to initial, unsorted data. The types of computations offered by the toolkit simply require sorted data. Rather than imposing those costs each time a utility is called, the toolkit requires ordered inputs, with which it keeps overhead minimal while also producing ordered outputs. This means that results generated from one BEDOPS utility can be used directly as an input to any subsequent BEDOPS operation without any overhead due to further sorting.

By default, *sort-bed* consumes as much system memory as needed to sort its input data. A person can temper this behavior by specifying a maximum allowed memory size with the *--max-mem* option, at the cost of a longer run-time.

```
$ sort-bed --max-mem 1G really_large_unsorted_dataset.bed > sorted_dataset.bed
```

4.3 Compression

BEDOPS introduces an archive format called Starch, which provides two useful features: lossless compression and indexing. Efficient compression is obtained by removing redundancy in BED coordinate data before applying compression with a choice between two common, open-source algorithms (Fig. 6). An index is created to provide random access to the start of each chromosome of the archive upon data extraction. The index additionally stores per-chromosome statistics.

The *starch* and *unstarch* tools, respectively, compress and extract data. The *starch* tool allows the person to choose the back-end compression method, as well as append useful metadata to the archive. An annotation note may be as simple as a sentence to describe the file's provenance or purpose, or as complex as a barcode or other machine-readable key that can be useful in identifying the file in an automated pipeline. On the other end, the *unstarch* tool extracts archive data and prints metadata to its standard

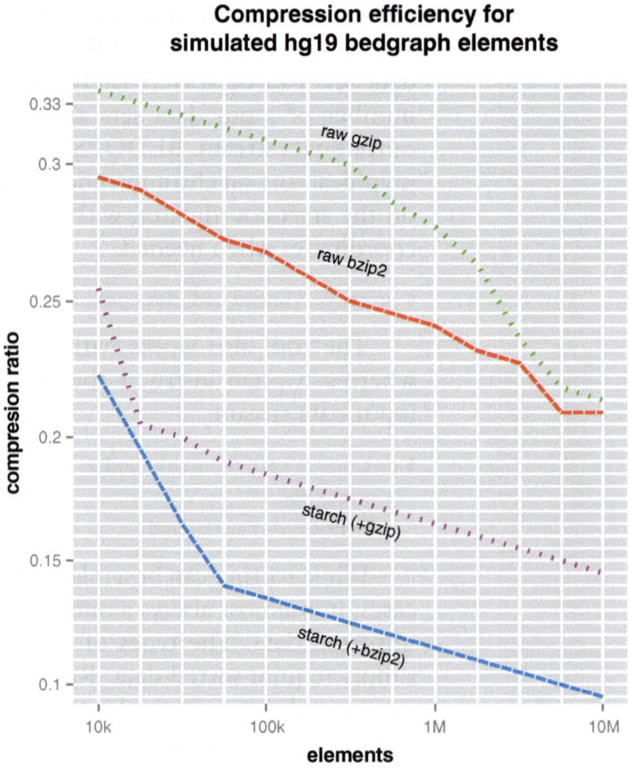

Fig. 6 Compression of 10k to 10 million simulated hg19 bedGraph (UCSC BED3 with a floating-point value) elements, with efficiency measured by comparing file sizes of raw *bzip2*, raw *gzip*, and *bzip2*- and *gzip*-backed Starch archives against the original dataset. Use of *starch* to reduce larger datasets provides considerable space savings over the associated raw compression approach

output. The metadata includes the base coverage and how many elements are contained in an archive, for each chromosome separately or altogether, as well as any note annotation.

Compressing sorted BED data with *starch* is very simple:

```
$ starch dataset.bed > dataset.starch
```

As with other tools in the BEDOPS suite, however, *starch* also accepts data from an upstream Unix process and writes a compressed archive to standard output, which facilitates integration with analysis pipelines.

```
$ awk '$6 == "-"' TSSs.bed \
  | bedops --range 0:1000 --everything - \
  | bedops --everything 1kb_windows_upstream_of_forward_stranded_TSSs.bed - \
  | bedmap --echo --echo-map-id-uniq - motif_predictions.bed \
  | starch --note "Experiment 27B/6 | JASPAR CORE 5.0_ALPHA | GENCODE v19 TSS" - \
  > motif_IDs_1kb_upstream_padding_TSSs.starch
```

Answers to metadata queries with *unstarch* are returned immediately.

```
$ unstarch --note motif_IDs_1kb_upstream_padding_TSSs.starch
Experiment 27B/6 | JASPAR CORE 5.0_ALPHA | GENCODE v19 TSS
$ unstarch --elements motif_IDs_1kb_upstream_padding_TSSs.starch 529
```

While *unstarch* is used to extract information from a Starch-formatted archive and put it on a standard output stream as BED, other BEDOPS utilities work directly with the archive files. Operations can be applied to the entire contents of the archive, or to just one specific chromosome, which can facilitate faster, more focused analyses and parallelization.

Compressed Starch archives can be merged efficiently into one larger archive with *starchcat*. This is highly useful for parallelized analysis pipelines, where results for individual per-chromosome Starch files can be concatenated into one final archive. The *starchcat* tool can also upgrade older Starch archives to add newer metadata annotation features as they become available.

5 Further Reading

The BEDOPS toolkit may be downloaded from its main site (*http://bedops.readthedocs.org/en/latest/*), which includes extensive documentation on all programs and their options, basic and advanced pipeline examples, and performance tips. Additional help from authors and community members is available through an online forum (*http://bedops.uwencode.org/forum/*).

References

1. Kahn SD (2011) On the future of genomic data. Science 331:728–729
2. ENCODE Project Consortium (2012) An integrated encyclopedia of DNA elements in the human genome. Nature 489:57–74
3. Neph S, Kuehn MS, Reynolds AP et al (2012) BEDOPS: high-performance genomic feature operations. Bioinformatics 28(14):1919–1920
4. Kent WJ, Sugnet CW, Furey TS et al (2002) The human genome browser at UCSC. Genome Res 12:996–1006
5. Li H, Handsaker B, Wysoker A et al (2009) The Sequence alignment/map (SAM) format and SAMtools. Bioinformatics 25 (16):2078–2079
6. McIlroy MD, Pinson EN, Tague BA (1978) Unix time-sharing system foreword. Bell Syst Tech J 57(6)
7. Maurano MT, Humbert R, Rynes E et al (2012) Systematic localization of common disease-associated variation in regulatory DNA. Science 337:1190–1195
8. Stergachis AB, Neph S, Reynolds A et al (2013) Developmental fate and cellular maturity encoded in human regulatory DNA landscapes. Cell 154(4):888–903
9. Neph S, Vierstra J, Stergachis AB et al (2012) An expansive human regulatory lexicon encoded in transcription factor footprints. Nature 489:83–90
10. Thurman RE, Rynes E, Humbert R et al (2012) The accessible chromatin landscape of the human genome. Nature 489:75–82
11. Neph S, Stergachis AB, Reynolds A et al (2012) Circuitry and dynamics of human transcription factor regulatory networks. Cell 150 (6):1274–1286
12. John S, Sabo PJ, Thurman RE et al (2011) Cell-specific chromatin landscapes determine cell-selective glucocorticoid receptor occupancy. Nat Genet 43:264–268
13. Pollard KS, Hubisz MJ, Rosenbloom KR et al (2010) Detection of nonneutral substitution rates on mammalian phylogenies. Genome Res 20:110–121

Chapter 15

GMAP and GSNAP for Genomic Sequence Alignment: Enhancements to Speed, Accuracy, and Functionality

Thomas D. Wu, Jens Reeder, Michael Lawrence, Gabe Becker, and Matthew J. Brauer

Abstract

The programs GMAP and GSNAP, for aligning RNA-Seq and DNA-Seq datasets to genomes, have evolved along with advances in biological methodology to handle longer reads, larger volumes of data, and new types of biological assays. The genomic representation has been improved to include linear genomes that can compare sequences using single-instruction multiple-data (SIMD) instructions, compressed genomic hash tables with fast access using SIMD instructions, handling of large genomes with more than four billion bp, and enhanced suffix arrays (ESAs) with novel data structures for fast access. Improvements to the algorithms have included a greedy match-and-extend algorithm using suffix arrays, segment chaining using genomic hash tables, diagonalization using segmental hash tables, and nucleotide-level dynamic programming procedures that use SIMD instructions and eliminate the need for F-loop calculations. Enhancements to the functionality of the programs include standardization of indel positions, handling of ambiguous splicing, clipping and merging of overlapping paired-end reads, and alignments to circular chromosomes and alternate scaffolds. The programs have been adapted for use in pipelines by integrating their usage into R/Bioconductor packages such as gmapR and HTSeqGenie, and these pipelines have facilitated the discovery of numerous biological phenomena.

Key words Genomic alignment, Genomic mapping, Bioinformatics algorithms, Next-generation sequencing, RNA-seq, DNA-seq, Sequence analysis, Transcriptome analysis

1 Introduction

The programs GMAP and GSNAP embody substantial experience in aligning sequences to genomes, with over 12 years of ongoing development and improvement. Changes in the programs have been driven largely by advances in both molecular biology and computing over the years. Sanger-based sequencing of single genes and expressed sequence tags (ESTs) have given way to massively parallel, array-based sequencing of short reads [40]. These reads initially provided only 30 or so bp but now routinely span 200–300 bp. Other modern sequencing technologies can produce

Ewy Mathé and Sean Davis (eds.), *Statistical Genomics: Methods and Protocols*, Methods in Molecular Biology, vol. 1418, DOI 10.1007/978-1-4939-3578-9_15, © Springer Science+Business Media New York 2016

very long reads of thousands of bp (e.g., [10]), which require additional considerations in alignment.

Concomitant with these advances in laboratory science have come changes in computing hardware, including 64-bit architectures that have become the norm over 32-bit architectures; random access memory that has expanded from a few gigabytes to hundreds of gigabytes typically available per computer; massive parallelization using large clusters of nodes and multiple cores per node; and special computer instructions for performing single-instruction multiple-data (SIMD) operations. In addition to more sophisticated hardware, several developments have been also made in bioinformatics algorithms and data structures.

To handle the wide range of read types produced by various sequencing technologies, we have introduced and employed numerous algorithms and data structures, each of which excels in particular aspects of alignment (Table 1). Therefore, GMAP and GSNAP represent dynamic entities, evolving as needed since they were introduced in their initial journal articles of 2005 and 2010, respectively [45, 46]. This article provides an opportunity for us to describe the enhancements in speed, accuracy, and features made since then.

Despite our many changes and improvements to GMAP and GSNAP, we have attempted to maintain certain objectives. Foremost has been a desire to achieve a high level of accuracy in detecting sequence differences, such as mismatches, indels, and splicing.

Table 1
Algorithms used in GMAP and GSNAP

Algorithm	Prog	Structure	Characteristics
Greedy match-and-extend	S	ESA	Multiple indels and splices, with sparse mismatches
Spanning set analysis	S	Gen HT	Dense clusters of mismatches or mismatches at ends
Complete set analysis	S	Gen HT	Very dense clusters of mismatches
Segment combinations	S	Gen HT	Single indel; one or two splices
Segment chaining	S	Gen HT	Multiple indels and splices, with dense mismatches
End pairing	M	Gen HT	Genomic localization
Diagonalization	M,S	Seg HT	Identification of exons
Oligomer chaining	M,S	Seg HT	Approximate alignment
Standard DP	M,S	Linear	Resolution of mismatches or an indel
Sandwich DP	M,S	Linear	Resolution of splice
End DP	M,S	Linear	Resolution of ends of alignment

Abbreviations for programs: M denotes GMAP; S, GSNAP. Abbreviations for data structures: *ESA* enhanced suffix array, *Gen HT* genomic hash table, *Seg HT* segmental hash table

These differences are more difficult to detect than reads that match identically to the reference genome, but they provide the means to discover novel biology. They need to be detected with high accuracy, because errors and omissions at the alignment stage will be carried over to subsequent analysis steps. Certain combinations of sequence differences can be particularly hard to detect and require special attention. For example, the RGASP evaluation of RNA-Seq alignment programs found that only GSNAP among the comprehensive set of programs tested could reliably identify indels near splice sites [12].

In general, the most difficult issues arise in aligning of transcriptional reads to a genome (RNA-Seq), rather than aligning genomic reads (DNA-Seq). RNA-Seq alignment must handle the phenomenon of splicing, and GSNAP has been shown by others to have the highest accuracy among available programs in finding such splices [16]. However, other sequencing phenomena, such as mismatches and indels, are present in both RNA-Seq and DNA-Seq. Therefore, although GMAP and GSNAP were designed for the complexities of RNA-Seq, they are well equipped to handle alignment of DNA-Seq reads as well.

Accurate detection of sequence differences can require additional computational time and effort, so a secondary consideration has been to achieve speed without sacrificing accuracy. The quest for speed has motivated additional changes in GMAP and GSNAP, typically in new data structures and leveraging computational techniques. For example, the original releases of the programs have supported multithreading, and subsequent enhancements have modified many algorithms to use SIMD programming. Another consideration in our development process has been to provide computational tools for practical use. That means both that the programs need to be robust to various types of input, and that they provide auxiliary features that help users or downstream pipelines in the analysis process.

In the remainder of this article, we present a brief overview of the programs' evolution. Then we describe computational features available in GMAP and GSNAP that facilitate accurate alignment, and provide some examples of their usage. We then discuss the underlying algorithms and data structures that implement these features. Finally, we outline the utilities and other user features provided by the program that facilitate the complete analysis of sequencing datasets.

1.1 Brief History

Development of GMAP began in 2002, when the draft human genome was first released [46]. The program was designed to take advantage of the newly available genomic sequence, by allowing users to align genes and ESTs rapidly to the genome and thereby reveal their intron–exon structure. These alignments also provided measurements of gene expression, which could be used to help identify cancer-specific genes and the amplification of genomic regions in cancer [48].

Aligning these transcript-derived sequences requires the ability to identify splicing, in which introns are removed from the pre-mRNA transcript before producing the final mRNA product. These introns, which can span 200,000 nucleotides or more, result in alignments that are characterized by continuous alignments of exon regions separated by large intronic gaps. Existing alignment tools, such as BLAST [2], and standard dynamic programming algorithms, such as Needleman–Wunsch [33] or Smith–Waterman [41], are not well suited for determining this type of exon–intron gene structure. At the same time, some programs at the time, such as SIM4 [14], were designed to find exon–intron structure for a transcriptional read, but only when the corresponding genomic segment was provided.

Therefore, GMAP was introduced to perform both the genomic localization of a read and the alignment of the read to the corresponding genomic segment. In addition, GMAP was designed to be extremely robust to sequence differences, to the point where it could even handle transcripts that had up to 20 or 30 % sequence deviation from the reference genome. In contrast, other programs for genomic localization and alignment, such as BLAT [21], generally had their alignments affected adversely by small amounts of deviation of only 1 %.

The robustness of GMAP makes it one of the few tools that can perform cross-species alignment, where the transcriptional reads of one species (e.g., mouse) are aligned to the genomic sequence of another species (e.g., human). This robustness was enabled by a hierarchical alignment strategy that used *end pairing* for genomic localization, *oligomer chaining* for approximate alignment, and specific types of nucleotide-level dynamic programming procedures for filling in gaps in the alignment.

The state of sequencing technology changed significantly with the advent of next-generation sequencing (NGS). In particular, array-based methods allowed millions of short reads to be sequenced in parallel. The resulting large volumes of sequence data placed a premium on speed, rather than the ability to handle the complexities of cross-species alignment. Therefore, we introduced a new program called GSNAP to align short reads that were expected to have relatively few differences from the reference genome, and could therefore use a faster algorithm [45].

GSNAP used the same basic data structure as GMAP, which was a genomic hash table of 12-mers in the genome sampled every 3 bp. Instead of having separate localization and alignment steps, GSNAP generated candidates directly from the hash table lookups, through two algorithms: *spanning set analysis* and *complete set analysis*, with the former designed to handle relatively few sequence differences in the read and the latter designed to handle more sequence differences. GSNAP also found indels and splicing with high accuracy, taking advantage of the fact that reads were relatively short, originally 30 or so bp and then lengthening to 75 bp. With such short reads, one could reasonably expect a single indel or one or two introns. Therefore, GSNAP employed specific algorithms for handling such cases, which were not given an explicit name in the original paper, but in retrospect could be labeled as *segment combination*.

As NGS technology produced more and more data, changes were made in GSNAP to improve its speed. A significant change involved making the hash table more specific, by cataloging 15-mers instead of 12-mers. To make this feasible, we initially introduced a compression technique based on Elias gamma encoding [11]. Recently, we accelerated the decompression process significantly by modifying a recent SIMD bitpacking compression scheme [25]. Another change for the sake of efficiency was to change the way in which the linear genome is represented, using a vertical format rather than a horizontal format for storing the bits for each nucleotide. The linear genome is used throughout GSNAP for determining the number and location of mismatches for a given candidate genomic segment. In addition to using a vertical format, we also now represent the genome in 128-nt blocks rather than 32-nt blocks, to enable the use of SIMD procedures that can perform computations on 128 bits in parallel.

More recent generations of NGS machines are producing longer reads of 200 or more bp. These longer reads pose additional challenges, such as multiple introns, which could not be handled by the specific procedures for segment combination. Therefore, we incorporated parts of the GMAP alignment algorithm into GSNAP to solve complex alignments when a simple alignment could not be found.

Alignment in GSNAP can be considered as triage, where fewer computational resources are needed for the vast majority of alignments that are straightforward, and more intensive computational resources are needed for the minority of alignments that are complex. In other cases, a feasible alignment may not exist due to major differences between the query read and the reference genome, which calls for an alignment program to recognize a futile effort and to abandon its efforts. Whereas the standard segment combination algorithm in GSNAP handles moderately difficult cases, and the GMAP algorithm handles very complex cases, there was a need within GSNAP to solve simple alignments quickly. For that need, we have borrowed upon the experience of other alignment programs, which use suffix arrays or its compressed analog, the Burrows–Wheeler Transform (BWT), as their genomic data structure [4, 23, 26, 27, 30, 43].

We have therefore added suffix arrays to our own hash table data representation, first using a standard suffix array and then an enhanced suffix array (ESA) [1] that includes some of our own technological innovations. The ESA allows GSNAP to handle the most straightforward alignments rapidly, giving it time to concentrate on more complex alignments using hash table methods. However, our initial use of ESAs merely supplemented our use of hash tables to find segments and combine them in pairs. To handle longer reads, we have more recently generalized our suffix array algorithm to handle multiple indels and splices, and therefore a larger proportion of cases. Our latest suffix array algorithm is a *greedy match-and-extend* method that can handle multiple introns in long reads. Nevertheless, rather than relying solely on a suffix array method, which can miss alignments, GSNAP uses it only as an initial triaging step, having its more general algorithms available as secondary methods when needed for solving more difficult alignments when needed.

Recent work on GSNAP has refined this triaging process, by reducing the duplication of effort involved in applying the suffix array, hash table, and GMAP algorithms in succession. To do this, we have streamlined the integration of the GMAP algorithm into GSNAP, by starting with the partial solutions obtained from the suffix array and hash table methods, rather than applying the GMAP algorithm from scratch, which previously required a time-consuming step of building a segmental hash table. These efforts have greatly increased the speed of GSNAP from previous versions, while still maintaining its high level of accuracy.

We have also generalized our original hash table methods in GSNAP to allow them, instead of the more time-consuming GMAP algorithm, to handle more cases. Rather than using segment combination, which joins pairs of segments with a single indel or splice in a read of 75 bp or so, we now have a more general segment chaining algorithm that can find multiple indels and splices in a read alignment.

At the same time, other NGS technologies, such as those from Pacific Biosciences, have been producing very long sequence reads of thousands of bp. Such reads can contain a higher number of sequencing errors, and so GMAP has been finding use as one of the recommended programs for aligning these reads. Therefore, we have also introduced enhancements to increase the speed of GMAP, which also benefits GSNAP when it calls upon that algorithm. One of the enhancements has been to introduce a step called *diagonalization* to accelerate oligomer chaining. This step essentially finds candidate exons in the alignment and therefore bounds the search process in the oligomer chaining step. Both diagonalization and oligomer chaining require a segmental hash table of small oligomers from a given segment in the genome. To increase the efficiency of this step, we have developed SIMD-based methods that generate segmental hash tables quickly.

2 Computational Features

GMAP and GSNAP are characterized by various computational features that aim to improve alignment results and find biological phenomena with high accuracy. Some of these features are general design strategies, such as an attempt to use global information across the read. Other features are specific tasks that GMAP and GSNAP apply in the alignment process, such as trimming and the detection of ambiguous splicing. We discuss both types of computational features in this section.

2.1 Using Global Information

A fundamental strategy in GMAP and GSNAP is to use global information across a read. This strategy differs from methods in other programs that typically try to identify a candidate quickly based on part of the read and then extend that alignment. In many programs, this strategy is called seed-and-extend [2], in which an oligomer seed is found that matches between the query sequence and genome, and then the alignment is extended outward from the seed.

In contrast, starting with its original implementation, GMAP takes a hierarchical approach that successively refines the alignment through a multiple-stage process, as shown in Fig. 1. In stage 1, GMAP looks at long oligomers (such as 24-mers or 30-mers) from both ends of the query sequence to identify candidate genomic regions that contain oligomers from both ends. Such long oligomer matches can be found from the genomic hash table by combining

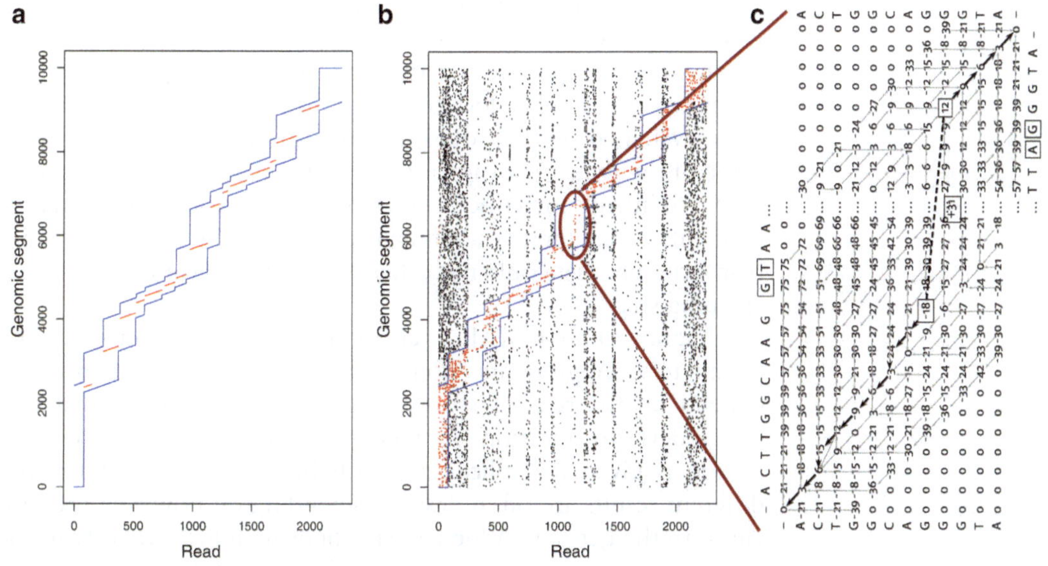

Fig. 1 Hierarchical approach in GMAP. (**a**) Diagonalization procedure with putative exons in *red* and bounds in *blue*. (**b**) Oligomer chaining with all *k*-mer matches shown as *black dots* and the dynamic programming solution as *red dots*. (**c**) Sandwich dynamic programming used to resolve a splice

information from lookups of 15-mers, and give rise to an end pairing algorithm in GMAP. In end pairing, GMAP scans from both ends of the read in order to find a match, skipping forward 1 nucleotide each time. Because one end of the read may be relatively nonspecific in the genome, GMAP may need to advance on the nonspecific end faster than the specific end. GMAP keeps track of this uneven scanning by keeping track of a cumulative specificity score that is based on the reciprocal of the number of distinct matches in the genome. Therefore, a single unique match in the genome counts for much more than a nonspecific oligomer that matches several places in the genome. Once GMAP finds one or more candidate genomic locations, it continues to scan for a certain number of cycles, to guard against the possibility of a false positive match. These additional cycles may lead to further support for the first candidate, or generate additional candidate locations.

In stage 2, GMAP attempts to align the read against each of the candidate regions found in stage 1. The candidate region is first analyzed by extracting a segmental hash table showing where all 7-, 8-, or 9-mer matches exist between the read and that genomic segment. This segmental hash table is then used by GMAP in two algorithms, diagonalization (Fig. 1a) and oligomer chaining (Fig. 1b), to generate an approximate alignment. This alignment contains blocks of ungapped and identical alignment between the query and genome, separated by gaps that are due to mismatches, indels, or introns. Finally, in stage 3, GMAP fills in these gaps using a variety of dynamic programming algorithms at the nucleotide level (Fig. 1c).

We have found that an iterative approach is sometimes necessary for very complex situations, including cross-species alignment. In an iterative approach, a dynamic programming routine that changes a single splice location will cause all of the intron gaps to be re-evaluated by their dynamic programming procedures. Since some of the dynamic programming routines are designed to combine two introns into one or to introduce an exon into the middle of an intron, an iterative alignment can lead to the correct exon–intron structure that could not otherwise be found from the oligomer chaining step.

Alignment in GMAP is also controlled by adaptive scoring parameters. Alignments from dynamic programming routines depend on the relative scores assigned to matches, mismatches, and indels. In addition, in sandwich dynamic programming, a reward is given for splice sites with high probability. However, different results are often desired depending on how well the read aligns to the genome. When a read aligns with high identity to the genome, we are more likely to trust the nucleotide alignments as a guide for finding intron boundaries. However, when a read aligns with lower identity, then the probability evaluations of splice sites provide a more reliable guide to intron boundaries than the nucleotides themselves. Therefore, GMAP evaluates its initial approximate alignment of a read as having low, medium, or high identity to the genome, and adjusts its dynamic programming scores accordingly. This adaptive scoring strategy is another example of how GMAP uses global information across the read in its alignment.

For GSNAP, information is also used globally across the read in the form of its spanning set and complete set analyses, as described in detail in our original paper on GSNAP. Each of these analyses is designed to look up oligomers in the hash table over the entire read. For the spanning set analysis, these oligomers are non-overlapping; for complete set analysis, they are overlapping. In both cases, we avoid the problems inherent in relying upon a single seed to determine the set of candidate genomic locations. The spanning set analysis is designed to find alignments with relatively few mismatches and is faster, so GSNAP tries that method first. That analysis yields a set of complete or partial segments that can be defined by a starting and ending position on the read, as well as a starting and ending position in the genome. Each segment can be compared against the genome to count its mismatches. If a single continuous match cannot be found, then GSNAP relies upon a set of segment combination procedures to join segments that are close in the genome and therefore represent candidate indels or local splicing events. If these procedures fail to yield a solution, then GSNAP employs its complete set analysis, which allows for a higher proportion of mismatches, to yield a larger set of segments, which are also compared against the genome and combined, if necessary, to yield alignments with indels or splicing.

2.2 Known Splicing Information

Another key philosophy underlying GMAP and GSNAP is their use of biological information to facilitate alignment. Rather than insisting that our algorithms be generalizable to other domains, we have been open to using as much domain knowledge as possible. For RNA-Seq, the most critical aspect is accurately identifying where splicing occurs within the alignment. Much is known about the domain of splicing and the patterns of nucleotides that surround donor and acceptor splice sites [5]. For example, the dinucleotide pairs at the ends of an intron are almost always GT–AG (or canonical splice sites) and the exceptions are generally GC–AG or AT–AC (semi-canonical splice sites). Probabilistic models have been developed to evaluate the probability that a given genomic location is a donor or acceptor splice site. GSNAP uses the MaxEnt model [47], which assesses 11 nucleotides around a proposed donor splice site and 23 nucleotides around a proposed acceptor splice site. GSNAP can then scan a candidate genomic segment for probable splice sites to determine whether to introduce an intron.

Although probabilistic models are accurate most of the time, there are some known splice sites that do not fit the models well and generate false assessments. Therefore, GSNAP pioneered the use of known splicing to assist with RNA-Seq alignment. A user can provide GSNAP with a database of known splicing, which can be specified at either the individual splice site or paired splice site (or intron) level. When known splicing is provided at the individual splice site level, GSNAP is free to splice between any pairing of donor and acceptor splice sites, subject to other limits the user may impose. For known splicing at the intron level, GSNAP can introduce splicing only between the given pairs of splice sites. In general, because of the phenomenon of alternate splicing, specifying information at the level of individual splice sites is generally advisable.

With known splicing, a user may still decide to allow GSNAP to find novel splice sites with the aid of its probabilistic model, and this is also advisable, given our incomplete knowledge of known splice sites and the phenomenon of cryptic splice sites. However, even when both known and novel splicing are allowed, known splice sites can still provide GSNAP with clues about the correct location of introns.

2.3 Tolerance to SNPs, Bisulfite Sequencing, RNA-Editing, and PAR-CLIP Sequencing

With GSNAP, we introduced the idea of SNP-tolerant alignment, which is another example of using domain knowledge in alignment. Human beings largely differ from another genetically by a set of ten million or so SNPs, or single-nucleotide polymorphisms, that occur commonly across the entire population [31]. The majority of sequence differences in a given sample will therefore differ from the reference genome at these SNP locations. However, a typical alignment program will consider such a difference as a mismatch, which is typically evaluated less favorably than a match. Therefore, most genomic alignment programs will have an inherent bias against aligning reads that have the non-reference SNP allele, manifesting as incorrect alignments, errors in copy number or allele-specific expression, and false associations between SNPs and phenotypes [7].

In contrast, with SNP-tolerant alignment in GSNAP, the user can provide a database of known SNPs, which is then used to construct an alternate genomic index. The resulting hash table is designed to allow both the reference-based and alternate-based oligomers to yield the same genomic location. GSNAP also uses both the reference genome and the alternate genome when comparing the read against a genomic segment to count mismatches. As a result of these modifications, GSNAP is able to treat the reference and alternate alleles equally and eliminate biases against the alternate allele. However, the utility of SNP tolerance is probably greatest for very short reads, where genomic localization can be affected by a single nucleotide. For longer reads, where sufficient evidence for localization is likely to exist elsewhere on the read, SNP tolerance alignment may be less important, as long as the aligner allows for a sufficient number of apparent mismatches in the alignment.

Although SNP-tolerance was implemented in our original release of GSNAP, we have expanded the concept to allow for wildcard SNPs. A wildcard SNP is one in which the minor allele is not defined, either because it is unknown or because more than one minor allele is allowed at that position. In any case, a wildcard SNP permits a match regardless of the read nucleotide at that position.

In addition to SNP tolerance, GSNAP is able to handle widespread substitutions of nucleotides, which occur in both bisulfite sequencing and in RNA-editing tolerance. In bisulfite sequencing, the library preparation changes all methylated Cs to Us, which appear as Ts in sequencing [15, 28]. Therefore, we must treat cases of read-T and genomic-C as matches, rather than mismatches. On the other hand, cases of read-C and genomic-T would constitute legitimate mismatches. GSNAP handles this alignment situation by creating special hash tables with only three possible nucleotides, and by using specialized procedures that compare read segments against corresponding genomic segments in the context of widespread C-to-T changes. Since our original paper on GSNAP, these comparison procedures have been made faster using SIMD-based bitwise operations, as we discuss in Sect. 4.1.

An analogous situation holds if we desire to align reads with tolerance to RNA editing, a feature that has been introduced after the initial implementation of GSNAP. In the majority of cases of RNA editing, certain occurrences of A are changed to Inosine, which appears as G in sequencing [3]. Although these changes are not as widespread as for bisulfite sequencing, we can still handle them with the same mechanism, with special hash tables for three possible nucleotides and specialized nucleotide-level comparison procedures. In addition, the hash tables needed for tolerance to RNA-editing can also be used by GSNAP to align cases where T-to-C changes are common, which occurs with data from the PAR-CLIP laboratory procedure used for studying RNA–protein interactions [17]. Therefore, GSNAP currently allows for alignment under a standard mode, or under tolerance to bisulfite sequencing, RNA editing, or PAR-CLIP sequencing. For the tolerant modes, GSNAP allows for library protocols that yield either strand-specific or nonspecific read sequences. For each of these modes, SNP-tolerance is available as an additional option.

2.4 Standardized Indel Positions

One problem in analyzing sequencing data for indels is the ambiguity of where to place the insertion or deletion. For example, in the case of repeated stretch of the same nucleotide, such as 10 A's, it is not clear whether an insertion or deletion of an A should be placed at the beginning, middle, or end of this homopolymeric region. Similar situations of ambiguous placement can hold for other combinations of genomic patterns and indels. When an aligner places an indel in different genomic locations across its alignments, it becomes more difficult to find patterns of repeated occurrences of an indel across a given dataset.

Therefore, GMAP and GSNAP attempt to standardize the placement of indels by shifting them to the lowest possible chromosomal coordinate that maintains the same number of mismatches. Of course, some individual reads will not cover the lowest possible coordinate, so they will report results other than the standardized position. Nevertheless, standardization still helps greatly in the analysis of indels. The standardization process is built into the algorithms that find indels, including the procedures for segment combination and for nucleotide-level dynamic programming. For dynamic programming, the direction of genomic alignment determines how tie scores are ranked. When the procedure generates an alignment with ascending genomic coordinates, then a subsequent tie score is considered a worse result than the first occurrence of that score. Conversely, for descending genomic coordinates, a subsequent tie score is considered a better result than the first occurrence of that score. These different policies for handling tie scores alter the pointers used in the traceback step, and thereby shift the indel to the lowest possible chromosomal coordinate.

2.5 Trimming Ends

Although GMAP and GSNAP attempt to align reads as completely as possible, there are situations when the full alignment of a single read cannot be determined in isolation. The most common situation occurs at the ends of reads, where there may be insufficient sequence to resolve complex phenomena, such as indels or introns. An indel or intron within the last few bp of a read can be difficult to distinguish from a series of mismatches, if no other information is available. Of course, subsequent analysis steps can make use of neighboring alignments to resolve these ambiguities, but GMAP and GSNAP do not have access to such information when aligning each read independently. Even if an aligner suspects an indel or intron may be present near the end of a read, there is often insufficient sequence to determine the location of the other end of the gap.

If an indel or intron at the end of a end cannot be resolved, then a straightforward extension of the alignment will result in calling false positive mismatches. Rather than make such errors of commission, GMAP and GSNAP have mechanisms for trimming alignments at their ends. In GMAP, Smith–Waterman dynamic programming is applied to each end to find the optimal score of matches, mismatches, and indels. The core GSNAP algorithm applies a simpler trimming procedure that assigns scores to matches, mismatches, and indels along the alignment. By default, a match receives +1 point, a mismatch receives −3 points, and an indel receives −2 points. Trimming is then determined by maximizing the cumulative score, as computed from the other end. The sensitivity for trimming at the ends can be controlled by the user by setting a different score for the amount of mismatches. If user sets the score to 0, then no trimming will be performed.

2.6 Ambiguous Splicing

The ends of an alignment pose challenges for detecting splices accurately. In particular, there may be several splices that are equally plausible, based on the distal sequence of nucleotides in the read. Therefore, rather than reporting either a series of mismatches or an incorrect splice, GMAP and GSNAP have a notion of ambiguous splicing. An ambiguous splice can be thought of as a set of equally good splicing possibilities. Since the correct splice cannot be determined, the alignment will need to report a soft clip at the medial splice site and not attempt to report the alignment of the distal exon. However, the alignment output will indicate that the soft clipping is due to an ambiguous splice and list multiple possibilities for the distal splice location. Internally, to compare alignments with ambiguous and unambiguous splicing fairly, GMAP and GSNAP give appropriate credit for the matches found on the other end of the gap.

Nevertheless, GSNAP will try to resolve ambiguous splicing whenever possible. For example, introns occurring near the end of a read can sometimes be found with the assistance of known splicing. A known splice site near the end can often provide an accurate splice even of only a few nucleotides are available for the distal end of the gap. This is because GSNAP can scan other known splice sites nearby to see if any match the distal nucleotides. To facilitate such matching, GSNAP builds a set of prefix tries at every given donor and acceptor splice site, based on the known splice site information. These prefix tries represent the possible substrings and their genomic locations within the allowable intron distance of the given site. Therefore, the distal substring at the end of the read can be translated quickly into one or more matching prefixes and genomic locations within an allowable intron distance. Mismatches can be allowed in this search process, with the goal to identify the distal splice end or ends with the fewest number of mismatches.

If only one known splice site matches, that is highly likely to be the correct distal end. However, if two or more known splice sites match, then an unambiguous alignment cannot be made. In that case, GSNAP reports a soft clip at the medial splice site, rather than guessing about the distal end. This notion of ambiguity holds for novel splicing as well. GSNAP has procedures for identifying ambiguous splicing at the ends and reporting them as such with soft clipping, rather than reporting an unsupported intron.

Another way in which GSNAP can resolve ambiguity in splicing is with paired-end reads. A paired-end read introduces a distinction

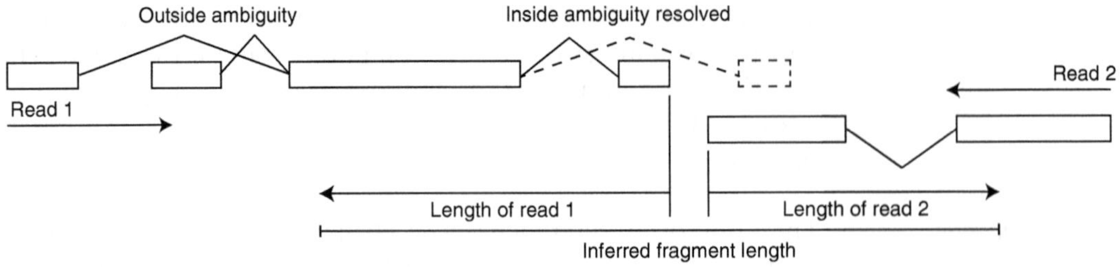

Fig. 2 Ambiguous splicing. The top gene structure represents the alignment for read 1 to the plus strand of the genome. The bottom gene structure represents the alignment for read 2 to the minus strand. The alignment for read 1 shows ambiguities on both of its ends. The 5′ end shows two equally good introns that cannot be resolved, whereas the 3′ end also shows two equally good introns. The 3′ introns can be resolved because one is inconsistent with the alignment for read 2, extending past its inside extent

between the inside and outside parts of the alignments. The inside parts are the 3′ regions of each end, which should be relatively close to each other in the genome. In contrast, the outside parts are the 5′ regions of each end, which represent the outermost bounds of the total extent of the sequence fragment. The inside parts are constrained because splicing should not extend beyond the mate end (Fig. 2). Therefore, GSNAP can often resolve ambiguous splicing when it occurs on the inside of paired-end alignments.

2.7 Chimeric Alignments and Distant Splicing

Another type of complex alignment handled by GMAP and GSNAP is distant splicing. Most splicing involves local excisions in an mRNA transcript of 200,000 bp or less, which is characteristic of typical mRNA processing. However, certain biological events can cause a splice to apparently span distant parts of the genome, such as different chromosomes (due to translocations) or large distances on the same chromosome (genomic deletions). Or the ends of the splice may occur in opposite directions on the chromosome (inversions) or in the same direction but in the wrong order, where the acceptor splice site is downstream of the donor splice site (scrambling events). Scrambling events can be indicative of circular transcripts or tandem genomic duplication.

GSNAP finds distant splicing from the partial segments obtained from its spanning set and complete set analyses, and used in its segment combination procedures. Typically, these partial segments are combined by GSNAP to find indels or local splicing events. But when such a local alignment cannot be found, GSNAP then tries to combine segments using distant splicing. For RNA-Seq, we expect that distant splicing should generally occur at exon–exon junctions with plausible splice sites. This is because introns are so large relative to exons that it is highly unlikely for a genomic rearrangement to occur in the middle of an exon. Instead, a distant splice with a true genomic origin should occur in an intron or intergenic region, and then distant exons should be joined through the normal splicing mechanism. In contrast, a joining of two transcripts in the middle of one or both exons is more likely to be due to a library artifact, in which a crossover occurs between transcripts that have already been spliced. Therefore, to generate distant splices with high specificity, GSNAP looks within partial segments for evidence of plausible splice sites, either from the user-supplied set of known splice sites or from high probability scores. The partial segments are then organized by the positions of the splice site within the read, and GSNAP searches for a pair of partial segments with matching splice site positions. If a single result emerges, GSNAP reports that alignment as a distant splice.

The algorithm described above for finding distant splices has been implemented in GSNAP since its original release. However, recent enhancements have been made to GSNAP for this type of alignment. One improvement has been to handle distant genomic joins in DNA-Seq, rather than in RNA-Seq. In DNA-Seq, because splicing is not involved, there is no requirement that the crossover point occurs at an exon–exon junction. Instead, the crossover point is most likely to occur between two intergenic or intronic segments. To allow for that situation, GSNAP organizes its partial segments by the range of read positions where the segment starts or ends. This range is defined by the locations of mismatches where the segment fails to align to the genome. Then GSNAP can combine distant segments that are consistent in their starting and ending positions in the read.

Another line of work has been to generalize the GSNAP distant splicing algorithm for longer reads. Currently, the algorithm depends on partial segments, which work ideally for short to medium-length reads. In our own simulations of paired-end 75 bp reads (unpublished), we have found that our algorithm is highly sensitive for finding splices, even with low coverage of only one or two reads that cross the distant splice junction. However, for longer reads that span multiple introns, a different approach is needed. We are therefore implementing an approach that can join partial alignments generated by the suffix array and GMAP algorithms.

GMAP handles the issue of chimeric alignments in a more general way than GSNAP. When an alignment does not cover the entire query, GMAP tries aligning the remaining piece or pieces of the read. GMAP then tries to combine the various partial alignment. Sometimes the partial alignments are near each other in the genome and can be merged into a single local alignment, possibly requiring the introduction of a splice. This situation can occur for very long reads or for very long introns, for which GMAP could not initially obtain a single continuous approximate alignment. When a local merge cannot be found, GMAP attempts to join partial alignments using a translocation, inversion, large genomic deletion, or other distant splice. The program attempts to find the best possible junction, which ideally involves an exon–exon junction, but non-canonical joins are also allowed. In the latter case, GMAP reports the range of possible crossover points that give equivalent alignment scores. This information can be used in subsequent analysis to detect a repeat at the site of a chimeric event.

2.8 Overlapping Alignments

Another complex situation in alignment occurs when paired ends overlap. This situation affects only GSNAP, since GSNAP accounts for paired-end relationships and attempts to find concordant alignments, whereas GMAP aligns each read independently. When a read fragment is sequenced at both ends for a total amount that exceeds the original fragment length, then the sequenced ends will overlap, which should result in alignments that overlap. The overlap may not actually become evident because of trimming or ambiguous splicing. However, when overlapping does occur in alignments, it can create problems for downstream analysis. In particular, algorithms for calling variant alleles often rely upon the number of major and minor alleles found. In an overlapping alignment, some nucleotides are counted twice, which can therefore skew results in regions of low coverage.

Therefore, GSNAP has mechanisms to handle overlapping alignments. We note first that GSNAP aligns paired ends independently, and then looks for concordance between these alignment results. This strategy differs from some algorithms, such as STAR, that try to find a continuous alignment across both paired ends, a method that will not work well for overlapping alignments. When the two ends overlap in their alignments, GSNAP attempts to find the midpoint of their overlapping region, and then hard clips the duplicate ends within the overlap (Fig. 3a). Alternatively, if the user

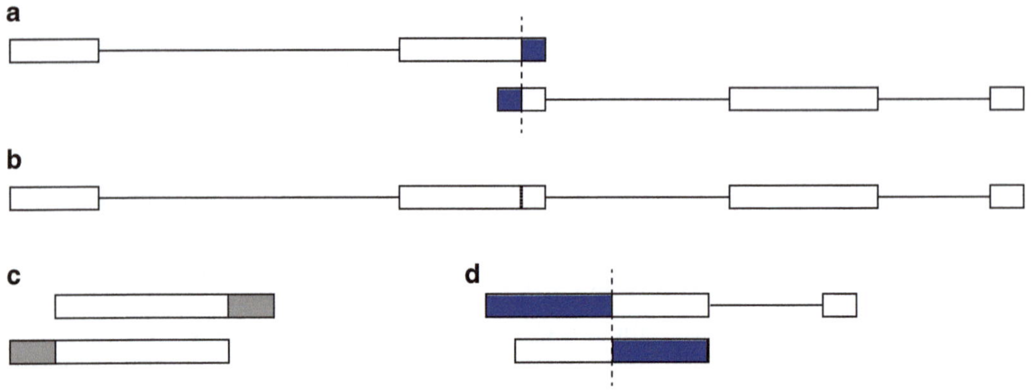

Fig. 3 Overlapping alignment. (**a**) Paired-end alignment for read 1 (*top*) and read 2 (*bottom*). The inside exons overlap, with the midpoint marked with a *dashed line*. The *shaded regions* can be hard-clipped to remove duplicate coverage. (**b**) The same alignment resolved by merging the two alignments at their midpoint. (**c**) A case of a fragment that is shorter than either of the sequenced ends. The sequences include adapters, shown in *gray*. (**d**) A case where the end of the alignment for read 1 extends past the beginning of the alignment for read 2. GSNAP resolves this by clipping the tails of each alignment, shown as *shaded regions*

desires, GSNAP can merge overlapping alignments to produce a single continuous alignment (Fig. 3b).

This mechanism works when the read fragment is longer than the amount sequenced at either end. However, for extremely short read fragments, the sequencer will continue past the end of the fragment into the adapters used for library preparation (Fig. 3c). Since these adapter sequences are unlikely to align to the genome, they will adversely affect the alignment results. Therefore, GSNAP tries to identify this situation before alignment, by analyzing every paired-end read for a high degree of identity between one end and the reverse complement of the other end. If such an identity is found, then GSNAP can infer that adapter sequence is present at the ends and remove that part of each read before performing its alignment.

Nevertheless, even when adapters are detected and removed, there can still be situations where the two alignments extend past each other. This generally occurs through imperfections in the alignment or differences in the sequence content of the overlapping regions. In such cases, GSNAP may be able to find a splice for one end, but not the corresponding splice for the other end (Fig. 3d). To handle such cases, GSNAP will attempt to hard clip the beginning parts of each end, rather than their ending parts, to preserve as much of the alignments as possible.

2.9 Suboptimal Alignments

Sometimes, a read may map to more than one place in the genome equally well (called multimappers), or almost equally well (called suboptimal alignments). A user may wish to find suboptimal alignments to compensate for possible alignment errors. Such errors can result from tandem duplications and paralogous genes in a genome, which can lead to some uncertainty about the correct genomic location of a read, especially when it is very short. Suboptimal alignments provide an indication about such uncertainty, and reads with suboptimal alignments can be excluded from downstream analysis in order to achieve high specificity in tasks, such as calling allelic variants.

Therefore, many aligners can report not just the best alignment, but also other alignments that are suboptimal. For example, Bowtie [23] is able to report all alignments with up to 2 mismatches. This behavior can be considered an absolute notion of suboptimality, since it reports a set of alignments up to a certain level, regardless of the best alignment found.

In contrast, GSNAP uses a relative notion of suboptimality. In this concept, the level of suboptimality is determined by the best alignment found. For example, if GSNAP is not able to find any exact alignments, but is able to find an alignment with one mismatch, then it considers that level to be the starting criterion for suboptimal alignments. In this example, if the user requests two suboptimal levels, then GSNAP will explore the space of possibilities involving one, two, or three mismatches (or actually points, if we consider the penalty scores for an indel).

2.10 Ranking Alignments and Eliminating Duplicates

An alignment program can generate multiple alignments for a single read. These alignments fall into two categories: duplicates and multimappers. A duplicate alignment occurs when two alignments, or slight variations of each other, are reported at the same genomic location. Such duplicates can occur in GSNAP, because of its use of multiple alignment methods, which can generate the same alignment more than once. In contrast, GMAP performs a separate alignment for each candidate genomic segment, so it does not have this issue. Duplicate alignments must be resolved by selecting the best representative among the competing possibilities at that genomic location, and GSNAP favors alignments that are more complete.

For multimappers, the issue is ranking alternative possibilities at different genomic loci, to try to determine the correct genomic location, if possible. These possibilities may differ in their amounts of completeness, especially in RNA-Seq where a splice may occur near the end of the alignment. Therefore, in comparing these different alternatives, GSNAP takes care not to penalize them for their differing amounts of trimming. Instead, GSNAP finds a common region among all alternative alignments, as defined by the maximum amount of trimming observed at each end, and then compares the alignments based on the number of mismatches in the common region. If the remaining multimappers still have equally good scores, then GSNAP will select one at random as being the primary alignment. (The user may also select an option for GSNAP to pick among multimappers deterministically, by selecting the one with the lowest genomic coordinate.)

2.11 Circular Chromosomes

Chromosomes can be either linear or circular, with circular chromosomes posing an issue in genomic representation. If a circular chromosome is represented linearly, then a read that crosses the circular origin will appear to be a two-part alignment, in which one part of the read aligns to the end of the chromosome and the other part aligns to its beginning.

GMAP and GSNAP provide a representation for circular chromosomes that makes it possible to find such alignments. They maintain two copies of a circular chromosome concatenated to each other, for use in building the genomic indices. Therefore, a read can be aligned without any further difficulty, even when it crosses from the end of the first copy into the beginning of the second copy. The difficulty arises only in reporting the alignment, when the coordinates of the second copy must be translated into coordinates that fit within the length of the circular chromosome. This translation process is called aliasing, and simply involves subtracting the total chromosomal length from the coordinates.

The second chromosomal copy leads to a complication for reads that do not cross the circular origin, in that such alignments will occur more than once. However, the process of aliasing resolves this problem easily, since that process will give the second alignment the same genomic coordinates as the first copy. Then, the mechanisms described above for removing duplicate alignments will discard the second alignment.

Circular chromosomes occur for many prokaryotes, so our representation provides a way to analyze those organisms properly. The mitochondrial genome in higher eukaryotes, including human beings, is also circular. It is important to represent the mitochondrion correctly, because the mitochondrion can generate high expression of transcripts in some tissues. In our experience, we have found that representing the mitochondrial genome properly as a circular chromosome leads to better alignment results, since some of the circular alignments would otherwise be reported elsewhere in the human genome, causing false results in downstream analysis.

2.12 Alternate Scaffolds

As our conception of genomes becomes more sophisticated, we are starting to identify and represent coordinated patterns of sequence differences among individuals. These patterns, or haplotypes, are being included in recent genomes as alternate scaffolds. The most recent version of the human genome, GRCh38, has 262 such alternate scaffolds. Each alternate scaffold has a corresponding primary genomic region that it is intended to replace in particular individuals. Alternate scaffolds pose an issue for genome alignment, because they complicate the notion of uniqueness in mapping. Most alignment pipelines ignore multimappers in their analysis and process only the unique alignments, because multimappers represent uncertainty about the correct genomic location of a read.

However, with alternate scaffolds, alignments to corresponding genomic alternatives are akin to duplicate alignments rather than multimappers. Therefore, they should be resolved as competing alignments, by selecting the single representative among them that is best. GSNAP can therefore handle alignments to alternate scaffolds by modifying its existing mechanisms for aliasing coordinates and eliminating duplicate alignments. Aliasing is more complex than for circular chromosomes, since we cannot simply subtract a single value to find the corresponding primary coordinate. This is because an alternate scaffold may be considerably longer or shorter than its corresponding primary segment. In fact, an alignment to an alternate scaffold may not have a matching set of coordinates in the primary genomic segment. Therefore, instead of applying aliasing to find an exact set of primary coordinates, GSNAP computes a range of possible primary coordinates, and then considers overlapping coordinates as a basis for resolving duplicates.

In resolving these duplicates, GSNAP will favor alignments with more matches, which will therefore result in reads being assigned to their closest haplotype. If there are multiple alternatives that are equally good, then preference is given to the primary chromosome.

3 Usage

We now provide an introduction for using the GMAP and GSNAP programs, which demonstrates basic usage plus some of the features described above. The GMAP package contains many parameters for configuration, and the alignment programs have numerous options, that cannot all be covered in this section. Additional details are provided in the README file included with the package, or by calling each program with the `--help` flag.

3.1 Downloading and Compilation

The GMAP and GSNAP programs are freely available for download as source files from the Web site http://research-pub.gene.com/gmap. They should run on any Unix system that has a C compiler and a version of the Perl program. The package should also be compatible with Windows machines that can simulate Unix using the Cygwin program.

The files are packaged in tape archive (tar) format and compressed with the gzip program. You will therefore need to decompress and extract the files, which can generally be done with the tar program by typing

```
tar xvfz <compressed tar file>
```

The source files then need to be compiled for your particular computer, by entering the directory containing the package and running the commands

```
./configure
make
make install
```

The "`./`" in front of the configure command ensures that you execute the configure program provided with the GMAP package and not some other configure command on your computer system that may take priority, based on your PATH environment variables. You may customize the configuration process by providing the flags `--prefix`, to specify where the programs will be installed, and `--with-gmapdb`, to specify where genome index files will be installed by default. Alternatively, you may store these and other parameters in a configuration file specified to the configure command using a `CONFIG_SITE` argument.

The configure program performs various tests on your machine to determine its characteristics, and then compiles the programs accordingly. In particular, one important set of characteristics is the type of SIMD instructions supported by your computer and compiler. Computer processors have been supporting increasingly sophisticated sets of SIMD instructions, with the earliest sets being MMX and SSE2, and the most recent sets being AVX2 and AVX-512. The GMAP package will attempt to take advantage of the range of SIMD instructions available for your computer. It will compile alternate versions of each alignment program, one for each available SIMD level, and then select the most appropriate version at run time, thereby allowing the program to be used in a cluster of mixed computer types.

The compilation process produces the binary programs GMAP and GSNAP, as well as versions of these programs called GMAPL and GSNAPL that can handle genomes larger than 2^{32} bp. Compilation also produces various auxiliary programs needed for pre-processing a genome for use by these alignment programs.

3.2 Pre-processing a Genome

GMAP and GSNAP are designed to align different read datasets applied to a common target sequence, such as a genome. The target sequence may also be a set of transcripts or other contigs, but for convenience, we will use the term "genome" to refer to any set of target sequences. To achieve speed, the programs depend on having the genome pre-processed into index files, such as suffix arrays and hash tables. This pre-processing step can be somewhat time-consuming, but needs to be done only once for any given genome.

The GMAP package contains a program called gmap_build that performs the necessary pre-processing of a genome in FASTA format, with each chromosome as a separate FASTA entry. This program determines the name of each chromosome or contig from the FASTA header, but can use a text file to substitute a different name for each chromosome or contig. Substitution can be useful when trying to add or remove a prefix such as "chr" from each chromosome name. The genomic build process also provides for flags to designate chromosomes that are circular, such as the mitochondrial genome for humans, and those that are secondary versions of the standard chromosomes.

By default, the gmap_build program puts the genome index files into the directory provided by `--with-gmapdb` to the configure program. However, the user may bypass this default with a flag.

The user may also prepare GSNAP for handling different alignment modes, such as bisulfite sequencing where C's in reads can be converted to T's. Likewise, GSNAP can align in a mode that is tolerant to RNA editing changes, where A's in reads are converted to G's. These alignment modes require additions to the genome indices, built using the cmetindex or atoiindex programs. The atoiindex program also allows for GSNAP to align with tolerance to T-to-C changes, which occur with PAR-CLIP sequencing.

The genomic build process can also handle large genomes having more than 2^{32} or 4 billion bp in length. At the present time, such genomes are represented by hash tables but not suffix arrays. Alignment to a large genome is handled by the GMAPL and GSNAPL programs, also provided in the GMAP package.

3.3 Auxiliary SNP and Splicing Information

The user may also provide information about known SNPs in building a known genome. This is done by using the snpindex program. SNP information is first put into a text format, for instance,

```
>rs62211261 21:14379270 CG
>rs62211262 21:14379281 CN
```

where each line provides the name of the SNP, followed by its chromosomal coordinate, and the two allowable alleles for the SNP. For a wildcard SNP, which allows any nucleotide to match at a given position, the reference allele and the letter "N" should be provided.

The GMAP package provides several programs for generating this SNP information from Variant Call Format (VCF) or GVF formats to generate a text snpfile. The text file can then be converted into an interval index tree (IIT) format using the iit_store program provided with the GMAP package. The IIT format is a way to represent intervals in a binary tree for fast retrieval.

```
cat <snpfile>.txt | iit_store -o <snpfile>
```

Finally, the snpindex program processes the IIT file to provide auxiliary information for a given genome, stored under a given name for the set of SNPs:

```
snpindex -d <genome> -v <snp_db_name> <snpfile>.iit
```

The snpindex program compares the two given alleles (or the one given allele for a wildcard SNP) against the reference genome and ignores any SNPs that are inconsistent with the genome.

Likewise, known splicing information can be specified in the following example text format

```
>NM_004448.ERBB2.exon1 17:35110090..35110091 donor 6678
>NM_004448.ERBB2.exon2 17:35116768..35116769 acceptor 6678
>NM_004448.ERBB2.exon2 17:35116920..35116921 donor 1179
>NM_004448.ERBB2.exon3 17:35118099..35118100 acceptor 1179
>NM_004449.ERG.exon1 21:38955452..38955451 donor 783
>NM_004449.ERG.exon2 21:38878740..38878739 acceptor 783
>NM_004449.ERG.exon2 21:38878638..38878637 donor 360
>NM_004449.ERG.exon3 21:38869542..38869541 acceptor 360
```

which can be generated from standard GTF or PSL formats using several utilities provided with the GMAP package. This format contains a name for the splice site, followed by the chromosomal coordinates straddling the exon–intron junction, the type of splice site, and optionally, an observed distance for the associated intron. The observed distances are used to indicate introns that are longer than the typical or given values for local splicing distance, so that splicing may be found correctly in those cases.

The text format is then converted into an IIT file

```
cat <splicing>.txt | iit_store -o <splicing>
```

and the IIT file can be placed into the maps subdirectory of the genome index, which is designed to store genomic annotations, such as SNPs and splice sites.

Rather than specifying splice sites, a user may elect to provide splicing information at the level of introns. In that case, the chromosomal coordinates denote the intron interval, rather than the position of an individual splice site.

3.4 Alignment

Once a genome has been indexed, a user may align reads against the genome using GMAP or GSNAP. Reads may be provided in either a FASTA or FASTQ format. For GMAP, a single FASTA file is generally used, with each read potentially extending over many lines, so the typical call would be

```
gmap -d <genome> <read_file>
```

where "⟨genome⟩" is the name of the genome database created by gmap_build. GMAP also allows a user to align against a genomic segment, without prior pre-processing by gmap_build. In that case, the genomic segment is also expected to be in FASTA format

```
gmap -g <genome_segment> <read_file>
```

For GSNAP, the FASTQ format is more common, with single-end reads input in a single file, and paired-end reads being input in two parallel files

```
gsnap -d <genome> <read1_file>
gsnap -d <genome> <read1_file> <read2_file>
```

However, GSNAP also allows for a FASTA format where a single-end read is provided in a single line, and a paired-end read is provided in two lines. Input files may be compressed with the gzip or bzip2 format, and decompressed during the alignment process with the `--gunzip` or `--bunzip2` flags.

If the genome is larger than 2^{32} or 4 billion bp, then the GMAPL and GSNAPL programs should be used instead. Each alignment program can detect whether it is being applied to a small or large genome, and will alert the user if it is being called inappropriately. Otherwise, the GMAPL and GSNAPL share the same source code as their GMAP and GSNAP counterparts, and use the same flags.

By default, GMAP is designed to find splicing, which was its original purpose, but this can be turned off using the flag `--nosplicing`. In contrast, GSNAP does not find splicing by default, but with the appropriate flags, it can find spliced alignments through novel discovery (`--novelsplicing`), by known splicing information (`--use-splicing`), or both.

As discussed above, GSNAP also allows for different modes of alignment, in which widespread substitutions of nucleotides are allowed in bisulfite sequencing (C-to-T), RNA editing (A-to-G), and PAR-CLIP sequencing (T-to-C). These alignment types can be handled with the `--mode` flag, which allows for the values cmet-stranded, cmet-nonstranded, atoi-stranded, atoi-nonstranded, ttoc-stranded, and ttoc-nonstranded. To obtain SNP-tolerant alignment, GSNAP should be given the flag "`-v ⟨snp_db_name⟩`", which matches the SNP database name provided to the snpindex program.

GMAP and GSNAP provide a variety of output formats. The most widely used format is the SAM format, which can be specified in GMAP by the flag "`-f samse`" and in GSNAP by the flag "`-A sam`". The SAM output includes some fields, all starting with the letter "X", that are specific to these programs. One field of note is "XO", which indicates the type of alignment found, depending on whether the paired ends were concordant and whether a single or multiple alignments were found.

The programs can also automatically place each alignment type into separate output files using the `--split-output` flag, which can be useful for some pipelines. The various output types are described in the README file provided with the GMAP package. Alternatively, the user can specify a single output file using the `--output-file` flag. If the user provides neither `--split-output` nor `--output-file`, then the programs send their output to stdout, and the user can then pipe that output into a file.

Besides the SAM format, GMAP and GSNAP also support various other output types, including PSL, GFF3, and BLAST m8 formats. They also provide output formats that are easier for humans to read and that can be useful for debugging purposes. In particular, the "`-A`" format for GMAP and the default output for GSNAP provide easily readable nucleotide-level alignments of the query reads against the genome.

In addition, GMAP and GSNAP allow for different ways to access the genomic index files. When sufficient memory is available on a given machine, each process can allocate memory for the files and read the file contents into that memory once. Most modern computers support the notion of shared memory, where multiple instances of a program can access a common store of memory for each file. Therefore, GMAP and GSNAP use shared allocations of memory when possible. They automatically keep track of the number of other instances of the program running on a given machine, and release the shared memory when all instances have finished. The user also has the option of keeping the genome indices resident in memory with a preload flag and explicitly releasing them with an unload flag.

For machines where physical memory is not sufficient to hold the entire set of genomic index files, the files can be accessed through memory mapping instead. In that case, the information in the files is kept on disk, and retrieved by page when needed. Cache memory on a computer keeps recently used pages for fast retrieval, since accessing disk memory is relatively slow.

The programs allow the user to decide between allocation and memory mapping through the `--batch` flag, which provides different choices for the various types of genomic files.

3.5 Parallel Computation

GMAP and GSNAP are multithreaded, and the number of threads can be specified with the `--nthreads` flag. Using multiple threads can speed up the program significantly on a computer that has multiple cores. In multithreaded mode, one thread is reserved for reading input and one for printing output, and these are not included in the value for `--nthreads`. Therefore, one may wish to reduce the value of `--nthreads` to be two less than the number of available cores.

Recent versions of the GMAP package also support MPI (message passing interface), which is the most widely used paradigm for coordinating large-scale computations over large numbers of computer nodes. The configure program determines whether MPI capabilities are available on your machine, and if so, will build MPI versions of GMAP and GSNAP called mpi_gmap and mpi_gsnap. In our case, the MPI versions coordinate the input of query reads and the output of alignment results through a master node, with the computation being spread across multiple worker nodes. The MPI versions of the programs will need to be run within an MPI environment, such as the program mpirun. The MPI versions also have an option for specifying whether the master node should also allocate threads for performing alignment, or whether it should focus only on coordinating input and output.

Even without MPI, GMAP and GSNAP can be run as multiple processes in a computer cluster. However, without MPI, the user will need to instruct each node how to access a fraction of the input reads. A flag called `--part` is provided for this purpose, and provides both a block size and item number to a given process. For instance, if we have 100 nodes available, we might provide each node with one of the following commands:

```
gsnap -d <genome> --part=0/100 --nthreads=<n> <input>...
gsnap -d <genome> --part=1/100 --nthreads=<n> <input>...
...
gsnap -d <genome> --part=99/100 --nthreads=<n> <input>...
```

The value `0/100` to `--part` indicates that out of each block of 100 input sequences, that process should process only item 0, or the first sequence in that block. Likewise, the value `1/100` indicates that the process should process item 1, and skip the other 99 sequences in the block. As indicated, each process on a separate node can be run with multiple threads. The user should arrange for each process to send its output to a unique filename or filenames, using the `--split-output` flag, the `--output-file` flag, or piping the result to a distinct filename. These distinct portions of the analysis will then typically need to be combined into a single BAM file for subsequent analysis.

4 Data Structures and Algorithms

To perform the computational tasks discussed in the last section, GMAP and GSNAP use not only a variety of algorithms, as shown in Table 1, but also three basic data structures. These data structures represent a given genome in three different ways: (1) as a linear representation, (2) as a hash table, and (3) as an enhanced suffix array (ESA). To generate these data structures, a user must pre-process the genome, which can be time-consuming, but this procedure needs to be done only once for each genome, and the GMAP/GSNAP package provides a utility program called gmap_build for this task. The resulting indices can then be used by both GMAP and GSNAP, although the ESA is currently used only by GSNAP. Each of these data structures has evolved over time, as shown in Table 2, to improve their functionality and efficiency. In this section, we discuss the changes made to the data structures, as well as new algorithms added since the initial releases of the programs. To save space, we do not discuss any algorithms that were presented in the original papers on GMAP and GSNAP, such as end pairing in GMAP, or the spanning set and complete set analyses and segment combination procedures in GSNAP.

4.1 Linear Genome: Vertical Storage Format

The GMAP/GSNAP package stores genomes linearly in order to rapidly compute the number of mismatches in a given read segment relative to the genome. Our linear representation compresses the genome using 3 bits per nucleotide. Representing the four nucleotides A, C, G, and T requires only 2 bits, so the extra bit allows us to represent uncertainty or lack of knowledge at any position, expressed as the symbol N, or unknown base. Unknown bases are often present in the centromeres or telomeres of genomes, where the genomic sequence is highly repetitive and either not fully determined or not generally of interest for the purposes of alignment.

Table 2
Evolution of each of the three types of genomic data structures in GMAP and GSNAP

Version	Linear genome	Hash table/ compression	Suffix array
2005	3 × 32-bits, horizontal	12-mers/none	
Sept 2011		15-mers/Gamma	
Oct 2013			Standard
April 2014	3 × 128-bits, vertical	15-mers/SIMD bitpack	Enhanced

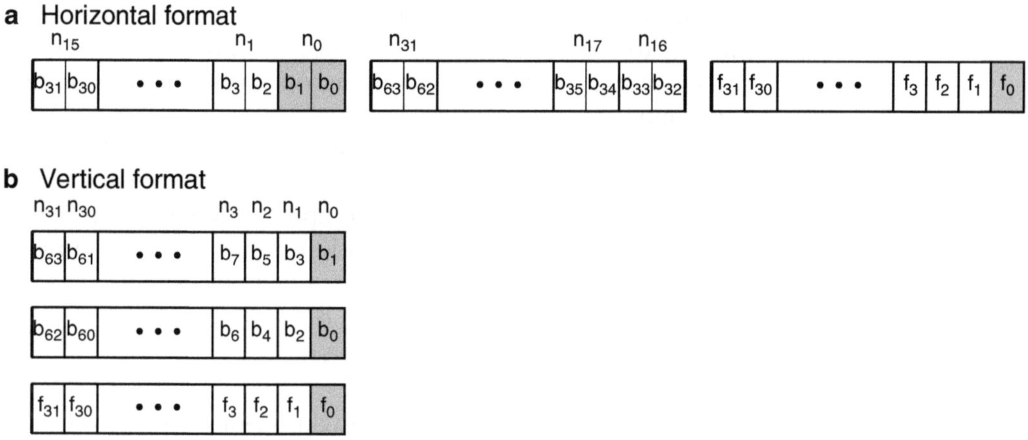

Fig. 4 Different bit arrangements for linear genomes. (**a**) Horizontal format, which stores the bits for each nucleotide next to each other. (**b**) Vertical format, which stores the bits for each nucleotide in separate words. Bits for nucleotide 0 are *shaded in gray*

The arrangement of these nucleotide bits has evolved over time. Since computer integers are represented using 32-bit or 64-bit words, which is not divisible evenly by 3, the linear genome divides its storage into blocks of words. The original implementation of the linear genome used three 32-bit words for each block of 32 nucleotides. The concatenation of the first and second words in each triplet represented the 2 bits representing the A, C, G, or T nucleotide, and the third word contained the third bit, indicating whether the position is an N.

This representation, at least for ACGT bits, can be considered a horizontal format, where the first and second bit for each nucleotide are located adjacent to each other in the same word (Fig. 4a). The current version of the linear genome introduces a vertical format. In a vertical format, we store the bit for all 32 nucleotides into the first word, and all second bits into the second word (Fig. 4b). We call these the high bits g_H, low bits g_L, and flag bits g_F. Actually, in our latest format, we store the genome not in blocks of 32 nucleotides, but as 128 nucleotides at a time. Therefore, consider that each of the words g_H, g_L, and g_F in the figure to be replaced by four words. The reason for the larger block size is for GMAP and GSNAP to exploit large-scale bitwise operations, such as those provided by SIMD instructions in modern computers, which can perform bitwise operations on 128 bits simultaneously.

The horizontal and vertical formats each have their own benefits. The horizontal format is useful for algorithms that deal with oligomers. This is because an oligomer is represented numerically as an integer in base 4, with A having the value 0, C 1, G 2, and T 3, and a horizontal representation represents that integer in the same way. As we show later, GMAP and GSNAP retain the horizontal format for computing segmental hash tables quickly. Although it is possible to convert between the vertical and horizontal formats on the fly, doing so requires a shuffling operation that can be time-consuming.

The vertical format is useful for comparing a read sequence with its corresponding genomic segment. With the vertical format, we can use bitwise operations to compute differences between the two sequences in parallel. The basic formula for computing the differences d between a read r and a genomic segment g, when both are stored in a vertical format, is

$$d_{\text{Std}} = (r_H \oplus g_H) \vee (r_L \oplus g_L) \quad (1)$$

which is computed simply as two exclusive-or operations plus one logical-or operation. These operations can be performed across 128 bits in parallel using the SIMD instructions `_mm_xor_si128` and `_mm_or_si128`, respectively.

To combine the read and genome bits, the read r must be represented in the same vertical format as the genome, with 3 bits per nucleotide, but shifted so that its bits are aligned with those of the corresponding genomic segment. The amount of shift necessary can be computed from the genomic position for the beginning of the read modulo 128. The resulting difference d_{Std} contains a 1 for every position in the block that has a difference in nucleotides between r and g. Masking of the difference may be needed to quantify mismatches over a portion of a read. The difference may also be converted into a count of mismatches using the popcount operation or into a set of mismatch positions by repeatedly applying operations to find and remove the lowest or highest bit set.

The vertical format offers significant advantages over the equivalent operation using a horizontal format. Although we could take an exclusive-or between a read and genomic segment in the horizontal format, we would then require a step that combines adjacent bits into a single difference bit, depending on whether either of the adjacent bits is 1. There is no standard bitwise operator for this step, so older versions of GMAP and GSNAP used lookups in a table of pre-stored differences for all possible bit patterns. With the vertical genome format, these lookup tables are no longer needed.

The difference quantity we have computed so far does not take into account the flag words g_F or r_F, which signify the location of unknown bases. To include this information into our computations, we need to decide whether an unknown base in the read represents a mismatch or match, which determines which of the following two formulas to apply:

$$d_N = d_{\text{Std}} \vee r_F, \text{ if read } N \text{ is a mismatch} \quad (2)$$

$$d_N = d_{\text{Std}} \wedge \neg r_F, \text{ if read } N \text{ is a match} \tag{3}$$

By default, GSNAP uses the match formula, but the user may opt for the other behavior. Likewise, we can treat unknown bases in the genome as either a match or mismatch:

$$d_{\text{Ref}} = d_N \vee g_F, \text{ if genomic } N \text{ is a mismatch} \tag{4}$$

$$d_{\text{Ref}} = d_N \wedge \neg g_F, \text{ if genomic } N \text{ is a match} \tag{5}$$

The difference quantity d_{Ref} accounts for the comparison between a read and the reference genome. However, GMAP and GSNAP allow for SNP-tolerant alignment. To provide this feature, we assume that we have an alternate genome, where the nucleotide bits represent the minor allele at SNPs but otherwise have the reference allele at non-SNP positions. If the non-flag bits are represented in words a_H and a_L, then we can compute SNP-tolerant mismatches as follows:

$$d_{\text{Alt}} = d_{\text{Ref}} \wedge ((r_H \oplus a_H) \vee (r_L \oplus a_L)) \tag{6}$$

In other words, a difference exists at a position only if the read nucleotide differs from the reference genome and it differs from the alternate genome.

In representing the alternate genome, we have not yet assigned bits to the flag word, a_F. We adopt the convention that all SNP positions contain a 1 in the corresponding bit of that word. This convention allows us to handle wildcard SNPs, in which the minor allele is not defined at a SNP and therefore any nucleotide in the read is counted as a match. For a wildcard SNP, since a single minor allele is not defined, we store the reference nucleotide into words a_H and a_L, but mark a SNP in the flag word a_F.

To allow for wildcard SNPs, we note that we have a wildcard SNP if the reference allele is equal to the alternate allele and the SNP flag a_F is true, as defined in our convention. Therefore, we compute the inverse quantity W, which marks all positions that are not wildcard SNPs:

$$W = (g_H \oplus a_H) \vee (g_H \oplus a_H) \vee g_F \vee \neg a_F \tag{7}$$

In other words, the quantity W is the logical inverse of our definition for wildcard SNPs, where the first three terms compute the concept that the reference allele differs from the alternate allele, assuming that a genomic unknown base is a mismatch.

To use this quantity W, we note that we have a match if we lack a difference between the read and genome or if we have a wildcard SNP. Conversely, we have a mismatch if we have both a difference and a non-wildcard-SNP position, meaning that our final equation for a SNP-tolerant differences is

$$d_{\text{SNP}} = d_{\text{Alt}} \wedge W \tag{8}$$

To allow for tolerance to bisulfite sequencing and RNA editing, recall that we wish to reduce both the read and the genome sequence to a 3-nucleotide space, to allow for the equivalence of C and T for bisulfite sequencing or A and G for RNA editing. Actually, since only the plus strand of the genome is represented linearly, we need separate equivalences depending on whether the read or its reverse complement aligns to the plus strand of the genome. We find that we do not need to store separate 3-nucleotide representations of the linear genome, but can compute the reduction on the fly, as follows:

$$d_{CT} = (\neg r_H \wedge r_L) \wedge (g_H \wedge g_L) \tag{9}$$

$$d_{\text{Met}+} = d_{CT} \vee ((r_H \vee r_L) \oplus (g_H \vee g_L)) \vee (r_L \oplus g_L) \tag{10}$$

$$d_{GA} = (r_H \wedge \neg r_L) \wedge (\neg g_H \wedge \neg g_L) \tag{11}$$

$$d_{\text{Met}-} = d_{GA} \vee ((r_H \wedge r_L) \oplus (g_H \wedge g_L)) \vee (r_L \oplus g_L) \tag{12}$$

The first two equations represent the calculation when the read aligns to the plus strand of the genome, whereas the second two equations are used when the reverse complement of the read aligns to the plus strand. In each case, we require a separate calculation to mark the case when a read-C and genomic-T is present, which represents a true mismatch rather than a conversion of a genomic C to T. On the plus strand, the first formula simply identifies when a read C is encoded as $r_H r_L = 01$ and a genomic T as $g_H g_L = 11$. The next formula converts C to T for both the read and the genome. To perform this conversion, we note that changing C to T can be done by changing the high bit to be 1 for C, and retaining it as 0 for A and as 1 for G and T. Therefore, the converted high bit is 1 as long as the nucleotide is not C, G, or T, which is the condition $r_H \vee r_L$ for the read or $g_H \vee g_L$ for the genome. Similar reasoning is used to derive the remaining formulas for the reverse complement of the read. The final values $d_{\text{Met}+}$ or $d_{\text{Met}-}$ represent tolerance to bisulfite sequencing that can substitute for d_{Std}. To consider unknown bases and SNP-tolerance, we can then apply the corresponding formulas above for those calculations.

Likewise, analogous calculations can be performed for tolerance to RNA editing, for both the read and its reverse complement:

$$d_{AG} = \neg(r_H \vee r_L) \wedge (g_H \wedge \neg g_L) \tag{13}$$

$$d_{\text{RNA}+} = d_{AG} \vee ((\neg r_H \wedge r_L) \oplus (\neg g_H \wedge g_L)) \vee (r_L \oplus g_L) \tag{14}$$

$$d_{TC} = (r_H \wedge r_L) \wedge (\neg g_H \wedge g_L) \tag{15}$$

$$d_{\text{RNA}-} = d_{TC} \vee ((r_H \wedge \neg r_L) \oplus (g_H \wedge \neg g_L)) \vee (r_L \oplus g_L) \quad (16)$$

In these formulas, the combinations of logical-and and logical-not can be performed efficiently using the single SIMD instruction `_mm_andnot_si128`.

Finally, for tolerance to PAR-CLIP sequencing, we apply the formulas for tolerance to RNA editing, except that we apply the formula for T-to-C conversion to the read, and the formula for A-to-G conversion to its reverse complement.

4.2 Genomic Hash Tables: Increased Specificity Through Compression

A hash table precomputes a list of genomic positions for each possible oligomer of a pre-determined length *k*, or a *k*-mer. The hash table used originally in GMAP was based upon the schema described for SSAHA [34]. Essentially, that scheme consists of an offsets table and positions table, where the offsets table contains starting indices in the positions table, one for each possible *k*-mer. The starting index indicates where the list of genomic positions begins for that *k*-mer. The end of the list can then be determined from the starting index for the next *k*-mer in the offsets table.

A hash table is used by GMAP for its end pairing algorithm and by GSNAP for its spanning set and complete set analyses. Because these algorithms may require many hash table lookups for a given read, we have explored methods for improving the speed of hash table-based algorithms. One key insight has been that larger *k*-mers make hash tables more specific. There are 4^{12}, or about 16 million, possible 12-mers, meaning that for a human genome of 3 billion bp, each possible 12-mer maps on average to 180 different genomic locations. However, a hash table of 15-mers would map to only 3 genomic positions each on average. This sharp increase in specificity helps to speed up hash table-based algorithms.

However, the difficulty with using larger *k*-mers is that the offsets table increases exponentially in size with *k*. An offset table for 12-mers requires 4^{12} entries, each of which requires 32 bits or 4 bytes, totalling 67 MB. However, a 15-mer table would require 4.3 GB. Although disk and memory resources are now commonly adequate to handle such large tables, savings in memory are still potentially valuable for other computational needs.

Therefore, our solution to the exponential growth of the offsets table has been to compress it. It would appear that each entry in the offsets table requires a 32-bit word of storage. However, because values in the offsets table point to a sequence of lists in the positions file, each value is guaranteed to be larger than or equal to the previous value. Also, since position lists are generally small, especially with large *k*-mers, each value in the offsets table is generally larger than the previous value by a small amount. Therefore, we can take the differences between adjacent values in the offsets table, which are generally small integers, and apply a compression technique that stores these small integers in fewer than 32 bits each.

This method of allocating fewer bits for small integers is called bitpacking, and the problem of compressing an increasing sequence of values is called differential bitpacking. In our application, we want to access individual offset values randomly, rather than decompressing them all in sequence, so we have the problem of differential bitpacking with random access. To allow for random access, we divide the increasing sequence of values into relatively small blocks, in our case, of size 64. Pointers for the blocks leads to another table in our hash table schema called the metainfo array, which also contains a prefix sum for each block to serve as a basis for adding the difference values.

In 2011, our first implementation of differential bitpacking used Elias gamma compression. Three years later, we changed the compression scheme to one we developed called columnar vectorized bitpacking. This scheme is designed to take advantage of SIMD instructions for the random access pattern needed for our purposes. The SIMD-based approach greatly increases the speed at which we can decompress the offsets table. However, details of this compression scheme are beyond the scope of this article, and will be presented elsewhere.

4.3 Handling Large Genomes

Genomic data structures have generally relied on the fortuitous circumstance that 32-bit words in computers are sufficient to represent 2^{32} or about 4 billion distinct genomic coordinates. Therefore, 32-bit words are sufficient to represent most genomes of interest, including the human genome, which has 3 billion bp. However, situations arise in which we wish to align to larger genomes, or perhaps collections of genomes. For example, the largest known genome to date is 152 billion bp, in the plant *Paris japonica*. Likewise, in metagenomics, in which samples can potentially contain any microorganism, it can be useful to align against a database comprising all known genomes.

To handle these cases, we have extended our hash table representations to handle large genomes. Accomplishing this requires addressing two challenges. First, we need to represent more than 4 billion genomic coordinates. Genomes that exceed this limitation can be considered *large* genomes. To address this issue, we essentially require more bits to represent each genomic coordinate. We could allocate 64-bit quantities, since 64-bit words are now standard in computing hardware. However, storing coordinates as 64-bit quantities would waste disk space, because 64 bits are overkill in practice. Instead, 40 bits should suffice for any foreseeable application, since they could represent up to 2^{40} or 1 trillion distinct genomic coordinates.

Therefore, our solution to representing large genomes is to allocate one 32-bit word and an additional 8-bit byte for each

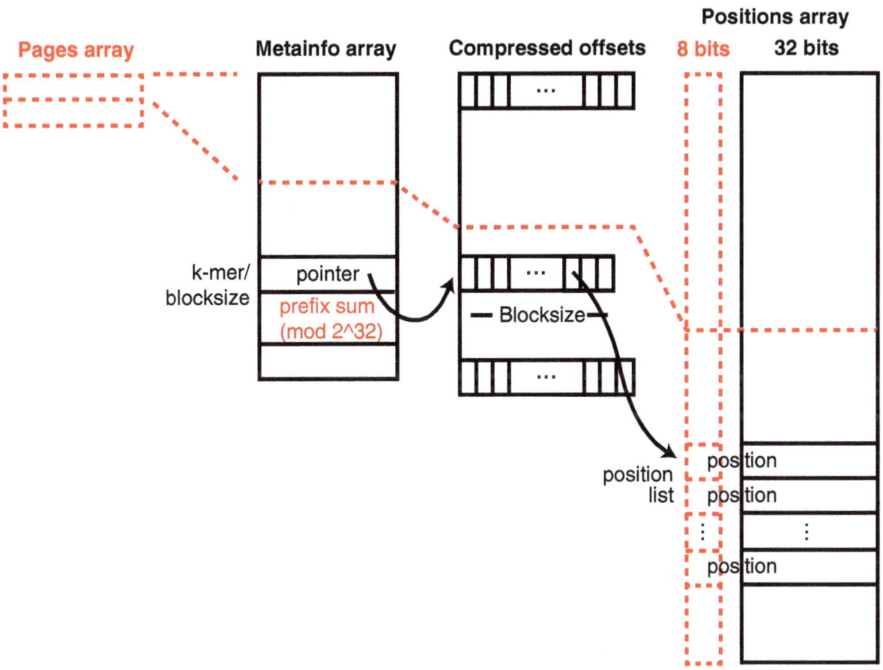

Fig. 5 Hash table for large and huge genomes. The compression scheme for a standard genome is shown in *solid lines*. The modifications needed for a large or huge genome are shown as *dashed lines*

coordinate (Fig. 5). The positions list in the standard hash table is therefore replaced with one list for the 32-bit words and a corresponding list for the 8-bit bytes. The two lists are needed only for the disk representation of genomic positions; for computation, when a list of genomic positions is needed for a given *k*-mer, the two lists are combined and placed into 64-bit quantities for subsequent computations.

The usage of 40-bits effectively widens the width of the positions list, but our second challenge arises when the length of the positions list exceeds 2^{32} entries. That is because the prefix sums in the metainfo array have a width of 32 bits, and can therefore represent at most 2^{32} entries in the positions table. This problem has been addressed somewhat in the earliest implementations of GMAP, since by default, only a sample of genomic positions are represented in the hash table. The sampling is performed by indexing every third position in the genome, rather than every position. This sampling interval therefore reduces the length of the positions table by a factor of 3. With that sampling interval, the representational limit of offsets table indices occurs at about 12 billion bp. Therefore, genomes that exceed that limitation can be considered *huge* genomes.

Although we could also widen the width of the metainfo array, as we did with the positions table, we have solved this problem instead by introducing the concept of pages with pages, the prefix sum can be reduced by representing them as the remainder modulo

2^{32}, and a page number can indicate the number of 2^{32} multiples that are represented by a given entry.

The pages table is just a list of pointers to the metainfo array, indicating where each new page begins. To look up a *k*-mer x, with its pointer and prefix sum at entry [x/64] in the metainfo array (where 64 represents the block size), we scan through the pages table to see how many of its entries are lower than [x/64]. For a small pages table, this scan can be performed linearly. For larger pages table, a binary search could be implemented. However, for genomes currently in practical use, only a few pages are typically needed. Finally, the prefix sum stored in the metainfo array is incremented by 2^{32} times the page index.

The utility program in the GMAP/GSNAP package that builds genomic indices can handle normal, large, and huge genomes, detecting each situation automatically. For large and huge genomes, which require modifications to the hash table format, alignment can be performed with specialized versions of GMAP and GSNAP, called GMAPL and GSNAPL, respectively.

4.4 Segmental Hash Tables

Previously, we described hash tables that catalog *k*-mers over the entire genome. Such genomic hash tables are constructed once for each genome and can take a fairly long time to build. However, certain steps in GMAP alignment require a hash table for small oligomers of 9 or fewer bp over a candidate genomic segment needed for a given read. Such segmental hash tables need to be constructed dynamically for a given read, and are needed by both diagonalization and oligomer chaining in GMAP.

Since a segmental hash table is built specifically for a given read, it can be specialized even further to contain only the *k*-mers that occur in both the given read and the given genomic segment. This contrasts with the genomic hash table, which catalogs all *k*-mers over the genome, albeit with a possible sampling interval. Such sampling is not needed for segmental hash tables, because of the limited number of positions that need to be represented. Likewise, the special considerations of compressing offset tables and handling large genomes need not concern us for segmental hash tables, because of their limited scope. In particular, instead of genomic coordinates, we can use chromosomal coordinates, which will not exceed 2^{32} bp in length.

We have developed an efficient, SIMD-based procedure for constructing segmental hash tables, which involves three steps. First, the query sequence is scanned to determine which *k*-mers are present and therefore relevant for this hash table. Second, we scan the genomic segment to count the number of genomic positions for each relevant *k*-mer. We use these counts to allocate memory for a position table, along with corresponding values for the offsets table. Finally, we scan the genomic segment again to fill in the position lists.

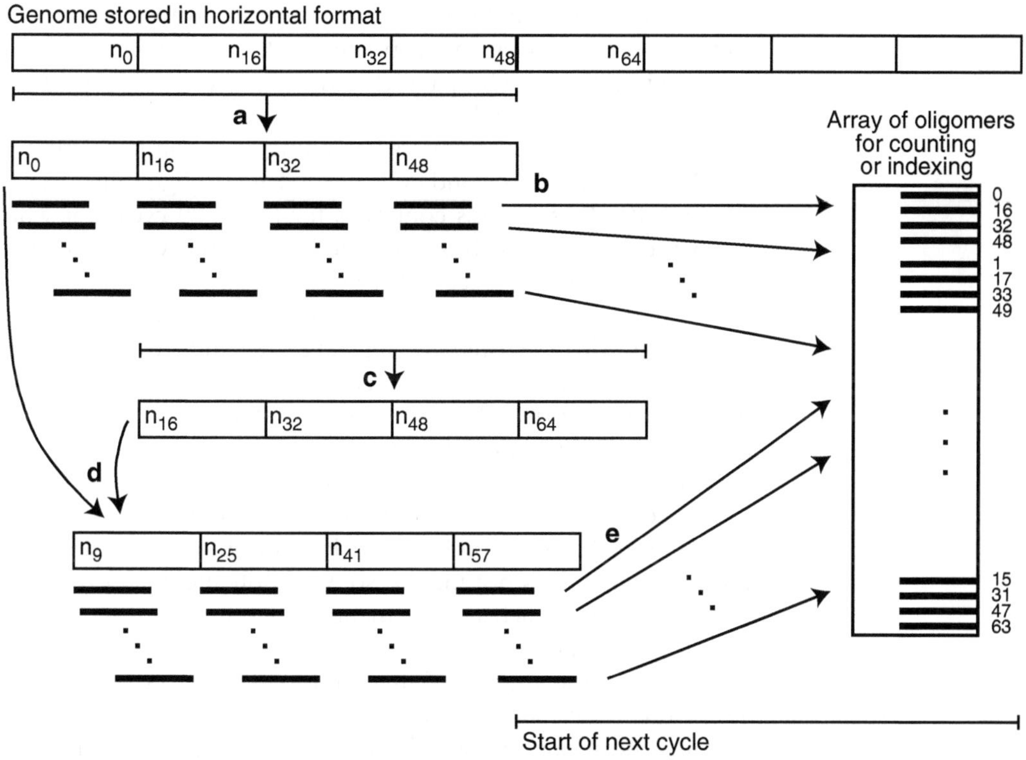

Fig. 6 SIMD algorithm for building a segmental hash table. (**a**) Load words and reverse nucleotides in each word in parallel. (**b**) Shift and mask 9-mers in parallel and store in array. (**c**) Load next higher words and reverse nucleotides. (**d**) Shift and combine words. (**e**) Shift and mask 9-mers in parallel and store in array

For scanning the genomic segment, we use the linear genome in its horizontal format, which uses three 32-bit words to represent each block of 32 nucleotides. To use 128-bit SIMD instructions, we process the genome in groups of 64 nucleotides, which occupy 128 bits for their ACGT representation. We then apply the SIMD-based procedure shown in Fig. 6.

In this procedure, the first step is to convert bits from the horizontal format into an order needed for generating *k*-mers. For alignments of the read to the plus strand, this is equivalent to reversing the order of nucleotides, Since the horizondal genome format orders nucleotides from least significant bit (lsb) to most sigificant bit (msb), while *k*-mers order them from msb to lsb. Algorithms are available for reversing the order of bits using shift and mask operations, and we can reverse two-bit quantities, rather than each individual bit, by omitting the step that operates on individual bits. We apply this reversal modified algorithm using SIMD operations so that each 32-bit word in the 128-bit register is processed in parallel. For alignments involving the reverse complement of the read, the desired conversion is to logically inverting the bits.

Then, within the SIMD register, we can perform all possible shifts, followed by an appropriate mask, to obtain each non-overlapping *k*-mer. The parallel nature of the SIMD operations means that we extract *k*-mers in a cyclic pattern. In other words, the first cycle yields the *k*-mers beginning at genomic positions 0, 16, 32, and 48. The second cycle yields *k*-mers beginning at 1, 17, 33, and 49. This process continues for $(17 - k)$ cycles, at which point the *k*-mers need bits from the next 32-bit word. To obtain these bits, we obtain a second register of 128 bits from one word higher in the genomic horizontal format. We then combine the bits in the two SIMD registers, by applying a left shift to the first and a right shift to the second, and then combining the registers with a logical-or operation. The shift-and-mask steps then continue until all 64 *k*-mers are extracted into an array.

The *k*-mers in this array are then either counted or used as indices for storing genomic positions, depending on the step being executed in the overall procedure. For counting, the order of the array elements does not matter, so we simply increment the count corresponding to each array element. In fact, for counting, we need not store *k*-mers into an array at all, but can use the values in the SIMD register as they are shifted and masked. However, for storage, order does matter, since we need a corresponding list of genomic positions for each *k*-mer in ascending coordinate order. Since the cyclic pattern of the *k*-mers in the array is known, we can easily write a series of instructions to access the array elements in the necessary order.

4.5 Diagonalization in GMAP

Once the segmental hash table is built, GMAP can use it to obtain an approximate alignment of the read against the genomic segment. For RNA-Seq, the alignment is characterized by ungapped strings of alignment between the read and the genome, potentially containing mismatches. These strings are essentially exons, although an exon with an indel will appear as two ungapped strings. Exons are interspersed with introns, which are jumps in the genome sequence.

In the original release of GMAP, we used oligomer chaining to build up strings using a dynamic programming procedure. However, oligomer chaining is potentially an $O(R^2G)$ procedure, where R is the length of the read and G is the length of the genomic segment. The dependence upon G occurs because each *k*-mer can appear with uniform probability over the length of the genomic segment.

In order to make this procedure more efficient, we have introduced a procedure called diagonalization that helps to limit the effect of G. In diagonalization, we try to find evidence for exons or alignment strings by accumulating evidence for all such strings in parallel. On a graph of *k*-mer matches between the read and genomic segment (Fig. 7), such evidence appears as a series of diagonals.

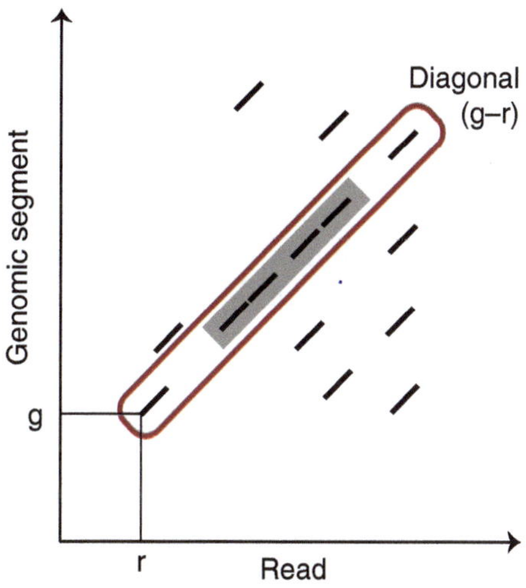

Fig. 7 Diagonalization. Matching *k*-mers between the read and genome are shown as *thick black line* segments. Matching *k*-mers for the diagonal ($g - r$) are surrounded by an oval. The best set of proximal *k*-mers is highlighted in *gray*

The diagonalization procedure maintains an array of all possible ($R + G$) diagonals. We can assign a distinct index to each diagonal from $-R$ to $+G$, and a *k*-mer that matches between read position r and genomic position g belongs to the diagonal with index ($g - r$). For each diagonal, we keep track of the read position for its last mapping; its current number of proximal mappings; the starting and ending read position for those mappings; and number and bounds for its best set of proximal mappings.

We scan the read sequence, and for every *k*-mer, we use the segmental hash table to find its list of genomic positions. For each genomic position, we determine its diagonal and update the appropriate entries in the list of diagonals if its last mapping was nearby, which is defined by some parameter along the read distance. This parameter allows for mappings that are a short distance away in the read, typically caused by a sequence mismatch that prevents a *k*-mer for that diagonal from mapping to the read or genome. If the mapping lies within the allowable parameter, the algorithm effectively extends the diagonal, and thereby tolerates the mismatch.

The result of this procedure is a set of diagonals, each of which could potentially correspond to an exon or alignment string. A diagonal with a large number of mappings provides strong support for an exon or alignment string. Our diagonalization procedure applies a dominance test that allows a diagonal to eliminate one more or diagonals from other same locations that it subsumes along the read. The remaining set of diagonals can then be used to place bounds on the *k*-mer mappings used for oligomer chaining. These bounds limit the range of genomic positions that are considered for each *k*-mer, and thereby eliminate the need to search positions over the entire genomic segment. The bounds can be extended from a given diagonal until the next diagonal in the set, so that a region without an explicit diagonal can still be bounded by nearby diagonals.

Our procedure appears to be similar to the anchoring algorithm in Pash [20], although that program is designed to divide a single problem among a multinode computer cluster, whereas ours is designed for a single thread. Moreover, our algorithm is designed to bound the oligomer chaining step, rather than serve as the final determination of the exon structure.

4.6 Resolving Gaps in GMAP: SIMD Dynamic Programming

After the diagonalization and oligomer chaining steps in GMAP are finished, we have an alignment at the nucleotide level with gaps, which represent mismatches, indels, and introns in the alignment. Gaps in the middle of the alignment essentially represent constrained dynamic programming problems, where the sizes of the missing read and genomic lengths indicate the underlying type of sequence difference. A mismatch or indel will give rise to a read gap and genomic gap of equal or approximately equal size. These can be aligned with a standard dynamic programming technique, such as Needleman–Wunsch, which constrains the alignment to connect the beginning and end of the matrix. A splice will create a genomic gap that is much larger than the read gap. That situation can be handled accurately with a technique called sandwich dynamic programming, which we introduced in our original paper on GMAP. In sandwich dynamic programming, we solve two dynamic programming problems, each of which is constrained at its origin, corresponding to the exon region, but otherwise open at the intron end of each matrix. We then find the best pair of scores that considers the probability of splicing at those positions. Finally, gaps at the end of the alignment can be solved with an end dynamic programming technique, where the solution is constrained on the medial origin of the dynamic programming matrix, but the distal end is allowed to be any point that maximizes the alignment score.

Since our original implementation of GMAP, we have improved the speed of our dynamic programming procedures by using SIMD operations. Such SIMD-based procedures have been implemented elsewhere, especially for homology searches against a database of proteins [9, 13, 37, 44]. However, we have had to modify such SIMD-based techniques in several ways. First, we have incorporated SNP-tolerance into our scoring procedures, by computing scores both for the reference allele and minor allele and taking the maximum.

Second, we encountered issues regarding the F-loop calculations in existing SIMD algorithms. The SIMD computations for dynamic programming define an E-direction and an F-direction, where E and F are terms in the recursion formulas for

Smith–Waterman dynamic programming. Only one direction, which can be labeled the E-direction, can be performed so that each SIMD bitwise computation is independent of its neighboring bits. The other, or F, direction requires each bit in a SIMD register to depend upon its preceding bit. Therefore, the F-loop calculation is inherently serial, which can slow down the dynamic programming calculations considerably. Existing algorithms provide a SIMD-based calculation to help speed up this calculation, but we have found that these calculations to be incorrect in some cases for nucleotide-based comparisons.

To circumvent this problem, we have designed SIMD dynamic programming procedures that avoid calculating the F-loop altogether. To do this, we need to add constraints to the allowed space of dynamic programming solutions. Such a constraint essentially prohibits the unusual combination of an insertion plus a deletion in a given dynamic programming path. With this constraint in place, we can divide dynamic programming problems into two triangular regions, on either side of the main diagonal (Fig. 8a). Within each triangular region, alignments are constrained to extend only outward from the main diagonal and not to return to the diagonal.

For dynamic programming at the end of the alignment, both triangular regions are considered, and the best path is chosen between the two. Likewise, for the two dynamic programming matrices involved with sandwich dynamic programming, both triangular regions are solved. However, for a dynamic programming problem that is constrained at both the origin and end of the matrix, only one triangular region needs to be solved, namely, the

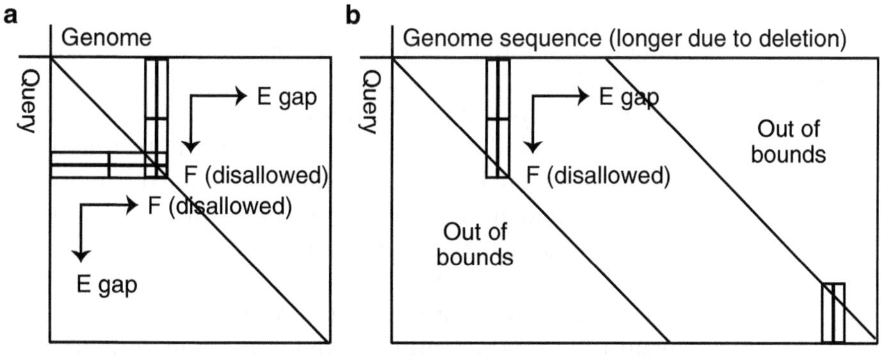

Fig. 8 SIMD-based dynamic programming without F loops. (**a**) An open-ended dynamic programming matrix with the SIMD algorithm divided into an upper and lower triangle. Within each triangle the F-gap is disallowed, which eliminates the time-consuming F-loop calculation, but still allows for a plausible local alignment ending somewhere in the matrix. (**b**) A fixed-end dynamic programming matrix that must traverse both read and genomic segment. The genomic segment is longer due to an apparent deletion. The calculation may be band-limited, with the far upper triangle and lower triangles considered out of bounds. Only vertical SIMD registers need to be considered, with the F-gap direction disallowed, to obtain a plausible alignment. Each small rectangle represents the matrix values contained in a SIMD register

one where the E-direction lies along the longer of the two dynamic programming axes (Fig. 8b). If both dynamic programming axes have the same length, then the constraint against the co-existence of insertions and deletions means that the main diagonal provides the only viable solution, so that dynamic programming is not even necessary.

4.7 ESAs: Greedy Match-and-Extend

Because GSNAP was derived from our work on GMAP, it originally used the same hash table data structure that GMAP used. In the meantime, other aligners have demonstrated that suffix arrays and related data structures, such as the BWT, can produce extremely fast alignments, provided that the read is identical or nearly identical to the genome sequence and mechanisms are provided for finding indels and splices [8].

Therefore, in our continuing effort to improve the speed of GSNAP, we have also incorporated suffix array methods into the program. However, rather than relying entirely on a suffix array, GSNAP uses that method only as an initial stage. If a single attempt at finding an alignment using suffix arrays fails, GSNAP then relies upon its standard algorithms and its backup GMAP algorithm. These backup algorithms are useful because BWT and suffix array algorithms are inherently greedy. They start from one end of the read and scan for a long segment that matches the genome exactly. If a given read is identical to the genome sequence, such a long segment can be found quickly and easily. However, mismatches or indels, especially those that are located near the initial end of the scan or that occur in clusters, can lead the scan astray, resulting in either a failure to align anywhere to the genome or alignments to the wrong genomic regions. To recover from such errors, suffix array algorithms try various methods, such as backtracking or scanning from multiple starting positions.

Similarly, the suffix array method implemented in GSNAP is a greedy match-and-extend procedure, in which the algorithm attempts to find a long match between the read and genome to serve as an anchor. Our method is applied to both the read sequence and its reverse complement simultaneously, and starts from a single starting position, namely, the 5′ end of each orientation. Each cycle of the method applies the suffix array to try to match the remaining part of the read and its reverse complement against the genome, until the end of the read or its reverse complement is reached. If the matching process fails, yielding only a limited substring match, then the next cycle continues at two positions after the last matching nucleotide. One position is skipped on the likelihood that failure to align at that position was due to a mismatch between the read and the genome.

When the scanning processes reaches the end of the read or its reverse complement, the algorithm proceeds with the orientation that completed its scan. It considers the longest matching substring to be an anchor, which divides the remaining matching substrings into those that are left and right (or upstream and downstream, respectively) from the anchor.

The anchor region can potentially match multiple locations in the genome. For each genomic location, the algorithm attempts to extend the alignment leftward and rightward using the other suffix array matches. A suffix array match extends the anchor if lies within an expected insertion, deletion, or splice distance from the previous genomic location. If no extension is possible, one likely explanation is that the scanning process was too greedy and extended the query match farther right than necessary. Therefore, the algorithm moves the boundary point between the failing substring and the previous successful substring as far as possible leftward, based on the nucleotides that do match against the genome. The algorithm then performs another suffix array search for the shortened substring to try to uncover a genomic location that can extend the alignment. If the algorithm is able to find genomic locations in each of the substrings that extend the alignment, it then refines the indel and splice endpoints between each substring. The refinement process uses the same procedures used by the GSNAP segment combining procedures. The size of the indel or intron can be determined from the difference between the genomic positions.

4.8 Integration of GMAP Algorithm into GSNAP

Alignment in GSNAP is a triaging process, in which an suffix array algorithm is tried initially to solve for straightforward alignments, and then segment combination procedures are tried for more complex alignments. To solve even more complex situations, such as those involving indels next to splice sites, GSNAP then applies its GMAP algorithm. In its standalone version, GMAP applies a stage 1 procedure to find candidate genomic locations, a stage 2 procedure to find approximate exon locations using diagonalization and oligomer chaining, and a stage 3 procedure to resolve nucleotide-level gaps using dynamic programming.

When we integrated the GMAP algorithm into GSNAP, we were able to skip the stage 1 procedure, since the initial attempts at using suffix arrays and genomic hash tables generally gave a set of candidate genomic locations containing the correct solution, especially for paired-end reads. Therefore, our initial integration of the GMAP algorithm began with stage 2.

However, stage 2 in GMAP is computationally very expensive, since it requires us to scan a genomic segment to build a segmental hash table. Furthermore, in most cases, the suffix array and hash table methods generate partial solutions that can then be refined by the stage 3 methods in GMAP, without having to find an approximate exon structure from scratch.

Therefore, the most recent work in GSNAP has involved streamlining the integration of the GMAP algorithm, by skipping the stage 2 procedure whenever possible, and converting partial solutions from the suffix array and hash table methods for direct application of the nucleotide-level dynamic programming procedures in stage 3. These procedures require the alignment to be represented as a linked list of individual nucleotide pairings between the query sequence and the genome, rather than as contiguous segments of alignments.

The stage 2 procedure of GMAP is still required when the suffix array and hash table methods do not yield any viable solution, especially when we are expecting a concordant alignment from a paired-end read. In that case, the alignment from one end of the pair is used as an anchor to determine the approximate genomic region for the other end, assuming the alignment is concordant. Then a segmental hash table is built for this genomic region and the stage 2 procedure is applied.

4.9 Segment Chaining

Both the partial results from the suffix array algorthim and the complete set algorithm in GSNAP can generate a set of segmental matches between the query sequence and genome, which can be used as building blocks to construct alignments involving indels and splices. For the complete set algorthim, segments are found by performing a multiway merging of the hash table lookups of each *k*-mer in the query sequence. In initial versions of GSNAP, designed for reads of 75 bp or so, these segments were then combined in pairs, which allowed GSNAP to find alignments involving a single indel or splice.

The most recent versions of GSNAP have generalized this process of segment combination into a segment chaining procedure that can concatenate multiple segments, thereby allowing for multiple indels and splices in a given alignment. Segment chaining starts with the most promising, or anchor, segments, which are those built from multiple *k*-mers in the query sequence. For each anchor sequence, the segment chaining procedure looks for segments nearby in the genome that are consistent with an indel or splice. The procedure then applies a dynamic programming algorithm to find the set of segments that gives the longest number of aligned nucleotides. Finally, these segments are converted into a linked list of individual nucleotide pairings, and given to the stage 3 algorithm of GMAP to identify the precise indel and splice locations and to fill in any gaps in the alignment.

5 User Features and Downstream Analysis

GMAP and GSNAP were designed for practical applications and in particular, for our own analysis pipeline at our institution. Many of their computational features were introduced based on our own experience in studying real-world sequencing datasets. In addition, GMAP and GSNAP provide user features that facilitate their use for analysis, and we discuss those features in this section.

5.1 Output Formats

GMAP and GSNAP provide a variety of output formats for various purposes. They can print alignments in a human-readable format, which is useful for understanding and debugging complex alignments. In particular, GMAP can produce a gene structure alignment showing the locations of mismatches, indels, and introns. For short reads, GSNAP has its own default alignment format that shows the original read, followed by the genomic segments that correspond to that read. GMAP can also divide a query sequence into its component exons, which can be useful for downstream processing.

The remaining output formats are largely designed for parsing by computer programs, and conform to community standards. For example, GMAP and GSNAP can produce output in BLAST m8 format, including scoring with an E-value for each alignment. For genome annotation, the GFF3 format is used widely. GMAP provides output in three different GFF3 output styles.

With the advent of NGS, the most widely accepted output format is the SAM format, which is interchangeable with its binary counterpart, the BAM format. GMAP and GSNAP provide standard SAM output, although in some situations, the programs generate alignment types that require some adaptation to be expressed in SAM format. For example, chimeric alignments, in which one end of a query sequence aligns to one part of the genome and the other end aligns at some distant location, require some special conventions. These alignments must be expressed on two separate SAM lines, with the distant ends being hard-clipped in each line. GMAP and GSNAP provide an XT field that provides information about the chimeric transition, including the location of the other end, and the dinucleotides and splice site probabilities for exon–exon junctions. For alignments with hard-clipping, the SAM format states that the original read and quality scores should be excluded from the alignment output. However, we have found it useful to include this information, so our programs include it in separate XH and XI fields. Similarly, circular alignments require a similar expression on two SAM lines, with a marker XC field to indicate this situation.

5.2 Alignment Types and Split Output

Alignment pipelines often make distinctions between different types of alignments. For example, many pipelines distinguish between unique alignments, which have only one genomic location, and multimapping alignments, which have more than one. Therefore, an early step in sequencing analysis is to characterize each alignment. GMAP and GSNAP make this process easier by assigning a type to each alignment, with the type declared in an XO field in SAM output. Furthermore, GMAP and GSNAP allow the output to be split into multiple files, with each file having a distinct alignment type. With split output, an analysis pipeline can process a subset of files or handle different files in different ways.

For single-end reads, an alignment can be classified as being non-mapping, unique, multiple, circular, or a translocation (which includes all types of distant splicing). For paired-end reads, the situation is more complex because of the different possible relationships between the ends. The most desirable relationship is concordance, in which the two ends align to the same genomic strand in the expected orientation and at the same local genomic region, as defined by a user-adjustable distance (default 200,000 bp for RNA-Seq reads and 1000 bp for DNA-Seq reads). Within the concordant category, the alignments may be unique, multiple (meaning two or more concordant alignments in the genome), circular, or a translocation. The circular and translocation categories hold if either of the two ends satisfy that attribute.

If the two ends align to the same local region, but do not align with a concordant strand and orientation, then the alignments fall into a category called paired alignments. Likewise, if they have the expected concordant strand and orientation, but their distance is greater than expected for a local alignment, then this is also considered a paired alignment. The critical distinction again is whether there are multiple paired alignments or only a single one. If there is a unique paired alignment, then the output files are further subdivided into the type of relationship. If the distance between the ends is greater than expected for a local alignment, then it is paired unique long. If they are local but strands are opposite, then it is an inversion. If they are local and same strand, but in the wrong order, then it is a scramble.

If not paired, then the relationship is considered unpaired, and the two ends are aligned independently of each other. The alignment types then depend on the number of alignments found for each end. If both ends cannot be aligned to the genome, we have a non-mapping alignment. If one end can be mapped, but the other cannot, then we have a half mapping alignment, either unique or multiple, depending on the number of alignments for the mappable end. If both ends can be mapped uniquely, the situation is considered half-mapping unique; otherwise, it is halfmapping multiple. Again, if either end contains a circular or translocation alignment, then the alignment is placed in a separate category for that condition.

5.3 Utility Programs

In addition to its alignment programs, the GMAP/GSNAP package includes several utilities that are useful in working with genomes. For example, the linear genome stored in a GMAP/GSNAP index is useful not only in alignment, but also for general-purpose retrieval of segments from the genome. Our package therefore includes a utility program called `get-genome`, which provides this functionality. Given a chromosome and start and ending coordinates on that chromosome, the get-genome program will print the nucleotides for the corresponding genomic segment. If the start coordinate is numerically greater than the ending coordinate, then the program understands this as a request for the contents of the minus strand, and will produce the reverse complement of the corresponding plus strand. The program is also able to incorporate information about SNPs if that information is provided in the genome building process.

Our programs also provide utilities for handling intervals and genome annotations. From its initial release, GMAP has had an implementation of IITs. An IIT is a tree structure that represents intervals, and can find all intervals in a given query region in logarithmic time. The package provides programs for converting a textual representation of intervals into a binary IIT format (`iit_store`), for performing the inverse conversion (`iit_dump`), and for retrieving all intervals that overlap a given query region (`iit_get`).

Since their initial release, these IIT formats have been generalized to allow coordinates that exceed 2^{32} in value and large amounts of data that collectively exceed 2^{32} bytes in size. The `iit_store` program automatically detects these situations and makes appropriate modifications to the IIT file automatically. Furthermore, the IITs have been enhanced with the concept of divisions, which can represent chromosomes, contigs, or other groupings of coordinates.

Nevertheless, our package retains its original type of IIT that lacks any divisions, now called a universal IIT. A universal IIT is used for representing the boundaries of the chromosomes, contigs, or divisions themselves. A genome can be considered as a concatenation of its chromosomes, and the universal coordinate for any genomic position is defined as its position in that concatenation. Therefore, the human genome has universal coordinates between 1 and approximately 3 billion. In order to handle large genomes, universal IITs also have the ability to store coordinates larger than 2^{32}. Since most alignment computations in GMAP and GSNAP are performed on universal coordinates, a universal IIT for chromosomes provides a translation between those coordinates and their corresponding chromosomal coordinates.

IITs are extremely useful for representing genome annotations. Such annotations can be characterized generally as being a directed (plus or minus) interval on the genome, together with annotation on that interval. Point annotations, such as SNPs, can be represented by intervals of length 0, where the start and end coordinates are the same. Although a splice site could also be represented by a single coordinate, it needs a direction, since splicing occurs on transcripts derived from either the plus or minus strand. Therefore, a splice site can be represented with an interval of length 1, where plus splice sites go from a lower coordinate to a higher one, and minus splice sites go from higher to lower.

To help represent genome annotations, intervals in GMAP/GSNAP can be enhanced further with a type, which is also built into our implementation of IITs. Such types are useful for representing certain biological concepts. For example, splice sites can be characterized as being either a donor or acceptor type. In universal IITs, chromosomes can be characterized as being of a linear or circular type.

Therefore, the IIT format can be used as a general representation for genome annotations, and the GMAP and GSNAP programs use them to represent information about SNPs, splice sites, and other domain information. Accordingly, our package provides several utility programs that can parse external annotation files and translate them into a format suitable for `iit_store`. External annotation formats that are currently supported include GFF3, GTF, and PSL.

5.4 Integration with R/Bioconductor

To make our alignment tools and utilities further accessible to researchers, we have integrated GSNAP and some of its utilities with the R platform for statistical computing [36]. The R package *gmapR* provides an R/GSNAP interface based on interoperable, domain-specific data structures standardized by the Bioconductor project [18]. Both R and Bioconductor are used widely within the bioinformatics community, and provide an alternative to the command-line execution of our programs.

The *gmapR* package allows a user to use the GMAP/GSNAP package within a Bioconductor workflow, including running the GSNAP aligner and constructing and querying genome indices. This package also generates statistical summaries of alignment data to support variant calling and other tasks. It computes these summaries from the output of `bam_tally`, an embedded utility that shares the same infrastructure as GSNAP and tabulates alignment data by genomic position.

To demonstrate *gmapR*, we outline a common workflow, which proceeds through constructing a genome index, aligning RNA-Seq reads to the genome, and summarizing the alignments for variant calling. This is a simplified sketch meant to illustrate the API. For a fully detailed, reproducible and up-to-date demonstration, please see the package documentation.

Let us assume the data are from the fruit fly, or *Drosophila melanogaster*. Bioconductor provides the *Dmelanogaster* reference assembly as a package. We can export the genome assembly object (`Dmelanogaster`) directly to a GMAP genome index, under a preexisting directory `genomeDir`:

```
library(BSgenome.Dmelanogaster.UCSC.dm3)
genome <- GmapGenome(Dmelanogaster, genomeDir, create=TRUE)
```

It is also possible to generate a genome index from a FASTA file, or an in-memory collection of sequences.

Next, we populate the index with known splice sites, which we derive from Bioconductor transcript annotations:

```
library(TxDb.Dmelanogaster.UCSC.dm3.ensGene)
spliceSites(genome, "ens") <- TxDb.Dmelanogaster.UCSC.dm3.ensGene
```

We could also incorporate known variations to enable SNP-tolerant alignment.

The next stage is alignment. The *gmapR* package supports every GSNAP parameter, of which there are dozens. For ease of use, we provide a formal container object for the parameters. In this case, we point GSNAP to the known splice sites, and also ask it to detect novel splicing. We assume the `fastqFiles` object points to FASTQ files on disk.

```
param <- GsnapParam(genome, splicing="ens", novelsplicing=TRUE)
result <- gsnap(fastqFiles, param)
```

By default, only the unique alignments are kept. The output from GSNAP is automatically converted to one sorted and indexed BAM file per FASTQ.

The final step is to summarize the alignments with the intent to call variants. First, we generate the tallies (for the first BAM), and then we summarize them. Again, the tallying tool supports dozens of parameters. Only the BAM file and genome index are required. We pass only one extra parameter, which enables indel tallying.

```
bams <- as(result, "BamFileList")
param <- BamTallyParam(genome, indels=TRUE)
tallies <- bam_tally(bams[[1]], param)
summaries <- variantSummary(tallies, high_base_quality=23)
```

The summaries include, for each position and nucleotide, the strand-specific count, average base quality, average read position, and so on. The Bioconductor packages *VariantTools* [24] and *VariantAnnotation* [35] support filtering and annotating the summaries, as well as exporting them to VCF format [6].

In summary, *gmapR* provides multiple integration points between GSNAP and the broader bioinformatics software ecosystem, all through a flexible, high-level programming language that is well suited for implementing *ad hoc* data pipelines.

5.5 Analyzing Datasets

At our own institution, we rely upon GSNAP for aligning reads and processing sequencing datasets for RNA-Seq, Exome-Seq, and whole genome sequencing data. We have developed an R/Bioconductor package called *HTSeqGenie* for analyzing datasets by aligning the samples using GSNAP, filtering reads, reporting the quality of the alignment results, and providing summary information for the user, including tabulations of coverage and splice junctions. Our filtering facility is flexible, allowing for a sampling of reads, filtering by read quality, filtering of ribosomal RNA, and filtering of adapter contamination. For quality reporting, our package uses the *ShortRead* package.

The behavior of *HTSeqGenie* is controlled by a cascading system of configuration files that allow successive refinement of analysis and output, starting from a standard set of parameters. In addition to the input and output file locations, the parameters control the set of computations to be done, as well as how they are to be done. Thus, for example, the pipeline can be told to align in chunks of one million reads, with duplicates marked and genomic features counted.

The output of the pipeline includes a set of files that describe computed counts for specific classes of genomic feature, such as exon, coding region, transcript gene, and so on. BAM files are also divided into useful categories, including the set of all uniquely mapped reads, read pairs whose ends are concordantly mapped, reads that are multiply mapped, reads that are unmapped, and so on. These subsetted files are then available to downstream processes for making biological inferences.

5.6 Biological Inference

Alignment by GMAP and GSNAP is limited because the programs process each read independently of other reads. However, the total set of alignments provides a global perspective that yields collectively more information than individual alignments. For example, an individual read may not be able to identify an indel at its end. But neighboring reads that cover the indel in their middle regions can identify the indel precisely. Therefore, inferring the indel is made easier when we have access to the entire set of alignments.

Accordingly, we have been developing a set of analysis tools, collectively called GSTRUCT, that can analyze sets of alignments. One of the basic programs in the GSTRUCT package is the `bam_tally` program for summarizing allele and indel counts. This program is designed to analyze BAM files quickly using minimal amounts of memory, and to group counts based on their quality scores, positions in the reads, or numbers of mismatches and indels relative to the genome. This program is already made publicly available through the *gmapR* package described previously.

The rest of GSTRUCT has undergone much development and we plan to make that publicly available as well. Among the other types of analysis available include realignment of reads based on global information about indels; calling of alternate splicing; and calling of distant splicing, such as gene fusions. A preliminary version of our alternate splicing program was assessed by the RGASP evaluation [42] as the GSTRUCT program, and was shown to be among the most accurate programs, ameliorating some of the errors inherent in the individual alignments generated by GSNAP. Our program for calling gene fusions has also been used internally to discover such biological phenomena as the R-spondin gene fusions in colon cancer [39].

6 Discussion

The ultimate goal of our work has been to make scientific discoveries, some of which may hopefully help meet unmet medical needs. Our own application of our program has given us valuable experience that has driven many of the computational and user features in GMAP and GSNAP. For example, our earliest analysis of single-end 33-bp reads [32] led to our implementation of SNP-tolerance as a necessary feature to avoid making false inferences about alternate splicing. Likewise, our development of variant calling routines demonstrated that it was critical to clip overlapping alignments and thereby avoid incorrect allele counts.

In general, we have found that aligning reads to the genome is largely straightforward, and most aligners generate similar results to one another. However, certain reads or situations pose difficult challenges in alignment; it is these reads that differentiate the behavior of aligners. Even though these reads represent only a small fraction of the whole, they can be concentrated in regions of the genome that can easily lead to incorrect biological inferences. At the same time, reads that are different from the reference are likely to provide interesting cases for study. Hence, difficult cases in alignments offer both the greatest challenges computationally and the greatest potential rewards biologically. Much effort in GMAP and GSNAP has gone into handling such difficult cases in alignment.

Development of GMAP and GSNAP has benefitted so far from continuous development by a single programmer, and support for the programs is ongoing. As sequencing technology advances, algorithms may need to change accordingly. Since our programs are not defined by any particular methodology, we have been willing to incorporate methods as needed to improve the behavior of our programs. At the same time, we are careful not to lose the accomplishments we have already made. To help with the development process, we have created a set of test cases for checking the correctness of our programs, as well as datasets for benchmarking their accuracy.

Our institution serves as one of the largest generators of sequencing data in the biomedical community, and our programs and downstream analysis tools have served a major role in making scientific discoveries in that data [19, 22, 29, 38, 39]. In addition, GMAP and GSNAP have found utility as components of other programs and pipelines, which have also shaped our programs' behavior. Our work has therefore benefitted from a large community of users, both at our own institution and worldwide. We are greatly appreciative of the feedback, suggestions, and bug reports they have provided, and it is our hope that our work will continue to help advance our understanding of biology and medicine.

References

1. Abouelhoda MI, Kurtz S, Ohlebusch E (2004) Replacing suffix trees with enhanced suffix arrays. J Discrete Algorithms 2:53–86
2. Altschul SF, Gish W, Miller W, Myers EW, Lipman DJ (1990) Basic local alignment search tool. J Mol Biol 215:403–410
3. Brennicke A, Marchfelder A, Binder S (1999) RNA editing. FEMS Microbiol Rev 23:297–316
4. Burrows M, Wheeler DJ (1994) A block-sorting lossless data compression algorithm. Technical Report Technical Report 124, Digital Equipment Corporation, Palo Alto, California
5. Burset M, Seledtsov IA, Solovyev VV (2000) Analysis of canonical and non-canonical splice sites in mammalian genomes. Nucleic Acids Res 28:4364–4375
6. Danecek P, Auton A, Abecasis G, Albers CA, Banks E, DePristo MA, Handsaker RE, Lunter G, Marth GT, Sherry ST, McVean G, Durbin R, and 1000 Genomes Project Analysis Group (2011) The variant call format and vcftools. Bioinformatics 27(15):2156–2158
7. Degner JF, Marioni JC, Pai AA, Pickrell JK, Nkadori E, Gilad Y, Pritchard JK (2009) Effect of read-mapping biases on detecting allele-specific expression from RNA-sequencing data. Bioinformatics 25:3207–3212
8. Dobin A, Davis CA, Schlesinger F, Drenkow J, Zaleski C, Jha S, Batut P, Chaisson M, Gingeras TR (2013) STAR: ultrafast universal RNA-seq aligner. Bioinformatics 29:15–21
9. Eddy SR (2011) Accelerated profile HMM searches. PLoS Comput Biol 7:e1002195
10. Eid J, Fehr A, Gray J, Luong K, Lyle J, Otto G, Peluso P, Rank D, Baybayan P, Bettman B, Bibillo A, Bjornson K, Chaudhuri B, Christians F, Cicero R, Clark S, Dalal R, deWinter A, Dixon J, Foquet M, Gaertner A, Hardenbol P, Heiner C, Hester K, Holden D, Kearns G, Kong X, Kuse R, Lacroix Y, Lin S, Lundquist P, Ma C, Marks P, Maxham M, Murphy D, Park I, Pham T, Phillips M, Roy J, Sebra R, Shen G, Sorenson J, Tomaney A, Travers K, Trulson M, Vieceli J, Wegener J, Wu D, Yang A, Zaccarin D, Zhao P, Zhong F, Korlach J, Turner S (2009) Real-time DNA sequencing from single polymerase molecules. Science 323:133–138
11. Elias P (1975) Universal codeword sets and representations of the integers. IEEE Trans Inf Theory 21:194–203
12. Engström PG, Steijger T, Sipos B, Grant GR, Kahles A, The RGASP Consortium, Rätsch G, Goldman N, Hubbard TJ, Harrow J, Guigó R, Bertone P (2013) Systematic evaluation of spliced alignment programs for RNA-seq data. Nat Methods 10:1185–1191
13. Farrar M (2007) Striped Smith–Waterman speeds database searches six times over other SIMD implementations. Bioinformatics 23:156–161
14. Florea L, Hartzell G, Zhang Z, Rubin GM, Miller W (1998) A computer program for aligning a cDNA sequence with a genomic DNA sequence. Genome Res 8:967–974
15. Frommer M, McDonald LE, Millar DS, Collis CM, Watt F, Grigg GW, Molloy PL, Paul CL (1992) A genomic sequencing protocol that yields a positive display of 5-methylcytosine residues in individual DNA strands. Proc Natl Acad Sci 89:1827–1831
16. Grant GR, Farkas MR, Pizarro A, Lahens N, Schug J, Brunk B, Stoeckert CJ Jr, Hogenesch JB, Pierce EA (2011) Comparative analysis of RNA-Seq alignment algorithms and the RNA-Seq Unified Mapper (RUM). Bioinformatics 27:2518–2528
17. Hafner M, Landthaler M, Burger L, Khorshid M, Hausser J, Berninger P, Borthballer A, Ascano M Jr, Jungkamp A-C, Munschauer M, Ulrich A, Wardle GS, Dewell S, Zavolan M, Tuschl T (2010) Transcriptome-wide identification of RNA-binding protein and MicroRNA target sites by PAR-CLIP. Cell 141:129–141
18. Huber W, Carey VJ, Gentleman R, Anders S, Carlson M, Carvalho BS, Bravo HC, Davis S, Gatto L, Girke T, Gottardo R, Hahne F, Hansen KD, Irizarry RA, Lawrence M, Love MI, MacDonald J, Obenchain V, Ole's AK, Pag'es H, Reyes A, Shannon P, Smyth GK, Tenenbaum D, Waldron L, Morgan M (2015) Orchestrating high-throughput genomic analysis with Bioconductor. Nat Methods 12 (2):115–121
19. Jiang Z, Jhunjhunwala S, Liu J, Haverty PM, Kennemer MI, Guan Y, Lee W, Carnevali P, Stinson J, Johnson S, Diao J, Yeung S, Jubb A, Ye W, Wu TD, Kapadia SB, de Sauvage FJ, Gentleman RC, Stern HM, Seshagiri S, Pant KP, Modrusan Z, Ballinger DG, Zhang Z (2012) The effects of hepatitis B virus integration into the genomes of hepatocellular carcinoma patients. Genome Res 22:593–601
20. Kalafus KJ, Jackson AR, Milosavljevic A (2004) Pash: efficient genome-scale sequence

anchoring by positional hashing. Genome Res 14:672–678

21. Kent WJ (2002) BLAT–the BLAST-like alignment tool. Genome Res 12:656–664
22. Klijn C, Durinck S, Stawiski EW, Haverty PM, Jiang Z, Liu H, Degenhardt J, Mayba O, Gnad F, Liu J, Pau G, Reeder J, Cao Y, Mukhyala K, Selvaraj SK, Yu M, Zynda GJ, Brauer MJ, Wu TD, Gentleman RC, Manning G, Yauch RL, Bourgon R, Stokoe D, Modrusan Z, Neve RM, de Sauvage FJ, Settleman J, Seshagiri S, Zhang Z (2015) A comprehensive transcriptional portrait of human cancer cell lines. Nat Biotechnol 33:306–312
23. Langmead B, Trapnell C, Pop M, Salzberg SL (2009) Ultrafast and memory-efficient alignment of short DNA sequences to the human genome. Genome Biol 10:R25
24. Lawrence M, Degenhardt J, Gentleman R (2015) VariantTools: tools for working with genetic variants. R package version 1.10.0
25. Lemire D, Boytsov L (2015) Decoding billions of integers per second through vectorization. Softw Pract Experience 45:1–29
26. Li H, Durbin R (2009) Fast and accurate short read alignment with Burrows-Wheeler Transform. Bioinformatics 25:1754–1760
27. Li R, Yu C, Li Y, Lam T-W, Yiu S-M, Kristiansen K, Wang J (2009) SOAP2: an improved ultrafast tool for short read alignment. Bioinformatics 25:1966–1967
28. Lister R, Ecker JR (2009) Finding the fifth base: genome-wide sequencing of cytosine methylation. Genome Res 19:959–966
29. Liu J, Lee W, Jiang Z, Chen Z, Jhunjhunwala S, Haverty PM, Gnad F, Guan Y, Gilbert HN, Stinson J, Klijn C, Guillory J, Bhatt D, Vartanian S, Walter K, Chan J, Holcomb T, Dijkgraaf P, Johnson S, Koeman J, Minna JD, Gazdar AF, Stern HM, Hoeflich KP, Wu TD, Settleman J, de Sauvage FJ, Gentleman RC, Neve RM, Stokoe D, Modrusan Z, Seshagiri S, Shames DS, Zhang Z (2012) Genome and transcriptome sequencing of lung cancers reveal diverse mutational and splicing events. Genome Res 22:2315–2327
30. Manber U, Myers G (1990) Suffix arrays: a new method for on-line string searches. In: Symposium on discrete algorithms, pp 319–327
31. Morin PA, Luikart G, Wayne RK, The SNP Workshop Group (2004) SNPs in ecology, evolution and conservation. Trends Ecol Evol 19:208–216
32. Nacu S, Yuan W, Kan Z, Bhatt D, Rivers CS, Stinson J, Peters BA, Modrusan Z, Jung K, Seshagiri S, Wu TD (2011) Deep RNA sequencing analysis of readthrough gene fusions in human prostate adenocarcinoma and reference samples. BMC Med Genomics 4:11
33. Needleman SB, Wunsch CD (1970) A general method applicable to the search for similarities in the amino acid sequence of two proteins. J Mol Biol 48:443–453
34. Ning Z, Cox AJ, Mullikin JC (2001) SSAHA: a fast search method for large DNA databases. Genome Res 11:1725–1729
35. Obenchain V, Lawrence M, Carey V, Gogarten S, Shannon P, Morgan M (2014) VariantAnnotation: a Bioconductor package for exploration and annotation of genetic variants. Bioinformatics 30(14):2076–2078
36. R Core Team (2015) R: a language and environment for statistical computing. R Foundation for Statistical Computing, Vienna, Austria
37. Rognes T, Seeberg E (2000) Six-fold speed-up of Smith–Waterman sequence database searches using parallel processing on common microprocessors. Bioinformatics 16:699–706
38. Rudin CM, Durinck S, Stawiski EW, Poirier JT, Modrusan Z, Shames DS, Bergbower EA, Guan Y, Shin J, Guillory J, Rivers CS, Foo CK, Bhatt D, Stinson J, Gnad F, Haverty PM, Gentleman R, Chaudhuri S, Janakiraman V, Jaiswal BS, Parikh C, Yuan W, Zhang Z, Koeppen H, Wu TD, Stern HM, Yauch RL, Huffman KE, Paskulin DD, Illei PB, Varella-Garcia M, Gazdar AF, de Sauvage FJ, Bourgon R, Minna JD, Brock MV, Seshagiri S (2012) Comprehensive genomic analysis identifies SOX2 as a frequently amplified gene in small-cell lung cancer. Nat Genet 44:1111–1116
39. Seshagiri S, Stawiski EW, Durinck S, Modrusan Z, Storm EE, Conboy CB, Chaudhuri S, Guan Y, Janakiraman V, Jaiswal BS, Guillory J, Ha C, Dijkgraaf GJP, Stinson J, Gnad F, Huntley MA, Degenhardt JD, Haverty PM, Bourgon R, Wang W, Koeppen H, Gentleman R, Starr TK, Zhang Z, Largaespada DA, Wu TD, de Sauvage FJ (2012) Recurrent R-spondin fusions in colon cancer. Nature 488:660–664
40. Shendure J, Ji H (2008) Next-generation DNA sequencing. Nat Biotechnol 26:1135–1145, 2008.
41. Smith TF, Waterman MS (1981) Identification of common molecular subsequences. J Mol Biol 147:195–197
42. Steijger T, Apbril JF, Engström P, Kokocinski F, The RGASP Consortium, Hubbard TJ, Guigó R, Harrow J, Bertone P (2013)

Assessment of transcript reconstruction methods for RNA-seq. Nat Methods 10:1177–1184

43. Trapnell C, Pachter L, Salzberg SL (2009) TopHat: discovering splice junctions with RNA-Seq. Bioinformatics 25:1105–1111
44. Wozniak A (1997) Using video-oriented instructions to speed up sequence comparison. Comput Appl Biosci 13:145–150
45. Wu TD, Nacu S (2010) Fast and SNP-tolerant detection of complex variants and splicing in short reads. Bioinformatics 26:873–881
46. Wu TD, Watanabe CK (2005) GMAP: a genomic mapping and alignment program for mRNA and EST sequences. Bioinformatics 21:1859–1975
47. Yeo G, Burge CB (2004) Maximum entropy modeling of short sequence motifs with applications to RNA splicing signals. J Comput Biol 11:377–394
48. Zhang Y, Luoh S-M, Hon LS, Baertsch R, Wood WI, Zhang Z (2007) GeneHug-GEPIS: digital expression profiling for normal and cancer tissues based on an integrated gene database. Nucleic Acids Res 35: W152–W158

Chapter 16

Visualizing Genomic Data Using Gviz and Bioconductor

Florian Hahne and Robert Ivanek

Abstract

The *Gviz* package offers a flexible framework to visualize genomic data in the context of a variety of different genome annotation features. Being tightly embedded in the Bioconductor genomics landscape, it nicely integrates with the existing infrastructure, but also provides direct data retrieval from external sources like Ensembl and UCSC and supports most of the commonly used annotation file types. Through carefully chosen default settings the package greatly facilitates the production of publication-ready figures of genomic loci, while still maintaining high flexibility due to its ample customization options.

Key words Visualization, Genomics, NGS, Annotation

1 Introduction

A typical genomics experiments these days produces massive amounts of data, which usually do not lend themselves to direct visual inspection. However, there is an intermediate level of processing that generates useful summaries, allowing to combine many different pieces of genomic information in a single graph. These visualizations are typically alignments of genomic features on a common horizontal scale, including such diverse feature types as gene or transcript models, CpG islands, transcription factor binding sites or other regulatory elements, chromosomal staining bands, Next Generation Sequencing read, ChIP measurements, and many more. Plotting these features together with the derived numerical data has proven to be tremendously useful for putting experimental results in context with genomic loci and to derive hypotheses. Relevant annotation data may be hosted in public repositories like the NCBI, Ensembl or at UCSC, or it may have been previously generated in-house. Many of the currently available genome browsers do a reasonable job in displaying genome annotation data, and there are options to connect to some of them from within R, for instance via the *rtracklayer* package [1]. However, none of these solutions offer the flexibility of the full R graphics

Ewy Mathé and Sean Davis (eds.), *Statistical Genomics: Methods and Protocols*, Methods in Molecular Biology, vol. 1418, DOI 10.1007/978-1-4939-3578-9_16, © Springer Science+Business Media New York 2016

system to display large numeric data in a multitude of different ways. The *Gviz* package aims to close this gap by providing a structured visualization framework to plot any type of data along genomic coordinates. It is loosely based on the *GenomeGraphs* package [2], however the complete class hierarchy as well as all the plotting methods have been restructured in order to increase performance and flexibility.

The fundamental concept behind the *Gviz* package is similar to the approach taken by most other genome browsing tools, in that individual types of genomic features or data are represented by separate tracks. A track in this context is really just a section within a composite visualization with a fixed horizontal scale. In the case of numeric data it may also contain a vertical scale to display their magnitude, however there are also more qualitative annotation tracks like transcript models for which the vertical axis does not convey any particular meaning. Within the *Gviz* package, each track constitutes a single object inheriting from a common base class, and there are constructor functions as well as a broad range of methods to instantiate, to interact with, and to plot these objects. When combining multiple track objects in a composite graph, the individual tracks will always share the same genomic coordinate system on the horizontal axis, thus taking the burden of aligning the individual elements from the user. A powerful customization interface provides ample opportunity to fine-tune almost every aspect of the visualization, while at the same time tries hard to come up with both appealing and meaningful default settings. In the remainder of this short article we want to highlight some of the features of the *Gviz* package in a real-world sample data set.

2 Materials

2.1 How to Install

Like any other Bioconductor package, *Gviz* should be installed using the *biocLite()* function. Mainly owing to its versatility it comes with quite a number of other package dependencies. The package can be loaded using the *library()* function.

```
source("http://bioconductor.org/biocLite.R")
biocLite("Gviz")
library(Gviz)
```

2.2 Data Set Description

For the purpose of this demonstration we will use a dataset from an RNA-seq experiment in which the authors studied the effect of RNAi depletion of the *Pasilla* (*PS*) gene on the transcriptome. *PS* is a known splice regulator and it is the *Drosophila melanogaster* ortholog of the mammalian RNA binding proteins *NOVA1* and *NOVA2* [3]. The authors used RNA interference to knock down the *PS* gene in S2-DRSC cells and compared their overall gene expression levels upon *PS* depletion to the expression in untreated cells. The experiment was performed with three biological replicates per condition. The data were deposited in GEO under the submission numbers *GSM461176-GSM461181*. Preprocessed data from this experiment are also directly available as the Bioconductor data package *pasilla*. In this data set, the original RNASeq reads were aligned using Tophat version 1.2.0 [4] with default parameters against the reference Drosophila melanogaster genome (version BDGP5.25.62 from Ensembl [5]). Similar as before, the *pasilla* data package can be installed using the *biocLite()* function and loaded using the *library()* function.

3 Methods

3.1 Track Objects

Before diving into the real world example we need to get a basic understanding of the fundamental concepts in the *Gviz* package, and to that end we first create the most simple track element there is: a genomic coordinates axis indicator. As briefly mentioned before, tracks within *Gviz* are represented by S4 objects inheriting from a common base class called *GdObject*. For the casual use it is not important to know all the details of this class. It mainly adds infrastructure for the very basic operations on a *Gviz* track like plotting and customization. The specific functionality of a particular track type is provided through a more dedicated child class and its methods. For our initial genome axis example we will need to instantiate an object of class *GenomeAxisTrack*. Conveniently, there is a constructor function available for this task with the same name.

```
axTrack <- GenomeAxisTrack() axTrack
## Genome axis 'Axis'
```

With our first track object being created we may now proceed to the plotting. There is a single function *plotTracks()* that handles all of this. As we will learn in the remainder of this chapter, *plotTracks()* is quite powerful and has a number of very useful additional arguments. For now we will keep things simple and just plot the single axis track using the default settings. The only two additional arguments we have to pass on are the start and the end coordinates of the plotting range.

```
plotTracks(axTrack, from=1e6, to=10e6)
```

The default values that have been chosen by the *Gviz* package for drawing the track object in Fig. 1 usually result in both visually

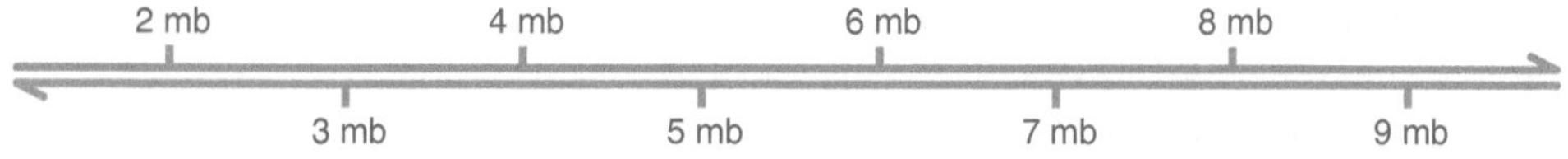

Fig. 1 A simple genome axis annotation track

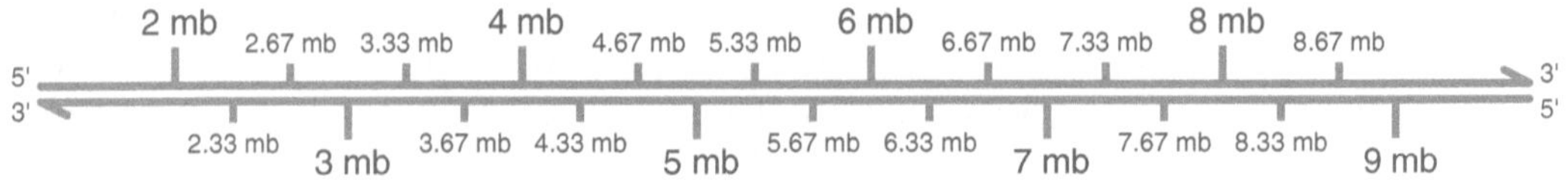

Fig. 2 A genome axis annotation track with some custom settings: direction indicators at and more fine-grained axis tick marks

appealing and also meaningful plots. However, visualization is an art in itself, and no software can ever be smart enough to guess the exact intention of a plot. To this end, *Gviz* offers a powerful customization framework to enable tweaking pretty much every aspect of the plot. We will discuss this in more detail in Subheading 3.2 below. As a simple example, we could decorate our genome axis with direction indicators and add additional tick marks at shorter intervals, as shown in Fig. 2.

```
plotTracks(axTrack, from=1e6, to=10e6, add53=TRUE, add35=
TRUE, littleTicks=TRUE)
```

So far we have not really plotted any genomic annotation data. One potentially interesting application would be to visualize the presence of repeats on a single chromosome. These data can be conveniently described by simple run-length encoding along genomic coordinates. All we need are the start and end positions for the repeats, and the chromosome to which these coordinates refer. The *Gviz* package provides the *AnnotationTrack* class to deal with exactly this sort of data. Its constructor function *AnnotationTrack ()* takes a variety of different inputs to digest the coordinate information, including *GRanges* objects, which are the recommended containers for genomic run-length encoded data in the Bioconductor world. However one can also create an *AnnotationTrack* object from quite generic inputs like *data.frames*. For this simple example we will download a RepeatMasker track for a single *Drosophila Melanogaster* chromosome and turn its first 50 entries into an *AnnotationTrack* object.

```
url <- "http://hgdownload.cse.ucsc.edu/goldenPath/dm3/database/chr3R_rmsk.
                         txt.gz"
con <- gzcon(url(url, open="r"))
repeats <- read.table(textConnection(readLines(con)), nrow=50)[, 6:8]
annoTrack <- AnnotationTrack(start=repeats[,2],
                         end=repeats[,3], chromosome=repeats[,1],
                              genome="dm3", name="Repeats", stacking="dense")
                         annoTrack
## AnnotationTrack 'Repeats'
## | genome: dm3
## | active chromosome: chr3R
## | annotation features: 50
```

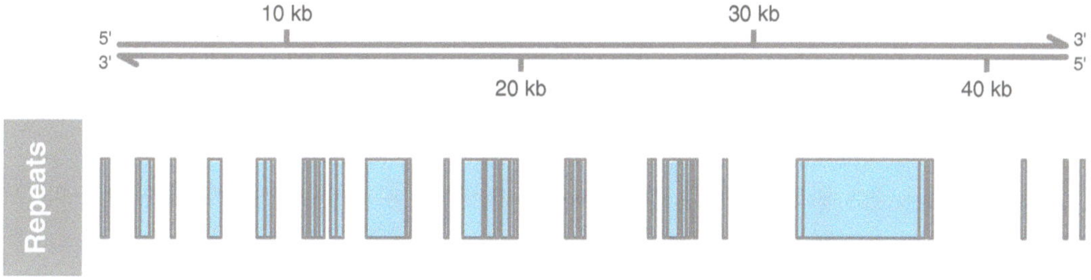

Fig. 3 Annotation track showing the distribution of the first 50 repeat elements on the *Drosophila melanogaster* 3R chromosome

Track objects in isolation are not particularly helpful, and we usually want to combine multiple tracks in a composite plot. In order to plot our new *AnnotationTrack* together with the *GenomeAxisTrack* from before, we need to provide them in the form of a list to the *plotTracks()* function (Fig. 3). We do not really have to explicitly provide the plotting ranges this time, since the *Gviz* package will try to extract the most extreme coordinates from the provided track objects if possible. Because the horizontal scales in all the individual tracks are synchronized, we automatically get the correct alignment of all the displayed features. We also see that the annotation track has been decorated with a title panel, displaying its name. Since names do not make too much sense for the genome axis, the title panel is hidden by default for that track. Again, this is an example where the package tries hard to select meaningful default settings. Later in this chapter we will see that the title panel is used to convey other information as well by some of the more complex track classes.

```
plotTracks(list(axTrack, annoTrack))
```

3.2 Display Parameters

Customization of track objects in the *Gviz* package is facilitated by so called display parameters. Even though we did not explicitly mention it, we have already made use of them in the previous examples. Display parameters are properties of individual track objects (i.e., of any object inheriting from the base *GdObject* class), and they can be adjusted at various levels:

1. During object instantiation as additional parameters to the constructor function
2. On existing track objects by using the *displayPars()* replacement method
3. Globally for all track objects in a composite plot by supplying additional parameters to *plotTracks()*

The parameter 'stacking' in the *AnnotationTrack* code chunk is an example for 1, and the *add53*, *add35*, and *littleTicks* parameter in the *GenomeAxisTrack* chunk are examples for 3. The following

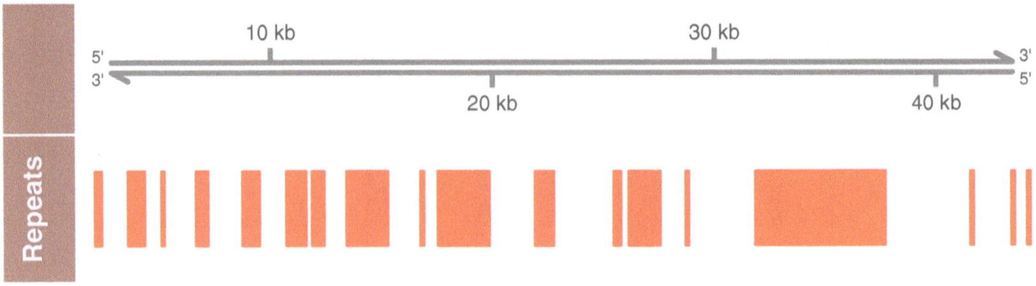

Fig. 4 Annotation track showing the distribution of the first 50 repeat elements on the *Drosophila melanogaster* 3R chromosome with customized plotting colors

code demonstrates the use of the *displayPars()* accessor and replacement methods and produces Fig. 4.

```
head(displayPars(annoTrack))
## $arrowHeadWidth
## [1] 30
##
## $arrowHeadMaxWidth
## [1] 40
##
## $cex.group
## [1] 0.6
##
## $cex
## [1] 1
##
## $col.line
## [1] "darkgray"
##
## $col
## [1] "darkgray"
displayPars(annoTrack) <- list(col="salmon", fill=
"salmon")
displayPars(annoTrack, "col")
## [1] "salmon"
plotTracks(list(axTrack, annoTrack), background.title=
"rosybrown")
```

In order to make full use of the flexible parameter system we need to know which display parameters control which aspect of a track class. The recommended source for this information are the help pages of the respective track classes, which list all available parameters along with a short description of their effect and their default values in a dedicated 'Display Parameters' section. Alternatively, we can use the *availableDisplayPars()* function in an interactive R session to print out the available parameters for a class as well as their default values in a list-like structure. The single argument to the function is either the name of a track object class, or the object itself, in which case the class is automatically detected.

```
dp <- availableDisplayPars(annoTrack)
tail(dp)
```

```
##
## The following display parameters are available for
'AnnotationTrack' objects:
##(see ? AnnotationTrack for details on their usage)
##
## rotation.item: 0
## rotation.title (inherited from class 'GdObject'): 90
## shape: arrow
## showAxis (inherited from class 'GdObject'): TRUE
## showFeatureId: NULL
## showId: NULL
## showOverplotting: FALSE
## showTitle (inherited from class 'GdObject'): TRUE
## size: 1
## stackHeight (inherited from class 'StackedTrack'): 0.75
## v (inherited from class 'GdObject'): -1
```

To avoid having to change the same parameters over and over again, the *Gviz* package supports display parameter schemes for registering settings in a central location. A scheme is essentially just a couple of nested named lists, where the element names on the first level of nesting correspond to track class names, and the names on the second level to display parameter names. The currently active scheme can be changed by setting the global option *Gviz.scheme*, and a new scheme can be added to the registry by using the *addScheme()* function. There is also a mechanism in place to register schemes upon package loading for even more permanent customization. The *getScheme()* function is useful to get the current scheme as a list structure, for instance to use as a skeleton when creating your own custom scheme. Please note that because display parameters are stored as integral parts of the individual track objects, a scheme change only has an effect on newly created objects.

3.3 A Real World Example

With all the necessary building blocks in place we can now proceed to a somewhat more applied biological example. The data originates from an RNASeq experiment, where the authors were interested in studying the effects of RNAi depletion of a known *Drosophila Melanogaster* splice modulator (*Pasilla*) on alternative splicing of its targets. They were able to identify several genes with effected splicing events, and we want to inspect one of them here in more detail: the *BMM* gene. The most obvious first thing to do is to take a closer look at the *BMM* gene locus, and the *Gviz* package provides the *GeneRegionTrack* class for this purpose. Objects can be created from Bioconductor-specific gene model abstractions (e.g., from *TranscriptDB* objects), or one can retrieve the necessary information from online resources like UCSC or Ensembl Biomart. In this case we select the latter by employing the dedicated *BiomartGeneRegionTrack()* constructor function. All we need to know is the genomic location of our gene of interested. To better put this location in the context of the whole chromosome, we also add our *GenomeAxisTrack* object from before and an *IdeogramTrack* object to the plot. Ideograms are schematic representations of chromosomes, showing their relative size and their cytostain banding patterns. When connected to the Internet, the *Gviz* package will automatically retrieve the necessary chromosome information for a large number of different organisms. The current relative plotting location on a chromosome is always indicated on the ideogram by a red box.

```
chr <- "chr3L"
from <- 14769250
to <- 14779800
geneTrack <- BiomartGeneRegionTrack(genome="dm3",
chromosome=chr, start=from, end=to,
transcriptAnnotation="symbol", name= "Genes")
ideoTrack <- IdeogramTrack(genome="dm3", outline=TRUE)
plotTracks(list(ideoTrack, axTrack, geneTrack), chromosome=chr, from=from,
                    to=to)
```

As we can see in the plot in Fig. 5, there are two transcript variants for the *BMM* gene that differ in the use of a single exon. Now the interesting question is how these exons are differentially used between the wild-type condition and the RNAi knockout of the *Pasilla* gene. The *DEXSeq* package [6] provides some nice infrastructure to make this inference. A full treatment of its features is beyond the scope of this article, and we refer to the package

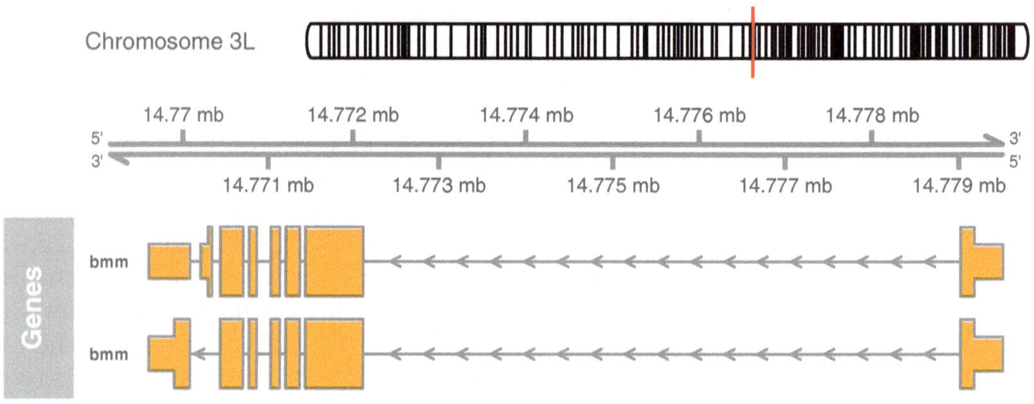

Fig. 5 Ensembl Biomart gene model track showing the Drosophila Melanogaster *BMM* gene locus

vignette for more details. In summary, we load the count data for six samples from the accompanying *pasilla* package, create an object of class *DEXSeqDataSet*, select only a subset of the available genes to speed up the computation and finally fit the differential exon usage coefficients in a generalized linear model.

```
library(pasilla)
library(DEXSeq)

## Loading required package: Biobase
## Welcome to Bioconductor
##
## Vignettes contain introductory material; view with
## 'browseVignettes()'. To cite Bioconductor, see
## 'citation("Biobase")', and for packages 'citation("pkgname")'.
##
## Loading required package: DESeq2
## Loading required package: Rcpp
## Loading required package: RcppArmadillo
## Loading required package: BiocParallel

## converting counts to integer mode
## using supplied model matrix
## using supplied model matrix
```

The resulting *DEXSeqResults* object holds all the information we need to create our visualization, but we have to extract it first in order to build a *DataTrack* object, which is the *Gviz* track class that handles displaying of numeric data. The general idea of the *DataTrack* class is that we can associated one or multiple numbers to a genomic location, representing values from one or several samples. Again, there is a multitude of different inputs for the constructor function, including the use of a *GRanges* object. All its numerical metadata columns will be automatically extracted and used for the track. Because the count data from an RNASeq experiment is often biased by the sequencing library size, we first apply a normalization using the library size factors that have been computed by the *DEXSeq* package. We also want to remove between-exon variation to place the focus on the group differences. Now we can assign these normalized values as metadata columns to the *GRanges* object, and also provide the sample grouping information. From the wide variety of different plotting options we choose a box-and-whisker plot because it nicely contrasts the data distribution between the two groups. We also ask the *Gviz* package to transform the count data to log2 space before plotting. The available horizontal space in this publication is fairly limited, so we will focus on the 3′ part of the *BMM* gene for now.

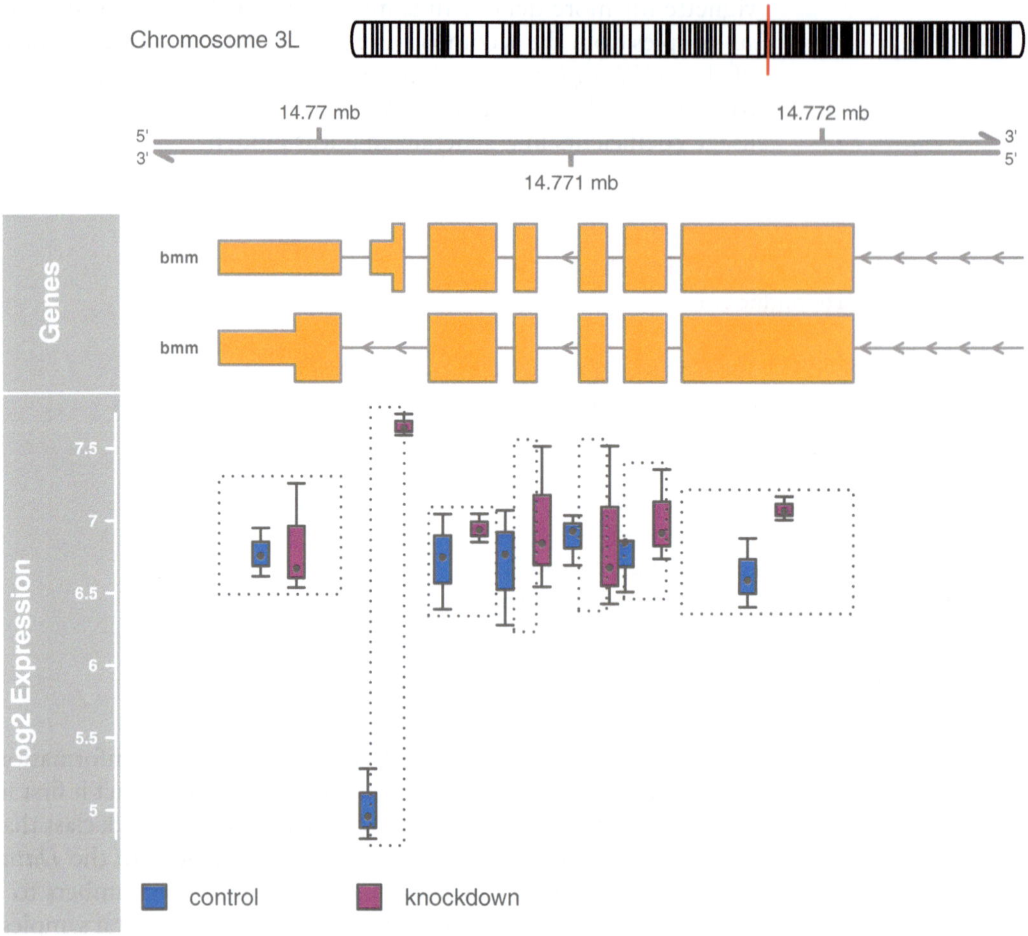

Fig. 6 Ensembl Biomart gene model track and box-and-whisker data track comparing the differential exon usage for the Drosophila Melanogaster *BMM* gene locus as a consequence of the RNAi knockdown of the *Pasilla* splice modulator

```
gr <- granges(dxd)
values(gr) <- (t(t(featureCounts(dxd)) / sizeFactors(dxd))) /
    rowMeans(featureCounts(dxd)) *
mean(featureCounts(dxd))
strand(gr) <- "*"
from2 <- 14769240
to2 <- 14772800
dataTrack <- DataTrack(genome="dm3", range=gr, groups=sampleTable$condition,
                     type="boxplot", box.width=10, size=6,
transformation=function(x) log2(x+1), name="log2 Expression")
plotTracks(list(ideoTrack, axTrack, geneTrack, dataTrack), chromosome=chr,
                    from=from2,
           to=to2, legend=TRUE)
```

The data values for the two groups at each exon in Fig. 6 have been automatically separated, placed at the correct genomic

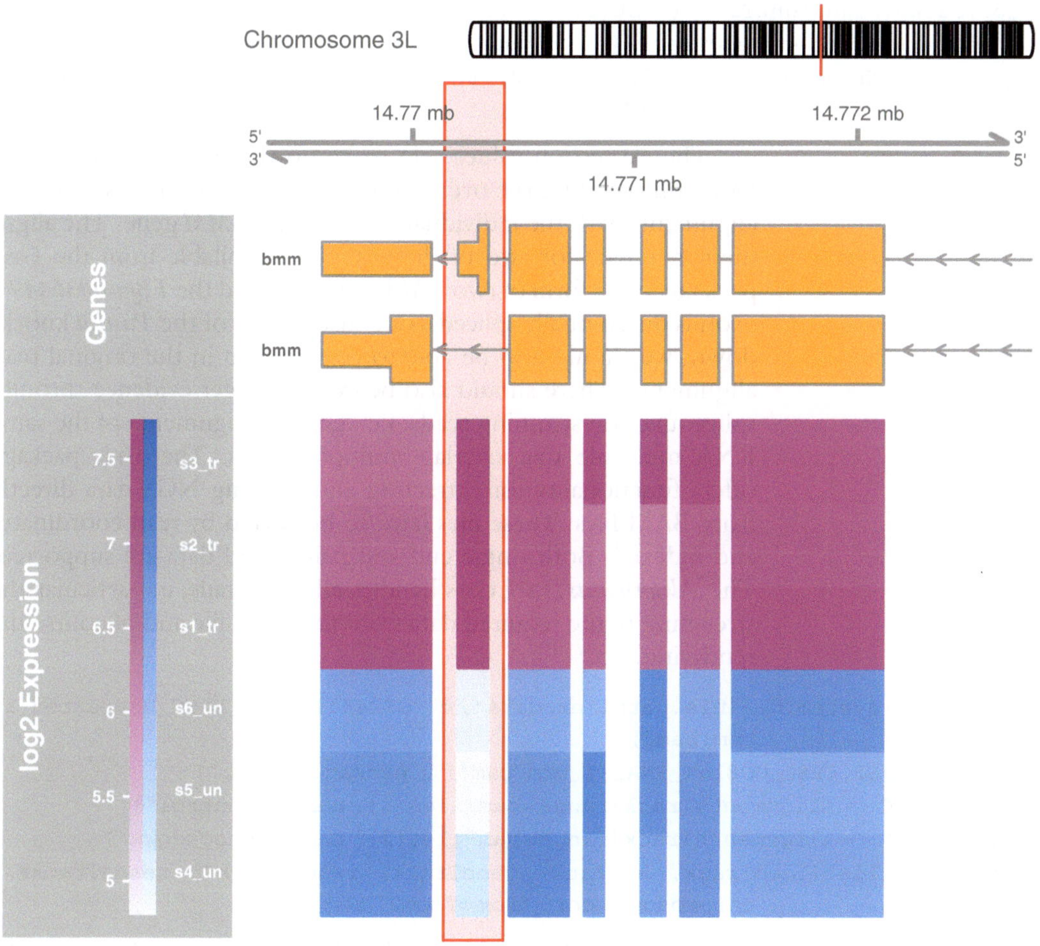

Fig. 7 Ensembl Biomart gene model track and heat map data track comparing the differential exon usage at the Drosophila Melanogaster *BMM* gene locus as a consequence of the RNAi knockdown of the *Pasilla* splice modulator. The location of the most differentially used exon FBgn0036449:2 is highlighted

location and differentially colored for easy contrasting. A vertical scale indicator has been included in the track title panel to show the magnitude of the normalized log expression values. We can immediately see that the second to last exon of the upper *BMM* transcript is used to a much larger extent in the *Pasilla* knockdown samples compared to the wild type. Creating a complex plot like this only took a few lines of code, which really is the fundamental idea behind the *Gviz* package. Another strength is its versatility when it comes to plotting numeric data. For instance we could choose from the long list of available plotting types a heat map instead of the box-and-whisker plot to give the visualization a slightly different spin. One can also draw the attention of the viewer to a particular plot region by wrapping our track objects into a *HighlightTrack* meta object which simply draws a colored box around the provided genomic locations (Fig. 7).

```
plotTracks(HighlightTrack(list(ideoTrack, axTrack, geneTrack, dataTrack),
range=range(ranges(geneTrack)[exon(geneTrack) == "FBgn0036449:2"]) * (-2)),
          chromosome=chr, from=from2, to=to2, type="heatmap", showSample
                    Names=TRUE)
```

The expression values that make up the box plot are actually based on counting the overlaps of a large number of RNAseq read alignments with the individual exons of the *BMM* gene. The alignment raw data for two of the samples is available from the *Gviz* package in the form of two BAM files. If indeed the *FBgn0036449:2* exon is differentially spliced as a consequence of the *Pasilla* knockdown, we should also be able to pick this up in the original read alignments. There should also be even stronger evidence through splice junction spanning reads. i.e., gapped alignments of the same RNA molecule that overlap multiple exons. The *Gviz* package offers functionality for extracting and plotting NGS data directly from BAM files. These files have to be sorted by read coordinates and indexed. Both single end and paired end data are supported. The *AlignmentsTrack* class handles all the details, in particular the streaming of the required data from the BAM file and of course the rendering.

```
bamFiles <- system.file(file.path("extdata", c("GSM461176_S2_DRSC_Untreated_1_
                    bmm.bam",
"GSM461179_S2_DRSC_CG8144_RNAi_1_bmm.bam")), package="Gviz")
readsTrackC <- AlignmentsTrack(range=bamFiles[1], name="Control")
readsTrackK <- AlignmentsTrack(range=bamFiles[2], name="Knockdown")
plotTracks(list(ideoTrack, axTrack, geneTrack, readsTrackC, readsTrackK),
                    chromosome=chr, from=from2, to=to2)
```

As expected, we do see the differential exon usage in the coverage plots in Fig. 8 in the top parts of the read alignments tracks, and also strong evidence for junction spanning reads in the pile-up views in the lower parts, indicated by the connecting lines between the read fragments. Even though it only make limited sense in this context, an *AlignmentsTrack* can also be used to show genotyping details based on the read alignments. It needs to know about the expected reference sequence, though. We can either provide that to the constructor function as a separate argument, or, even better, add a *SequenceTrack* object to our plot which holds an indexed version of the whole *Drosophila Melanogaster* genome. The package will detect the presence of this track and automatically extract the relevant sequence data. A convenient way to create the object is to start from one of the *BSGenome* packages that are provided as part of the Bioconductor suite.

```
library("BSgenome.Dmelanogaster.UCSC.dm3")
## Loading required package: BSgenome
## Loading required package: Biostrings
## Loading required package: rtracklayer
```

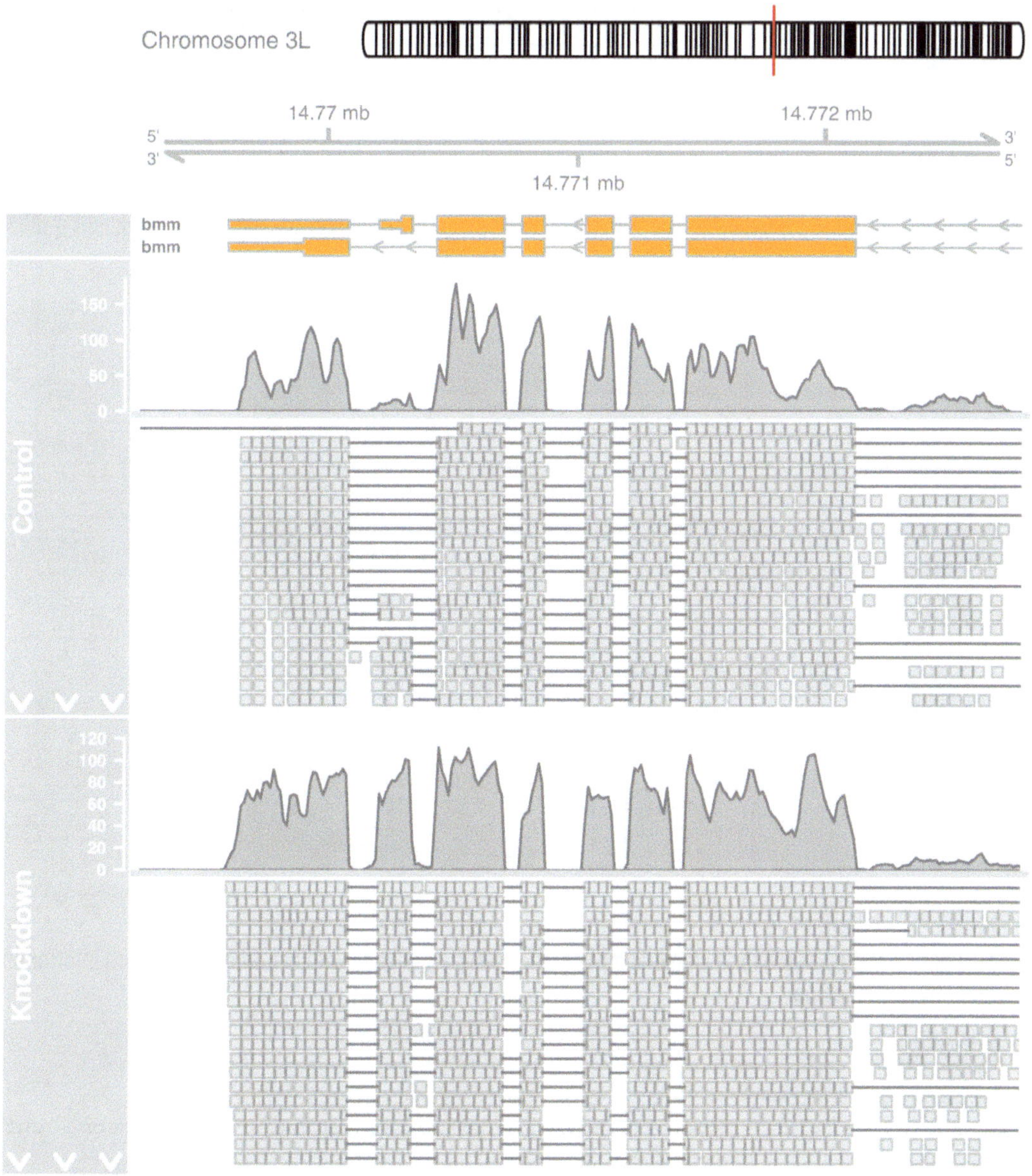

Fig. 8 Aligned RNASeq reads for two samples at the *BMM* gene locus

```
seqTrack <- SequenceTrack(Dmelanogaster)
plotTracks(list(ideoTrack, axTrack, geneTrack, readsTrackK, seqTrack),
                  chromosome= chr, from=from2, to=to2)
```

Mismatches in the read alignments are now highlighted by color-coding, both in the coverage section and in the pile-up section (Fig. 9). When zooming in a little bit we can actually make out locations showing two apparent homozygous SNPs (Fig. 10).

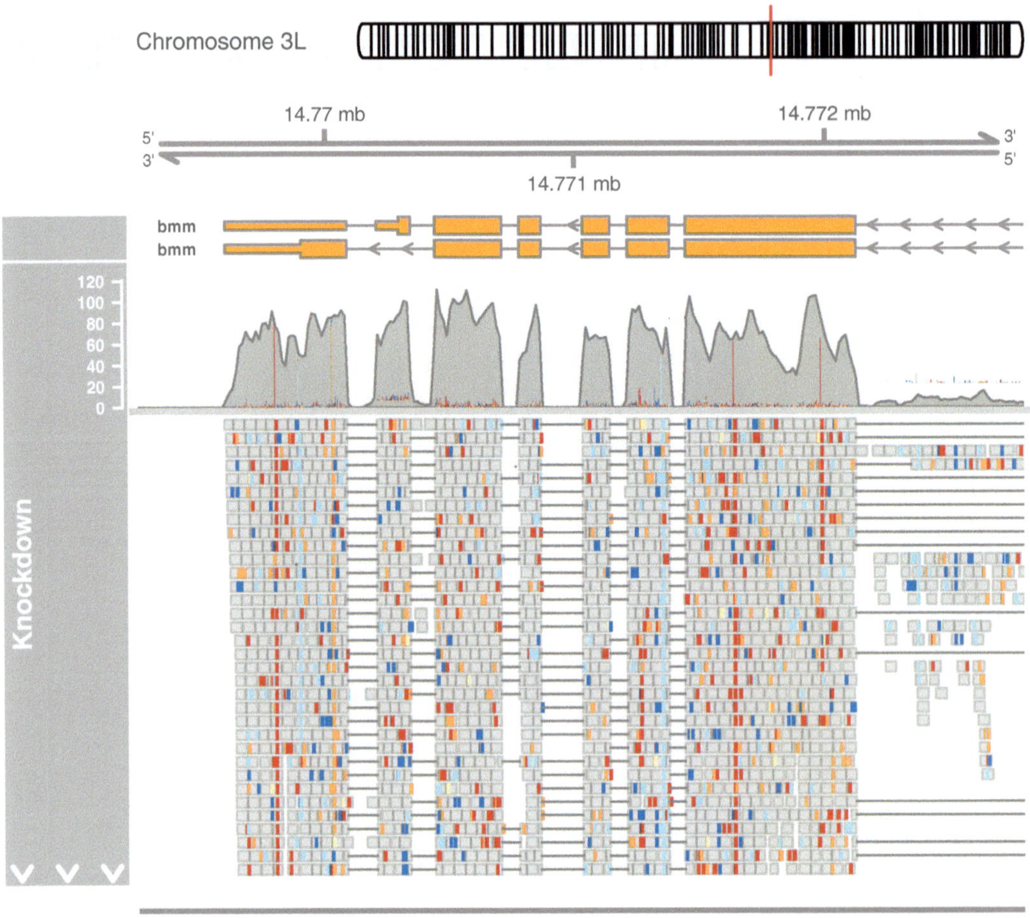

Fig. 9 Aligned RNASeq reads for a single sample at the *BMM* gene locus with alignment mismatch information indicated by color-coding

```
from3 <- 14769770
to3 <- 14769900
plotTracks(list(ideoTrack, axTrack, readsTrackK, seqTrack), chromosome=chr,
                          from=from3, to=to3)
```

Genomic variability is typically much higher in regions that are not highly conserved between closely related species. The UCSC database is a good source for interspecies conservation data, and we can make use of yet another track type in the *Gviz* package to quickly add this information to our plot. To be more precise, we will use the *UcscTrack* meta-constructor to automatically download the relevant data from UCSC and use it for building a *DataTrack* object. The rational here is that one can readily access the underlying UCSC data tables by means of the *rtracklayer* package, and only needs to provide a mapping between the retrieved table columns and their respective *Gviz* track constructor counterparts. An obvious prerequisite to this is that a user needs to know about the

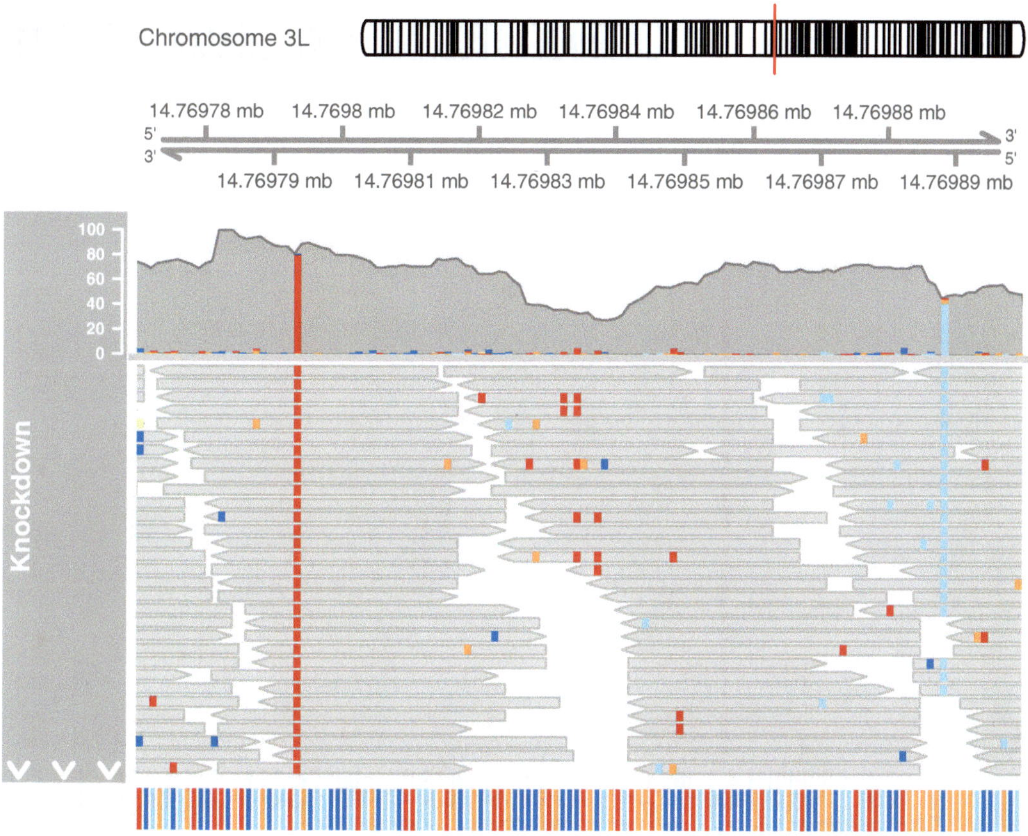

Fig. 10 Zoomed in view on aligned RNASeq reads for a single sample with alignment mismatch information indicated by color-coding showing two homozygous SNPs

underlying structure of the UCSC data table, which can easily be inspected via the online UCSC table browser. In the following code chunk we apply exactly this approach to the *phastCons15way* table within the UCSC *Conservation* track. The *trackType* argument tells the meta-constructor to build a *DataTrack* object, and the *start*, *end* and *data* arguments provide the mapping from the table columns to the *DataTrack* constructor arguments. We also add some smoothing by applying a running window summarization and slightly adjust the colors and the relative track size.

```
consTrack <- UcscTrack(genome="dm3", chromosome=chr, track="Conservation",
                 table="phastCons15way",
from=from3, to=to3, trackType="DataTrack", start="start", end="end",
                 data="score", type="histogram",
window="auto", col.histogram="darkgray", fill.histogram="darkgray", name="%
                 Cons", size=6)
plotTracks(list(ideoTrack, axTrack, readsTrackK, seqTrack, consTrack), chro-
                 mosome=chr, from=from3, to=to3)
```

As seen in Fig. 11, both SNPs do fall in relatively unconserved regions.

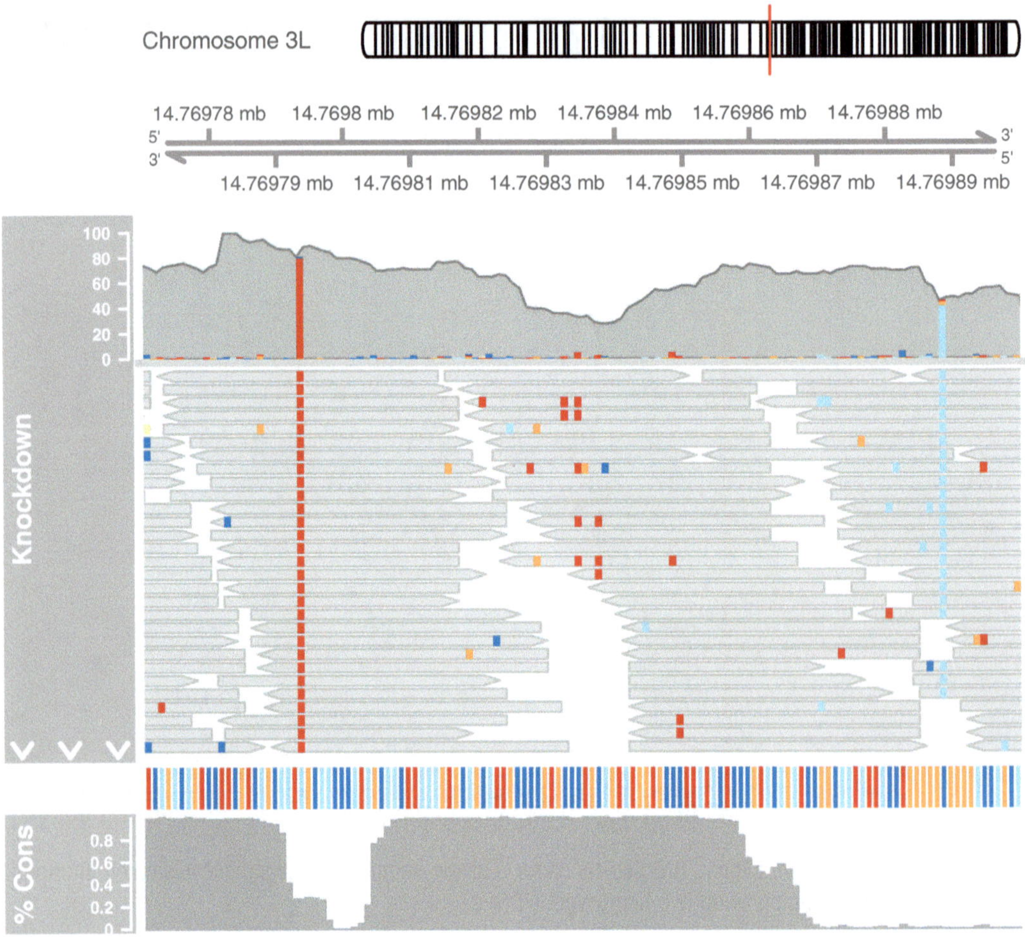

Fig. 11 Zoomed in view on aligned RNASeq reads for a single sample with alignment mismatch information indicated by color coding showing two homozygous SNPs. The interspecies conservation of the locus is shown in the additional data track

4 Notes

In this chapter, we described how to visualize different features along genomic coordinates using the *Gviz* package. Similar to other Bioconductor packages for genomic data visualization, *Gviz* is using the concept of aligned tracks representing different annotation and numerical data types which are combined to produce the final figure for a given genome region. *Gviz* can load data from both commonly used genome browsers (Ensembl Genome Browser [7], UCSC Genome Browser [8]) and various file formats, and offers a large collection of plotting types for numerical data. The package empowers users to generate figures almost identical to or often times better than the seemingly omnipresent screenshots from genome browsers in a much more flexible way. It even offers zoom-in capability to single-base-pair resolution similar to the Integrated Genome Viewer [9], which is often useful to highlight sequence variation identified in the analyzed samples. Compared to more "point and click-like" approaches by other tools, *Gviz* does not lend itself very much to interactive data browsing, even though there are smart caching and streaming mechanisms in place to speed up the plotting. That said, excellent new packages like *shiny* [10] and *rmarkdown* [11] provide some opportunity to create interactive documents for data exploration using *Gviz* plots. More importantly, though, in combination with tools like *Sweave* [12] or *knitr* [13], *Gviz* can be used to generate high quality figures in a scripted fashion, thus facilitating reproducible research. Its tight integration in R and Bioconductor environment allows users to perform the entire analysis from the very start to the final figure generation within a single software tool.

References

1. Lawrence M, Gentleman R, Carey V (2009) Rtracklayer: An r package for interfacing with genome browsers. Bioinformatics 25:1841–1842. doi:10.1093/bioinformatics/btp328
2. Durinck S, Bullard J. GenomeGraphs: plotting genomic information from ensembl. R package version 1.30.0
3. Brooks AN, Yang L, Duff MO, Hansen KD, Park JW, Dudoit S, Brenner SE, Graveley BR (2011) Conservation of an rNA regulatory map between drosophila and mammals. Genome Res 21:193–202. doi:10.1101/gr.108662.110
4. Trapnell C, Pachter L, Salzberg SL (2009) TopHat: discovering splice junctions with rNA-seq. Bioinformatics 25:1105–1111. doi:10.1093/bioinformatics/btp120
5. Ensembl (2011) Drosophila melanogaster genome version BDGP5.25.62
6. Anders S, Reyes A, Huber W (2012) Detecting differential usage of exons from rNA-seq data. Genome Res 22:2008–2017. doi:10.1101/gr.133744.111
7. Cunningham F, Amode MR, Barrell D, Beal K, Billis K, Brent S, Carvalho-Silva D, Clapham P, Coates G, Fitzgerald S, Gil L, Girón CG, Gordon L, Hourlier T, Hunt SE, Janacek SH, Johnson N, Juettemann T, Kähäri AK, Keenan S, Martin FJ, Maurel T, McLaren W, Murphy DN, Nag R, Overduin B, Parker A, Patricio M, Perry E, Pignatelli M, Riat HS, Sheppard D, Taylor K, Thormann A, Vullo A, Wilder SP, Zadissa A, Aken BL, Birney E, Harrow J, Kinsella R, Muffato M, Ruffier M, Searle SM, Spudich G, Trevanion SJ, Yates A, Zerbino DR, Flicek P (2015) Ensembl 2015. Nucleic Acids Res 43:D662–D669. doi:10.1093/nar/gku1010
8. Kent WJ, Sugnet CW, Furey TS, Roskin KM, Pringle TH, Zahler AM, Haussler D (2002) The human genome browser at uCSC. Genome Res 12:996–1006. doi:10.1101/gr.229102
9. Robinson JT, Thorvaldsdottir H, Winckler W, Guttman M, Lander ES, Getz G, Mesirov JP (2011) Integrative genomics viewer. Nat Biotech 29:24–26
10. Chang W (2015) Shiny: web application framework for r
11. Allaire J, Cheng J, Xie Y, McPherson J, Chang W, Allen J, Wickham H, Hyndman R (2015) Rmarkdown: dynamic documents for r
12. Leisch F (2002) Sweave: dynamic generation of statistical reports using literate data analysis. In: Härdle W, Rönz B (eds) Compstat. Physica-Verlag HD, Berlin, pp 575–580
13. Xie Y (2014) Knitr: a comprehensive tool for reproducible research in R. In: Stodden V, Leisch F, Peng RD (eds) Implementing reproducible computational research. Chapman and Hall/CRC, Boca Raton. ISBN 978-1466561595

Chapter 17

Introducing Machine Learning Concepts with WEKA

Tony C. Smith and Eibe Frank

Abstract

This chapter presents an introduction to data mining with machine learning. It gives an overview of various types of machine learning, along with some examples. It explains how to download, install, and run the WEKA data mining toolkit on a simple data set, then proceeds to explain how one might approach a bioinformatics problem. Finally, it includes a brief summary of machine learning algorithms for other types of data mining problems, and provides suggestions about where to find additional information.

Key words Machine learning, Data mining, WEKA, Bioinformatics, Tutorial

1 Data Mining and Machine Learning

A principal activity of all scientists is *trying to make sense of data*, and computers are often the primary means by which that is attempted. Spreadsheets and statistical packages are widely used tools for obtaining distributions, variances, and scatter plots, which can offer *some* insights; but there are other, much more difficult and important objectives of data analytics for which these metrics are of little utility. Biologists, for example, spend a good deal of time and effort looking for patterns in data that can be converted into an explanation of some phenomenon observed in nature. That is, the ultimate goal is to discover new, interesting, and potentially useful knowledge that furthers understanding of the natural world.

The process is one of first detecting regularities in data, then formulating some hypothesis as a characterization of those regularities, and finally testing the hypothesis against new data to see how robust it is. One might call such a process *data mining*—the search for valuable nuggets of insight in a slurry of observations. Done manually, this can be a tiring and frustratingly difficult thing to achieve; particularly when the amount of data involved is large. Fortunately, much of the process lends itself to automation, and computer scientists have devised a great many algorithms that can rapidly find and characterize patterns in data. Collectively, this type

Ewy Mathé and Sean Davis (eds.), *Statistical Genomics: Methods and Protocols*, Methods in Molecular Biology, vol. 1418, DOI 10.1007/978-1-4939-3578-9_17, © Springer Science+Business Media New York 2016

of computation is called *machine learning*, and it is increasingly a key tool for scientific discovery.

This chapter aims to provide the reader with a quick and practical introduction to data mining using machine learning software. It explains the different types of machine learning, provides an overview of some of the algorithms available to address these types of learning, shows how to download, install, and run one of the more widely used open-source data mining software tools—namely, WEKA [1]—and takes the reader through some examples demonstrating how to prepare data and load it into WEKA, how to run machine learning experiments on that data, and how to interpret the results. In short, by the end of this chapter the reader will be able to carry out machine learning.

Having said that, it is important to bear in mind that the depth of discussion here is still necessarily limited because the subject is very large. Like so many things, learning how to do data mining is not particularly difficult, but learning how to do it well is quite another thing. This chapter is thus a kind of *crash course* and a *small sampler*. Along the way, it attempts to alert the reader to some of the more common pitfalls that arise when carrying out machine learning experiments, and makes suggestions as to where more information can be obtained.

1.1 Structured vs. Unstructured Data

Before explaining how machine learning works, it is useful to get an idea of what its objectives are and what it is working with. There are many kinds of machine learning, and many different algorithms to choose from depending upon the data available and the goals of the study. We might start by first differentiating between *structured* data and *unstructured* data.

It is perhaps easiest to think of structured data as the kind of information one might record in a plain old spreadsheet. Each row of the spreadsheet is a single datum (i.e., an *instance*) and each column is a *feature* (i.e., or *attribute*) of the instances, such that each cell of the spreadsheet holds a value recorded for that feature for that instance—an *attribute-value* pair. The nature of the spreadsheet means that every instance is represented in the same structured way, even if some cells are empty or irrelevant or wrongly recorded. In contrast, unstructured data are things like documents (i.e., texts) or pictures (i.e., images) that have different dimensions, different formats, or other widely disparate characteristics.

Machine learning algorithms work well on structured data, and it is frequently the case that unstructured data simply has a structure imposed upon it prior to the actual learning. For example, a collection of images might be transformed into some regular set of attributes such that each image is generalized according to its colors, textures, shapes, intensities, dimensions, and so forth; and a collection of documents might see each text converted to something like a vector of counts for how often each word in the complete lexicon of the entire collection occurs in each text. Unstructured learning is generally less common than structured learning, and has overhead associated with it that requires a fair bit of explanation, so its peculiarities are not covered in this chapter. We focus on learning from data as it may be found in a spreadsheet or a database table; but we include some examples where unstructured data (e.g., primary sequence data) is converted into tabular form.

1.2 Supervised vs. Unsupervised Learning

Another way to characterize data mining paradigms is to differentiate between *supervised* and *unsupervised* learning. In supervised learning, the data used for learning are treated as exemplars that have in some way been annotated with *the thing* to be learned. In such situations, the goal is to find a characterization of these examples so that judgments can later be made for new instances. In the simplest case, the value of one attribute of each example is the thing we want to predict (i.e., the judgment) and the assumption is that values of the other attributes allow for that if only the right relationship can be found. The learning algorithm seeks out a model of the data that supports such prediction. By learning on a subset of the existing data, we can test predictive accuracy on the remaining data; hence the learning is directed by examples provided by an expert and performance is supervised through evaluation against existing judgments. Whether or not performance on that test sample is indicative of future performance on novel data is an important question, and one we will address later.

1.3 Prediction: Classification and Regression

Supervised learning itself comes in two principal forms: classification and regression. If we are trying to predict something that has a discrete value, it is called classification. For example, we may want to predict whether the molecular composition of a blood sample (e.g., mass spectrum data) is or is not indicative of cancer; thus we have a binary classification problem. Or, we may have a sample of mitochondrial DNA and we want to predict the genus or species of the organism from which it most likely came; giving us a potentially very large number of classes to choose from in a so-called multi-class problem. And, if it is possible for an instance to belong to more than one class then we have what is known as a multilabel classification problem. For all classification problems, the objective is to accurately associate an instance with one or more labels from a finite set of alternatives. In comparison, if we are trying to predict a numeric value then the learning objective is to find some formula for generating a good estimate of the true value. In this situation, accuracy is measured by how close the estimate is to the actual value provided by the expert, rather than whether the predicted value is precisely right or not. This is regression, and typically requires different algorithms than those used for classification.

1.4 Clustering and Associations

In comparison, unsupervised learning is where we do not know the right answer ahead of time for any of the data—there is no prior basis to judge how good our result is. The goal is not to be able to make judgments that are right or wrong, per se, but simply to find interesting and useful generalities within the data.

A common form of unsupervised learning is that of *clustering*—given a collection of data, separate instances into two or more groups (clusters) based upon their similarities. Several issues arise in clustering that greatly affect the output and, consequently, dash any hope of an objective assessment about success. For one thing, it is generally not clear how many groups there should be. Should there be two? Twenty? One for each instance? Is it possible for one group be a subset of another (i.e., hierarchical), or overlap with other groups? Can an instance belong to more than one group at the same time and, if so, can membership be fractional or probabilistic?

In addition to this, a clustering algorithm must have criteria to decide whether two things are similar or dissimilar. Should they be in the same cluster or not? Similarity is not always as obvious as one would hope. Consider the situation where a child has been given a bunch of building blocks of all different shapes, sizes, and colors. One might find, after a time, that the child starts sorting them into groups; perhaps putting all the cubes together into one group, all the pyramids in another, all the cones in another, and so forth. Or maybe the child puts all the big blocks in one group, all the small ones in another, and all the rest in another. Or one might instead see the child sort all the blue ones together and all the red ones in another group, and so on, without regard to shape or size. Or the blocks may end up being sorted based upon which ones are the child's favorites, or which blocks are the newest, or which ones were played with most recently, or any other criterion that is difficult to assess objectively by a third party. Similarity is often pretty subjective, and the measure utilized greatly affects the outcome.

Another kind of unsupervised learning is one where the objective is simply to discover any interesting correlations within data. This is called association mining, and a classic example is that of so-called *basket analysis*, where supermarkets analyze the contents of shoppers' baskets at the checkout to see if the purchase of one item appears to correlate with the purchase of another; a discovery with the potential to help target future buyers through careful product placement. As an example from genomics, one might use association mining to look for correlations between genotypes and phenotypes, identify codependent genes, or find potentially interesting variant calls that co-occur.

Similar to clustering, the value of the output for this kind of unsupervised learning is difficult to assess and often specious or uninteresting. For example, one might be excited to discover that the presence of six variants correlates highly with both a specific disease and the geographical area where someone lives, suggesting an inheritable genetic cause for that disease; but, one might be a little less excited to discover that the same data suggests a very high correlation between the town where someone lives and their nationality. Association mining algorithms can find correlations, but cannot judge their utility.

Clustering and association mining (i.e., unsupervised learning) are very different from classification and regression (i.e., supervised learning), and very different from each other, and specific algorithms have been devised to carry out these kinds of data mining tasks. Moreover, because classification and regression aim to predict something accurately, we can only use some of our data for learning and must withhold some for testing the result of that learning; and there are several important aspects to how one goes about partitioning the data for these two stages (i.e., training and testing). Unsupervised learning, on the other hand, has no objective measure of success, and therefore, all the data can be used as input to the algorithm.

Whatever kind of machine learning you are doing, it is important to remember that the algorithms are effectively heuristic. They rely on principled techniques (usually based upon statistical methods) to find patterns and express correlations, and they do this very well; but they cannot assess the true value of what they find. It is the scientist carrying out the machine learning study that must assess the quality and utility of the result. Put another way, machines do learning, but people do data mining.

1.5 Nominal and Numeric Data

One more important distinction to make relates to types of data. As mentioned earlier, some attributes have values that are discrete and chosen from a finite set. We can think of these as labels, or categories; but more generally we call them nominal. The color of a car, or the symbolic value of a nucleotide or amino acid in a primary sequence is an example. Nominal attributes typically only submit to tests of equality (e.g., "first amino acid in this peptide is Methionine, yes or no?", "base immediately upstream from this position is C or G, true or false?") and other comparisons like "greater than" or "has a value between X and Y" do not generally make sense. Values that are numbers can be tested for equality as well, but also submit to range tests and comparisons involving closeness (e.g., "how big is the difference between this value and some other one?"), while nominal values do not.

The *type* of a feature can place restrictions on which learning algorithms can be used, and what operations can be performed on that feature when formulating a generalization of it. For example, many algorithms construct a model of the data by considering how much knowing the value of a particular feature improves the probability of predicting the class of an instance. This is like the game of 20 questions, where the best question to ask next is the one that you think improves your chance of guessing the right answer quickly. So it is that an algorithm will compute, for each attribute, the improvement in the probability of guessing the correct result (i.e., formally, the information gain) when this feature's value is known; then choose the best feature and use it as a test to partition the data into subsets; repeating the process for each subset until a sequence of attribute tests ends with a prediction.

For nominal attributes, only tests for equality to one or more of the observed values are possible. For numeric values, a learning algorithm must compute some constant against which an observed value can be compared (e.g., greater-than, less-than) to maximize information gain and partition the data into subsets. These are different processes, therefore WEKA needs to know whether it is dealing with a number or not. There are two other attribute types available in WEKA (dates and strings, which have their own peculiarities), but the vast majority of data types are typically nominal or numeric.

2 WEKA: Getting Started

Many machine learning systems exist, but the one we use in this chapter is the WEKA Machine Learning Workbench. It is one of the more popular open-source machine learning toolkits, and contains a wide range of learning algorithms. It is easily obtained simply by typing "download WEKA" into a search engine, which will likely bring up a link to the corresponding SourceForge project site, or to the University of Waikato (where WEKA was developed), which itself will take you to the SourceForge site. At the time of writing, the latest stable version is WEKA 3.6.12, but any version close to this will be sufficiently similar. You will want to select the version of WEKA that corresponds to your operating system (Windows x86 or 64 bit platforms, Mac or Linux).

WEKA is written entirely in Java, which means it can run on any computer that has a Java virtual machine (JVM) installed. If you are not sure whether your machine has a JVM, you can simply download the version of WEKA that includes Java VM 1.7 and when you go to install it your machine will tell you if you already have it and ask if you want to replace or update your JVM. A virtual machine is a piece of software that pretends to be a specific computer architecture that runs code specifically written for it (in this case, Java) so that such code can run on any machine with the corresponding virtual machine without having to recompile the code or fiddle with compatibility issues. It is what makes this kind of code very portable and easy to install and run.

Once downloaded, click on the downloaded file to start the install wizard that will guide you through the installation process. You will need about 65 MBytes of disk space, and the installation

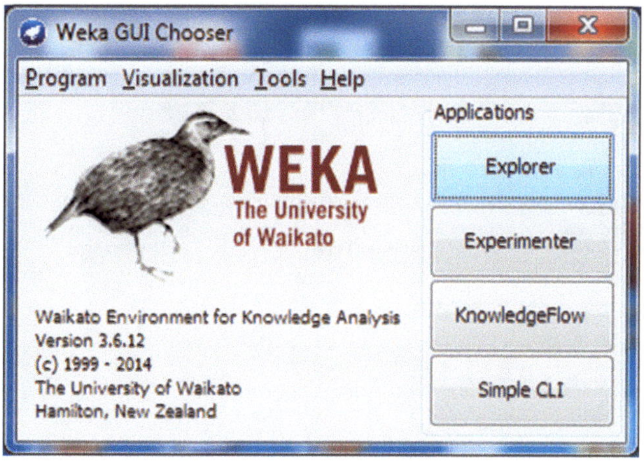

Fig. 1 Start screen for WEKA

should be very quick. Once completed, if you have installed everything correctly and launched WEKA, you should get a startup window that looks like the one in Fig. 1. Do not worry if the size or layout is a little different, or if the typeface is not quite the same, as sometimes individual preferences on a machine will affect how the JVM displays things. The important thing is that you are looking at the same set of options.

For this tutorial, we will use the *Explorer*, which is the most straightforward way to use WEKA interactively. The *Experimenter* is a way to set up a series of machine learning experiments so that they can be saved and repeated in the future; *KnowledgeFlow* offers a drag-and-drop interface to set up a machine learning experiment by piecing together data sources, learning algorithms, training and test procedures and so forth with corresponding icons linked with arrows (which some people prefer); and the *Simple CLI* is a way to invoke parts of WEKA with type-written commands as one might do if setting up machine learning as part of a process pipe (e.g., where the input comes from a running process, or the output is required by a subsequent program). That is to say, WEKA is not just a stand-alone application, but has a complete API that allows it to be incorporated into other software systems using just those bits that are needed—either as a serial or concurrent thread, or by importing the classes of WEKA into another Java program. Unless you are a programmer yourself, this might not make a lot of sense, and is anyway beyond the scope of this chapter. To learn more about it, simply type "WEKA Simple CLI" into your favorite search engine and many useful links to explanations and tutorials are easily obtained.

2.1 Loading Data

If all has gone well, clicking on the Explorer option (so named because it is a way to go exploring with WEKA) will produce a new window similar to the one pictured in Fig. 2. Note the six tabs

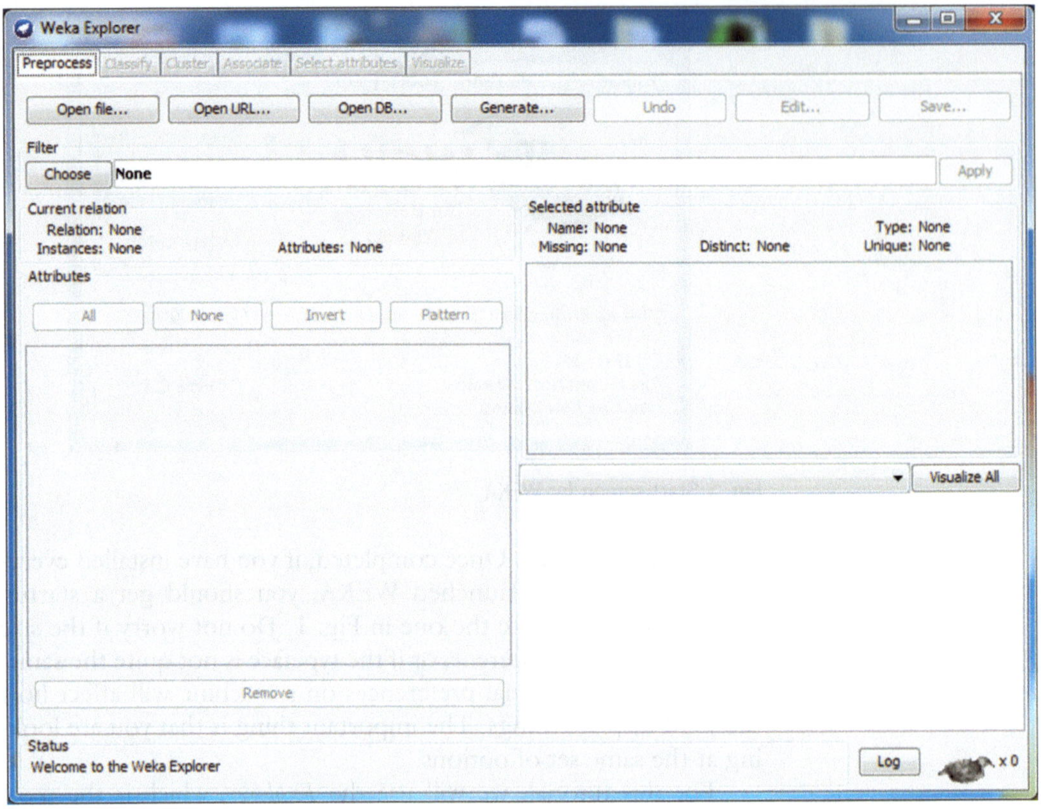

Fig. 2 Start screen for the WEKA Explorer

along the top, and you will see that you are in the Preprocess area. This is where interactive data mining begins. It is where you load data, examine it, perhaps modify it (more on that later) or otherwise get your data ready for input into a learning scheme. As suggested by the buttons along the top of the Preprocess window, data can be obtained from a file on your machine, or from a website via a URL, or it can be pulled out of a database if you are familiar with SQL, and you can even generate artificial data if you just want to experiment with the learning procedures more generally. We are only concerned here with loading data from a file, and the WEKA download actually includes a couple of dozen sample data files to experiment with.

Depending upon how search paths are set on your own machine, clicking on the *Open file …* option should take you directly to the WEKA folder (i.e., directory) where the sample data, documentation and log folders are located; or you may have to traverse your directory structure to get there (for example, on a Windows machine, you would likely go to C:\Program Files (x86) \Weka-3-6 to find the data folder).

Use the "*Open file…*" option to load the sample data file called "weather.numeric", and your WEKA window should end up

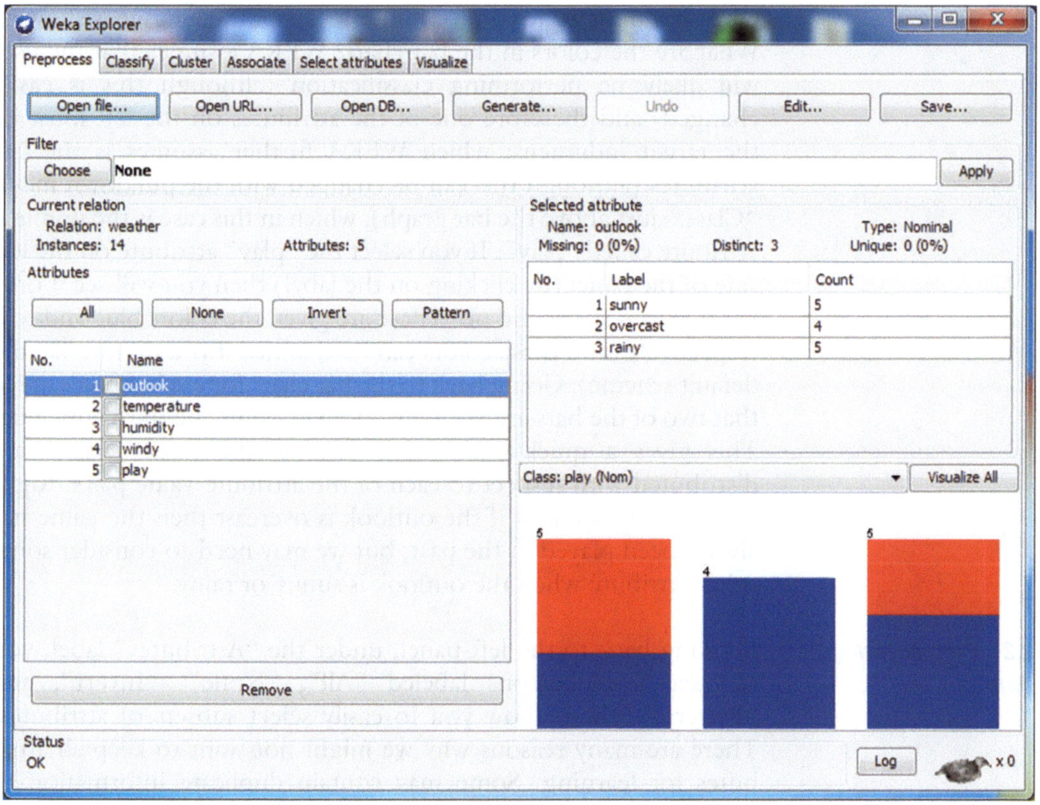

Fig. 3 After loading the weather.numeric sample data

looking like the one in Fig. 3. This data is contrived for demonstration purposes, supposedly recording weather conditions for 14 different days upon which some fictitious outdoor sport was either played or not played. The learning objective is to find a generalization of these past examples that helps decide if this game should be played on some future occasion given the prevailing conditions.

On the left side of the screen is information about the number of instances (i.e., examples) loaded from the file, how many attributes each instance has, and a checkbox list of those attributes. On the right hand side is some information about the specific attribute selected (i.e., highlighted) on the left, in this case the "outlook" attribute, and it includes information about its type (either nominal or numeric), how many distinct values for this attribute have been found in the data, what percentage of the instances are missing the value for this feature, and what percentage of the values only occur once (i.e., are unique). Below that is a table that lists the values observed for this feature and their frequencies (or, when a numeric attribute is selected, you will see the maximum and minimum values found for this attribute, and the mean and standard deviation), and below that is a colored bar chart showing the distribution for those values. Rolling the mouse over each bar brings up the corresponding value (or, for numeric attributes, a range is shown). What are the colors in the bar chart? WEKA assumes that the user will likely be performing classification (although this is easily changed) and therefore one of the attributes on the left must be the target judgment; which WEKA further assumes is the last attributes (although this can be changed with the pulldown menu "Class:" just above the bar graph), which in this case is the nominal attribute called "play". If you select the "play" attribute on the left side of the panel (by clicking on the label) then you will see it only has two values: yes and no. These are given the colors blue and red, respectively (and these colors can be changed if you do not like the default scheme). Going back to the bar chart for *outlook* you can see that two of the bars are both red and blue while the third is all blue. This gives a quick perspective of how the target classes are distributed with respect to each of the attribute-value pairs. Right away we can see that if the outlook is overcast then the game has always been played in the past, but we may need to consider some other attribute when the outlook is sunny or rainy.

2.2 Preprocessing Data

Looking back to the left panel, under the "Attributes" label, you can see four buttons labeled "All", "None", "Invert", and "Pattern". These allow you to easily select subsets of attributes. There are many reasons why we might not want to keep all attributes for learning. Some may contain duplicate information of others (e.g., the *windy* attribute here is either true or false, so if we happened to also have a "*not windy*" attribute with the complement value then one of these two attributes could and should be tossed out as redundant), or some attributes may have too many missing values, or be too tightly related to the concept being learned to be of general use (e.g., if each instance had a unique identification code as an attribute, then knowing that code would let us perfectly predict the class simply by looking it up, but if a future instance has its own unique code then no generalization of this attribute would be useful for prediction). On the left side, you can use the checkboxes to select attributes, and the Remove button at the bottom allows you to delete them from consideration in subsequent learning. The buttons above the list of attributes are shortcuts to quickly select "All" attributes, "None" of the attributes, "Invert" your selection, or use a regular expression pattern to select a subset. These come in handy when a large number of attributes are being manipulated. For example, if you wanted to see what could be learned from just a few attributes in a data set that included hundreds of features, it is much easier to select the few you want to keep, invert your selection and then hit *Remove* than it is to select *one-by-one* the hundreds of attributes you do not want. Note that anything we do to the data on this panel does not affect the original data file, and we can repeal any operation with the *Undo* button (near the top). If we like the changes we've made and want to keep them, we can always save them in a new file using the *Save* button, and we can also *Edit* individual values in the data, say, if we want to correct an error or fill in a missing value.

Near the top of the window, just below the top row of buttons, you can see a "Filter" section with a "Choose" button on one side and an "Apply" button on the other. This is where you can select from a very large number of filters that can transform data in a myriad of ways prior to learning. For example, you might want to take a nominal attribute whose value can be one of three possible labels and turn it into three new binary attributes that test each value individually instead. That is, if one attribute can have the value "A" or "B" or "C", we could turn it into three binary attributes called "Is A", "Is B", and "Is C" where one of these has the value "true" and the others "false" to reflect the original multi-valued attribute. This can occasionally be useful because it helps some learning algorithms search for characterizations more effectively.

Another common filter is to transform a numeric continuous valued attribute into a discrete valued one by binning values into a small number of categories defined by ranges. Many learning algorithms do this themselves, but do so heuristically and it may be the case that you know something about the data that would allow for better discretization, and this filter can be set up to do a better job of binning values prior to learning. In another situation, you might want to add a numeric feature that is the product or sum of two other numeric features (something that most learning algorithms do not do on their own because the possible combinations that could be tested are large), and this can be done with a filter. Indeed, just about anything you can think you might want to do, there is likely a filter for it in WEKA. Filtering can be extremely useful, but too complex to discuss in more detail in this chapter. Let us move on and get WEKA to learn something from the data we have loaded.

2.3 Choosing a Classifier

Predicting whether or not to play this game is a binary classification problem. Once we have prepared our data for learning, we select the *Classify* tab to move to the training and testing area. After selecting this tab you should see something similar to (but not exactly the same as) the screenshot shown in Fig. 4. There are two principal decisions to be made here that will most greatly affect the outcome of our machine learning experiment: what classifier to use, and what kind of testing to do for evaluation. The default classifier is ZeroR; also known as a *majority* classifier because it simply predicts the most frequent class found in the training data—which is often a good thing to check because if 90 % of your data belongs to one class and your machine learning experiment only gets 10 % wrong during testing, then you have not necessarily learned anything interesting because you are not doing any better than you would predicting the majority class.

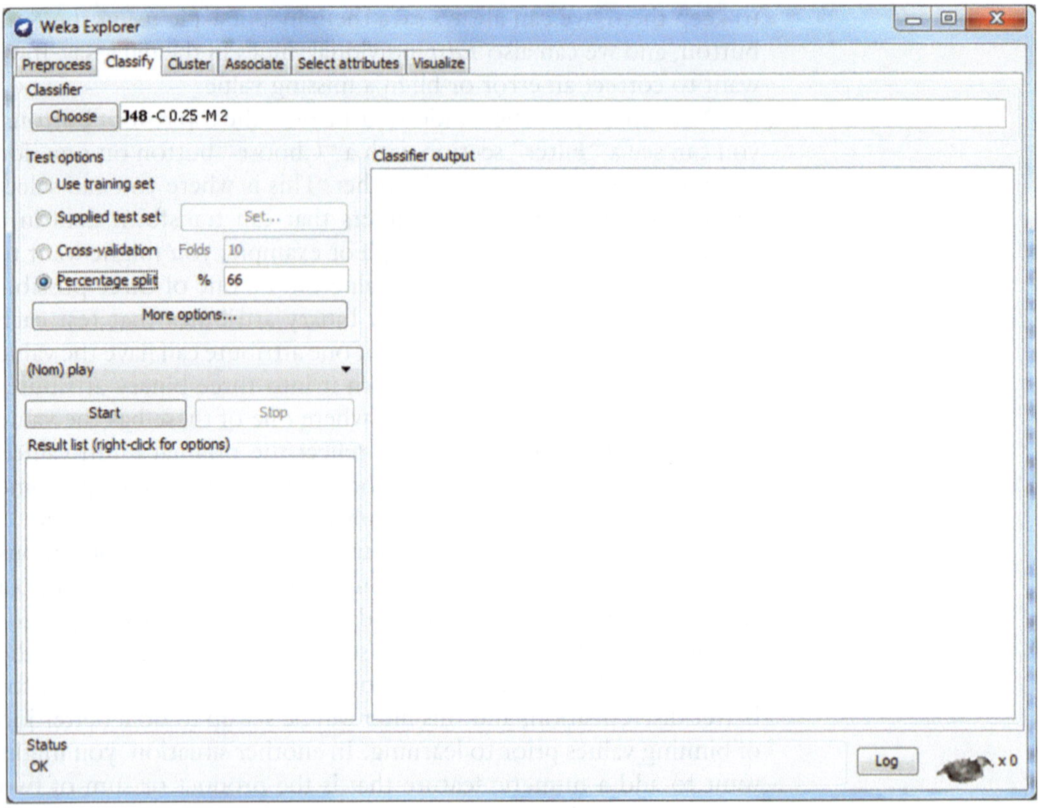

Fig. 4 Ready to learn with J48 and a 2:1 training-test split

For our first experiment, we want to use one of the most commonly used *decision tree* learners: J48 (an open-source version of Ross Quinlan's C4.5 algorithm [2]). Decision trees are easy to build and easy to use. They represent a simple top-down decision procedure, where one simply tests some condition first and, based on the result of that test, chooses another condition to test after that, and so on until you hit the end of a sequence of questions where a decision (i.e., judgment, or prediction) is finally made. It is a kind of *expert system*.

To choose the J48 classifier you click on the *Choose* button which brings up an expandable directory of classifiers. There are well over a hundred algorithms to choose from in WEKA, and they are organized according to the type of output they produce, the type of input they expect, and the type of decision process used to build the model. Some algorithms will be grayed out because they are not available for the kind of input you have prepared. For example, some algorithms cannot handle numeric input, and some produce a numeric prediction instead of a class label. J48 produces a tree, so it can be found under the *tree* subdirectory. Once selected it should appear in the classifier text box as shown in Fig. 4, along with a couple of default parameter settings (i.e., –C

0.25 −M 2, which set the confidence and minimum support required by the algorithm when building the tree). Almost every algorithm has parameters that can be altered to fine-tune or guide its behavior, and these can be accessed simply by clicking on the textbox where the current parameters are displayed. We have to stick with default parameters in this chapter, which are often satisfactory anyway.

2.4 Training and Testing

The next decision we have to make is how we are going to test the final model. You can see a list of *Test options* just underneath where you select the classifier, and the options are *Use training set*, *Supplied test set*, *Cross-validation*, and *Percentage split.*

Obviously, the learner needs examples to learn from. This is the training set. To assess performance of the final model, there needs to be some data that we can make predictions for and then compare those predictions against what is known for these data. This is the test set. If we use all of our data to train up a model, and then use the same data for testing, then we run the risk of over-estimating the value of the model because it has seen the test data during training. For example, if the model managed to memorize every training instance then it may likely get 100 % accuracy when testing on that same data. If the model has any generalization to it, then it may get a few wrong, but either way there are likely to be fewer surprises than if the test data had not been seen during learning. The result obtained by using the training data as test data gives rise to a thing called *resubstitution error*, and is often unjustifiably optimistic for predicting performance of the model on future novel data.

Another option is to have a completely separate test set stored in a separate file that is supplied only during testing. This is useful when the test data might be completely different than the training data in some fundamental way. For example, if one were trying to predict something about gram-negative bacteria based solely upon what is observed in gram-positive bacteria, the gram-positive training data could be loaded into WEKA at the outset to build the classifier, then the gram-negative data could be loaded from a separate file for testing; eliminating any risk the learner had a sneak peek.

Most often, we have one set of data and we must partition it into training and test sets. One way is to simply use a percentage split, using (say) 67 % of the data for training and test on the remaining 33 %. What percentage split to use is hard to generalize. The more data seen for training the better the chance of the learner finding a valid characterization for making predictions. On the other hand, the more data we have for testing, the more likely our success rate reflects the true virtue of the inferred model. Ideally we want an abundance of both training and test data, but in practice we just have whatever data we have.

A good way to make the most of our data is use all our data for training and all our data for testing, but not at the same time (as resubstitution error would ensue). To do this, we divide our data into a number of equal-sized subsets, called folds. For each fold, we remove it from the training set, build a model on the other folds and then test on the withheld portion. If we have *k* folds, then this is called *k*-fold cross-validation. That is, if *k* is 5 then we have fivefold cross-validation. We would withhold the first fold, train on the remaining four, then test on the first (which has not been seen by the learner). Next, we put the first fold back into the training set but take out the second fold and repeat training and testing. This is done once for each of the five folds, such that the test results are always for data not seen by the learner, and all the data is used for testing.

Cross-validation is widely regarded as the most reliable way to judge the results from machine learning when data are all in one set. How many folds should be used? Again, this is difficult to generalize. If too few folds are used then we may deprive the learner of enough data to build a good model. The maximum number of folds we can have would put one datum in each fold—so-called leave-one-out cross-validation (LOOCV)—and performance from this type of testing is on average closest to the true level of performance we can expect from the model on novel data. However, since each fold requires building a new model and evaluating with a new test set, too many folds with a large data set could take far too long to complete in a practical amount of time. A good compromise is to find a tractable value for *k*, then repeat *k*-fold cross-validation some number of times using different folds on each iteration, then average the results.

One final note about cross-validation: care should be taken to make sure that the distribution of classes in the test set is the same as in the training set. Imagine, for example, if 80 % of our data was in one class and 20 % in the other in a two-class problem. If we do fivefold cross-validation and all 20 % that comprise the minority class happen to end up in one fold then the learner would not have a chance to infer what differentiates the two classes. Keeping class distributions the same in training and test data is called stratification, and WEKA seeks to do this automatically. In a multi-class problem, however, if one class has very few exemplars then it may not be possible to maintain stratification if too many folds are used.

2.5 Running the Experiment

The weather data we loaded for this sample experiment has only 14 instances, so we set the test option to threefold cross-validation. Having chosen J48 (with default parameters) as our classifier, we can start the training and test phases in WEKA simply by pressing the START button in the middle of the left panel. For such a small data set, WEKA will complete these phases in the twinkling of an eye. For much larger and/or more complex data sets it could take a

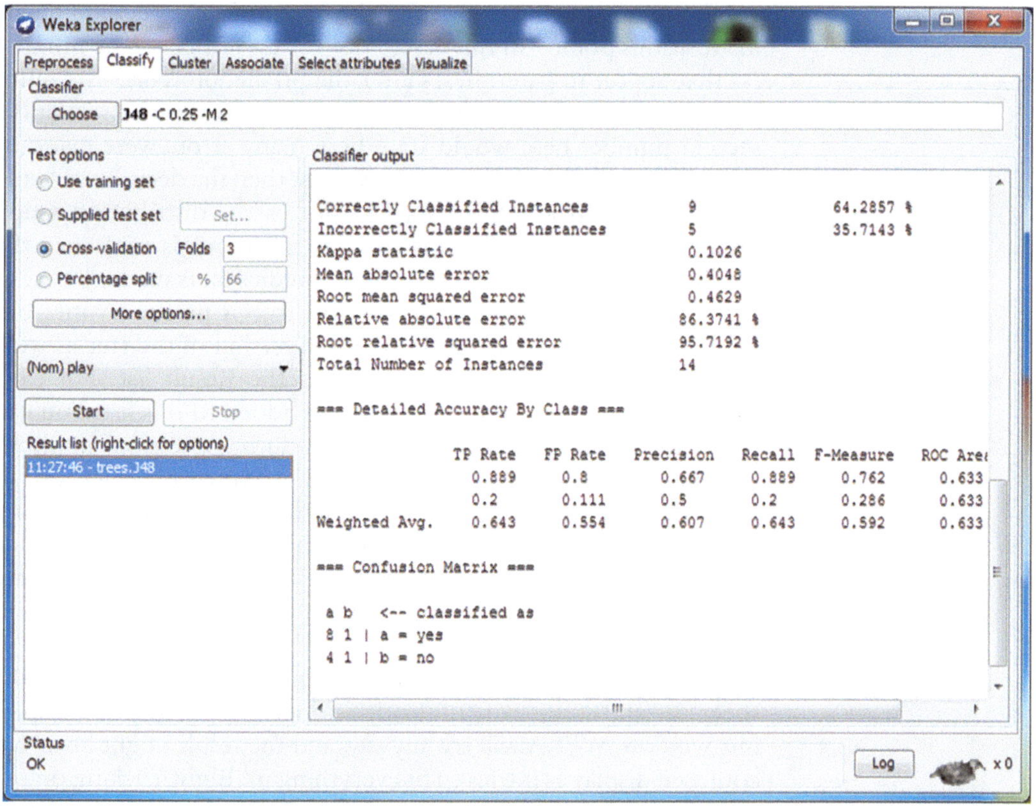

Fig. 5 Output after training and testing on weather data

lot longer. You can see if WEKA is still working because the icon of the bird in the bottom right corner (a picture of an actual Weka) will be in motion. If we notice progress going too slowly to complete in the time we have, we can simply interrupt it by hitting the STOP button, then rejig our parameters (or data) to try and make things tractable.

In this example experiment, WEKA completes in an instant and your display should look pretty much like the one shown in Fig. 5. The text window on the right side is now filled with details about the set-up of the experiment and the results from testing. At the top is a record of the classifier used and the parameter settings, some information about the data set, and the training and test procedure. Under that is a textual representation of the J48 tree. This is a bit obscure for people who are not computer scientists, but the basic method of interpretation is that each level of indentation (indicated by a vertical bar character) represents dropping a level in the tree. The decision criterion for the top level of the decision tree is flush to the left, and in this case it tests the "outlook" attribute. If the outlook is sunny, then evaluation proceeds down one level (indicated by one vertical bar) to test if humidity is greater or less than 75. If greater, then instead of dropping another level and testing another attribute, the colon after the test shows the decision that is made at this point—in this case, "no" . . . do not play. The number (s) that appear in parentheses after the prediction reflect the number of instances that reached this point in the decision tree (and a second number here would state how many errors were made by this prediction). If the outlook is overcast then the decision of "yes" is made without any further tests (just as the distribution graph under the Preprocess tab showed), and if outlook is rainy then the windy attribute is also tested before a prediction is made.

This way of presenting a decision tree is pretty cryptic, but WEKA allows you to see a nice graphic version of the tree as well. On the bottom left of this window is the Result list. For each experiment you run, a new item will be added to this list. You can revisit the result of any experiment run during a session simply by clicking on the one you want to see. If you right-click on an experiment in this list, some other options appear for visualizing the results. To see an actual decision tree for this experiment, right-click on the experiment and select Visualize tree and a separate window will pop up with a picture similar to the one shown in Fig. 6.

For many (if not most) data sets for real world problems, the sheer size and complexity of the decision tree is too much to display in a window. WEKA will try anyway, and the result might be a very confused display as it tries to fit everything in. Right-clicking on the visualization window provides some re-display options that sometimes help.

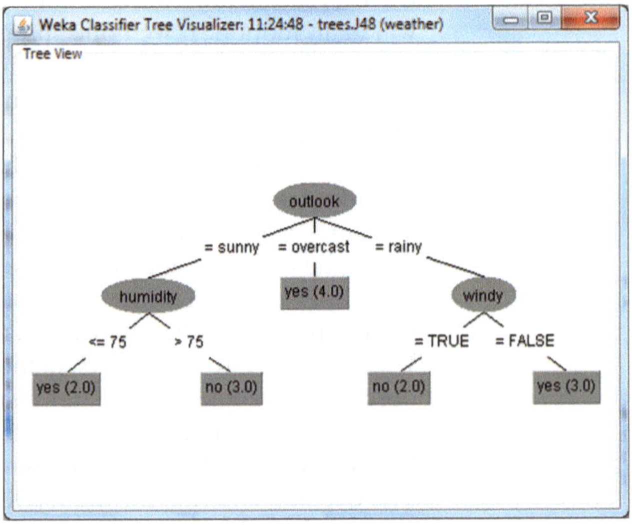

Fig. 6 The decision tree for the experiment from Fig. 5

2.6 Understanding the Results

Apart from a textual representation of the decision tree, the *Classifier output* area contains a number of performance statistics for the learning algorithm, all computed from the test sets used in the evaluation process. Note that the model that is shown—in this case, a decision tree—is built from the entire dataset loaded into the Preprocess panel. Thus, this model itself is not involved in the evaluation process at all (unless evaluation is performed on the training set or a separate test set). Rather, models built from subsets of the full data are used to establish values for the performance metrics. In a *k*-fold cross-validation, these metrics are aggregated across *k* test sets, which are used to evaluate *k* models built from the corresponding training (sub)sets.

Considering the values of the performance metrics, we can see that five instances in the data are misclassified in the cross-validation process. At the bottom of the output, in the *Summary* section, we can see how these errors are distributed across classes, in the so-called "confusion matrix". One instance has been incorrectly classified as class *no*, and four instances have been incorrectly classified as class *yes*. It is often useful to consider how the classifier compares to a random predictor that assigns instances randomly to classes (maintaining the same column totals in the confusion matrix). The kappa statistic, also shown in the *Summary* section, measures exactly this. A value of zero means that the classifier does not improve on the random straw man, which means that it has not learned anything useful from the data. One is the best possible value; it is achieved when all test instances are classified correctly.

WEKA also outputs error statistics, such as the squared error and the absolute error. These are the primary metrics for evaluating regression methods. In classification problems such as the example considered here, they measure the accuracy of the class probability estimates produced by the classifier. (Most algorithms in WEKA can produce a probability estimate for each class value when making a classification, and it can be useful to consider how accurate these probability estimates are.) WEKA also shows the relative absolute error and the root relative squared error. To compute the former, the mean absolute error of the classifier is divided by the mean absolute error obtained when simply predicting class probabilities based on the class frequencies in the training data, without considering any of the non-class attributes. The computation for the root relative squared error is performed analogously. These relative errors are expressed as percentages. Values close to 100 % mean that the classifier has not learned anything useful from the data.

The output area also shows a table of per-class performance statistics, with one row per class value. The corresponding target class for each row is called the "positive" class; all other classes are considered "negative". There is the *TP Rate* (true positive rate), which is the number of instances correctly assigned to the positive class, divided by the number of positive instances in the data; the *FP*

Rate (false positive rate), which is the number of negative instances incorrectly assigned to the positive class, divided by the number of negative instances; *Precision*, which is the number of instances correctly assigned to the positive class, divided by the total number of instances assigned to the positive class; and *Recall*, which is the same as *TP Rate*. The *F-Measure* is the harmonic mean of *Precision* and *Recall*. The last column contains the area under the ROC curve, an estimate of the probability that the classifier ranks a randomly chosen positive instances above a randomly chosen negative one, assuming its class probability estimates for the positive class are used for ranking. If the *ROC Area* is one, all positive instances are ranked above all negative ones—the ideal outcome—if it is zero, the classifier does not improve on randomly shuffling the data.

We have discussed the classification case here. When a regression method is evaluated, the same error statistics are output in the *Classifier output* area. Additionally, WEKA outputs Pearson's correlation coefficient, measuring the correlation between the predicted values and the actual values.

3 Approaching a Bioinformatics Problem

There are about as many bioinformatics problems as there are biologists, but a sizeable number start with primary sequence data. One class of problems has to do with characterizing some site in a set of sequences, such as glycosylation points [3] or inhibitor binding sites [4, 5]. For this example, we address the problem of identifying the cleavage point of the signal peptide.

Often, sequence data is obtained from a database in a FASTA format (or similar text-based form), such as the sample of three signal peptide cleavage records shown in Fig. 7. Each record has three lines where the first includes a description of the example sequence, the second is the primary sequence itself showing the complete signal peptide at the beginning of the nascent protein along with the first 30 residues of the mature protein, and the third line is a character-mask mapping one-to-one with the preceding primary sequence using the letter S to indicate the corresponding residue above is part of the signal peptide, M indicates a residue in the mature protein and C marks the cleavage site as the first residue of the mature protein. The problem is stated in this way: given a new sequence, identify the cleavage site. But, before we can begin mining this data, we must transform it into a format amenable as input to a learning algorithm, such as the spreadsheet form discussed earlier, where each row is an instance and each column is an attribute, one of which is the attribute to be predicted. We assume only the sequence itself will be available when predicting the cleavage point in a future novel instance.

```
  50 11S3_HELAN    20 11S GLOBULIN SEED STORAGE PROTEIN G3 PRECURSOR
MASKATLLLAFTLLFATCIARHQQRQQQQNQCQLQNIEALEPIEVIQAEA
SSSSSSSSSSSSSSSSSSSSCMMMMMMMMMMMMMMMMMMMMMMMMMMMMM
  51 11SB_CUCMA    21 11S GLOBULIN BETA SUBUNIT PRECURSOR.
MARSSLFTFLCLAVFINGCLSQIEQQSPWEFQGSEVWQQHRYQSPRACRLE
SSSSSSSSSSSSSSSSSSSSSCMMMMMMMMMMMMMMMMMMMMMMMMMMMMM
  54 1B39_HUMAN    24 HLA CLASS I HISTOCOMPATIBILITY ANTIGEN, BW-42 B*4201 ALP
MLVMAPRTVLLLLSAALALTETWAGSHSMRYFYTSVSRPGRGEPRFISVGYVDD
SSSSSSSSSSSSSSSSSSSSSSSSCMMMMMMMMMMMMMMMMMMMMMMMMMMMMM
...
```

Fig. 7 Signal peptide cleavage data

If we assume there are some physicochemical properties of the nascent protein that determine the cleavage point, then our goal is to use machine learning to find a characterization of those properties from known examples. Well, really our goal might be simply to predict the cleavage site of a new protein accurately; but, we might instead be *more* interested in the characterization itself (i.e., the model) as it may lead to understanding and new knowledge about the cleavage process. These are two different aspirations and typically suggest we might use different learning algorithms based on what we want at the end. For output with explanatory power, we might want a decision tree like the one in Fig. 6 as they are easy to understand. If we are more interested in predictive accuracy, a support vector machine or random forest is typically a better choice because they do not need to make a decision based on a linear sequence of tests; however, the models they produce are often difficult to interpret and thus not useful as "knowledge".

The next choice we have to make before we prepare the data for learning is how to frame the learning objective as an attribute of each instance. In the case of this cleavage problem, we could try to predict the length of the peptide, and therefore each instance is one complete sequence. This requires a learning algorithm that is capable of predicting a number from properties embodied in an entire sequence. One reason we might not take this approach is that, as mentioned earlier, numeric prediction algorithms tend to seek an estimate with a small error margin and do not often produce exactly the right value. Their performance is evaluated in terms of that margin of error. We need to be right or wrong, not close enough, here.

Another way to view the problem—one that is probably more useful—is to predict whether or not a given residue is the first of the mature protein, which is a binary classification problem. In this case, each instance is one residue at a given point in the sequence.

Having framed the objective, we now must generate a set of features to describe each instance. If we pursue the binary classification approach then we want to characterize each residue with attributes that we think are relevant to whether or not it marks the cleavage point. If we have no idea what properties might be relevant then we may start naively using whatever we can think of. For example, if we suspect that the residues on either side of the cleavage point play a critical role, then we might simply create (say) six features that record the residue at each of the three positions upstream and downstream from the cleavage site. As it happens, distributional analysis of residue frequency in known peptide data shows alanine and serine are commonly found in the -1 and -3 positions (i.e., one and three residues upstream of the cleavage site), and a couple of other residues are also unusually frequent at these positions, so some traction might be gained from these features. However, given 20 possible residues at each of these six positions, there are 20^6 combinations of values that could be seen, times two for the class attribute (i.e., YES or NO, this is a cleavage site), or 128 million possible instances in a comprehensive data set. In practice, one would be lucky to have a couple of thousand instances to learn from, presenting potential for a *data sparseness* problem, where so many combinations are not seen by the learner that it is almost trivial to find some characterization that is pretty accurate simply because there are no instances to disprove it. This is not to say that such a characterization is necessarily wrong. If the data we possess happens to contain the necessary information for inferring a correct generalization of the underlying principle, it is said to be a "characteristic set". The problem is that when data sparseness exists it is virtually impossible to tell whether an accurate model is correct or just lucky.

Biologists typically know *something* about the data that has a reasonable chance of factoring into a good generalization, and this is a good basis for formulating a set of features. For example, it is known that signal peptides are pretty short, having an average length of about 23 residues. Thus it might be useful to have a feature that records how far the residue in question is from the start of the nascent protein; but, this too might yield too many numeric values to generalize easily. Given that the vast majority of signal peptide lengths are within six or seven of the mean, it may be more fruitful to have a binary feature that simply records if the residue in question lies within this range; or, one might record how many standard deviations from the mean it is, thus reducing the range of possible values and mitigating possible sparseness.

Rather than using specific amino acid labels as feature values, one could reduce the combinatoric explosion by recording more general physicochemical properties for them. This is also helpful because the hypothesis we seek from the data mining process is one that generalizes according to such properties anyway. Moreover, existing research about signal peptides has identified a number of regularities based on such properties. For example, the region of about six residues upstream of the cleavage site is generally not very hydrophobic, while the region of about eight residues upstream of that is typically quite hydrophobic. Thus we might have two binary features, one for each of these regions, whose value is "true" if the region is hydrophobic or "false" if it is not. Or each could be a numeric attribute quantifying just how hydrophobic a region is.

Once a set of features has been devised to describe each instance in our dataset, there is the challenge of actually producing them. Functions and macros in spreadsheets can often do the job, or any programming language will suffice; but obviously some modicum of computer programming skill is required. Sometimes the best solution for a biologist is just to make friends with a good computer scientist.

One more issue needs to be addressed, however, before we can proceed to computing the final data set. Successful learning needs not only positive instances of the concept to be learned, but negative ones as well, in order to eliminate overly general models like "predict every residue marks the cleavage site", which will be 100 % accurate on our data. If negative instances are available, then great. But many biological studies end up with just samples of the thing demonstrating the characteristic of interest.

In the signal peptide problem, negative instances are simply every other residue in our sequences that does not mark the start of a mature protein. Given the lengths of the sample sequences, this gives us about 75 times as many negative instances as positive ones. If we give them all to the learning algorithm there is a good chance that it will have a difficult time inferring a model that does better than simply predicting the majority class (i.e., predict that every residue is NOT at the cleavage site and the prediction is wrong only 1 in 75 times). In a binary classification problem, it is generally good to have the same number of positive and negative instances.

In this specific case, we might want to create negative instances by choosing residues *near* the cleavage point so that the properties are similar to positive instances, forcing the learner to find a discriminating model. It is probably also a good idea to generate several samples of negative data and run the machine learning experiment with each to avoid putting too much stock in an outcome that only works well serendipitously for the particular sample of negative instances we selected.

Once all features have been generated for all the data, it can be saved in a comma-separated-values (CSV) file and this can be loaded directly into WEKA. When WEKA imports a CSV file, it assumes the first line of input is a list of feature names/labels (which is typically the case for a CSV file saved from a spreadsheet) that it uses to name each attribute in the "Preprocess" window and in the resulting output after learning. The rest of the lines are taken to be instances. Recall, however, that it is important to differentiate between nominal and numeric features. When reading a CSV file, WEKA attempts to parse every value as numeric, and if this succeeds for every value associated with a feature then that feature is assumed to be numeric, otherwise it is assumed to be nominal. There are many situations where this assumption can fail, such as when one feature-value that is supposed to be numeric has a stray character that prevents it from being parsed as a number, then the whole set of numbers associated with that feature are treated as nominal labels. Another situation is when a numeric value is not really being used as a number and is better treated as being nominal, as with an identification code. To solve the first problem, you're best to go back to the procedure that generated the data and fix the error. For the second type of problem, WEKA needs to be informed that what looks like a numeric attribute is actually a nominal one. WEKA's native file format for data is the AARF format, which is very similar to a CSV file but has some header information specifying types and ranges of data (and some other things). If loading a CSV file leads WEKA to misinterpret data types then a quick way to fix this is to save the data (using the "Save" button on the "Preprocess" tab) which generates an *ARFF* version that you can then edit directly to change type information.

4 Other Kinds of Learning

WEKA provides a uniform user interface to a large number of machine learning algorithms. This is useful because it is often difficult to determine a priori which algorithm is most suitable for a particular dataset and will produce the highest accuracy or most pertinent patterns. Decision tree learning is attractive because trees can be grown very quickly, even for large datasets, and may provide valuable insights. A closely related approach is to learn classification rules [6]: sets of simple if-then rules that describe the conditions (i.e., attribute-value combinations) under which a certain classification is to be made. Rule sets can be learned almost as quickly as decision trees but may provide a more compact representation of the salient relationships in the data. In WEKA, algorithms of this type can be found in the "rules" package, just as tree learners can be found in the "trees" package.

Trees and rules are not the only models that are intelligible and, thus, potentially able to provide useful insights. Another option is to learn a graphical model of the probability distribution underlying the data. A Bayesian network (or belief network) is a type of graphical model in the form of a directed graph [7]. Each node in this graph corresponds to an attribute in the data and directed edges connecting these nodes encode statistical dependencies between the attributes. Cycles are not allowed in this graph.

If there is a directed edge from node A to node B, node A is called a "parent" of node B. Each node has a conditional probability distribution attached to it that gives the probability of each of the node's attribute values given those of its parent nodes. Learning a Bayesian network amounts to determining suitable parent nodes for each node and estimating a conditional probability distribution based on the chosen parents. With nominal data, estimating the latter is as simple as counting how often a particular attribute value occurs for each combination of values of the parent nodes. The trickier part is to determine appropriate connections between nodes; various search methods are employed for this. The goal is to find an accurate model with as few edges as possible because experience has shown that simpler networks are more likely to reflect causal relationships between attributes in the data.

Bayesian networks can be used for supervised learning as well as for unsupervised learning. WEKA employs them for the former task, where they are used to estimate probabilities for the class attribute. A particularly simple form of Bayesian network is the naive Bayes model [8], which has a predetermined structure and only requires estimation of conditional probabilities. In this model, each attribute, excluding the class attribute, has exactly one parent, namely the class attribute. This corresponds to the assumption that the values of the other attributes are statistically independent given a particular value for the class attribute (i.e., considering the data for a particular class, knowing the value for one attribute does not give us any information regarding the values of other attributes). Learning a naive Bayes model is thus extremely fast. Moreover, in spite of its simplicity, it often yields accurate classifications, making it a very popular choice amongst practitioners. General Bayesian networks, as well as naive Bayes, are available in the "bayes" package for WEKA.

The above techniques can provide valuable insight, but in some applications the goal is to simply maximize predictive accuracy on new data. In this case, learning algorithms from WEKA's "functions" package are worth investigating. This package provides algorithms for learning basic linear models, such as linear regression and logistic regression, but also contains several more sophisticated methods. Two very prominent models implemented in this package, which often yield high accuracy, are support vector machines [9] and multilayer perceptrons [10]. Both are models that can be expressed as standard mathematical functions. Given an input instance, they produce an output by applying a sequence of mathematical operations. In the regression case, this output can be directly used as the prediction; in the classification case, it needs to be compared to a threshold to determine what the classification should be. A multilayer perceptron is a type of artificial neural network, where the component functions—the nodes of the network—mimic the behavior of neurons in the brain. The number of nodes in the network, their arrangement into layers, and the connections between them, are predefined by the user. Learning the network amounts to determining appropriate parameter values for the component functions. This is done by applying a gradient-based optimization technique to minimize the error on the training data, starting with randomly chosen parameter values.

A conceptual drawback of multilayer perceptrons is that the final model found using gradient-based optimization depends on the initial parameter values because the error function contains local minima. Support vector machines are theoretically appealing because their optimization process converges to a unique solution. Just like multilayer perceptrons, they can be used to represent nonlinear relationships in the data. This is possible because the model can be expressed in terms of dot products between instances. The training instances that form part of the solution expressed in this way are called the "support vectors". Support vector machines are able to model nonlinear relationships by replacing the simple dot product with a so-called "kernel function." A kernel function implicitly maps the instances into a higher-dimensional space and then takes the dot product in this space, yielding nonlinear behavior in the original instance space. It is best to think of kernel functions as similarity functions with this particular property. Different kernel functions exist and can be plugged in when configuring the learning algorithm in WEKA. A popular kernel function that often works well is the Gaussian radial basis function kernel.

The idea of using similarity functions, or distance functions, for machine learning has been around much longer than support vector machines. One of the oldest machine learning methods is the nearest neighbor classifier [11]. The basic version simply stores the training data; to make a prediction for a test instance, it simply finds the most similar training instance and predicts its class value. This can be made more robust by combining class values from the *k* most similar instances, yielding the *k* nearest neighbor classifier. Because the basic version of this method simply stores the training data and only does serious computation when predictions are to be made, it is called a "lazy" learning method. Implementations of this method, and closely related methods, are to be found in WEKA's "lazy" package. WEKA supports various distance functions to measure similarity and also provides advanced data structures such as KD trees [12] to speed up the search for the nearest neighbor. A drawback of all lazy learning methods is that they do not generate a model that provides insight into the data.

The biggest variety of supervised learning algorithms is located in WEKA's "meta" package. It contains algorithms that wrap around user-specified base learners, which are themselves supervised learning algorithms. For example, it contains algorithms for bagging [13], boosting [14], and randomization [15] that all generate a so-called "ensemble" classifier by repeatedly invoking a base learner, e.g., a decision tree learner. The resulting ensemble classifier is often significantly more accurate than the base learner by itself. However, any kind of interpretability will be lost.

Apart from ensemble classifiers, the meta package also provides several useful wrappers for tasks such as performing cost-sensitive learning, combining multiple different learning algorithms by a simple voting process or a more powerful "stacking" approach [16]—where the process of combining their predictions is phrased as another learning problem—and performing attribute selection prior to learning. Cost-sensitive learning is useful in practice when different types of classification errors incur different costs. This can be accommodated by adapting the model itself [17], or its predictions [18]. Attribute selection [19] is useful when the data contains a large number of attributes and only a small subset is likely to be of predictive value.

The vast majority of algorithms in WEKA perform supervised learning (i.e., classification or regression), but the workbench also contains algorithms for clustering and association rule mining (performed interactively in WEKA by selecting the appropriate tab at the top of the window). Classical algorithms for clustering are *k*-means [20], which yields a flat set of clusters represented by cluster centroids, and hierarchical clustering [21], which yields a dendrogram of clusters that can be displayed graphically. A probabilistic alternative to k-means is to fit a Gaussian mixture model to the data [22], where each Gaussian represents a cluster. The seminal Apriori algorithm for association rule mining [23] is also available in WEKA, with several other, more recent, algorithms for mining such rules efficiently by applying sophisticated data structures to speed up the search for frequent patterns in the data.

5 Concluding Remarks

Biological research yields tremendous amounts of data; machine learning provides tools that can help to make sense of this data and turn it into actionable models. The content of this chapter provides a brief, high-level introduction to basic concepts and algorithms in machine learning, using the WEKA workbench to illustrate how these can be applied. WEKA makes it relatively easy to enter the world of machine learning because it provides graphical user interfaces that do not require any scripting or serious programming. Readers who do not mind getting their hands dirty by writing scripts may also want to take a look at machine learning libraries for the statistical computing environment R [24] or the scikit-learn library for the programming language Python [25]. Just like WEKA, they come with extensive documentation.

References

1. Witten IH, Frank E, Hall MA (2011) Data mining: practical machine learning tools and techniques, 3rd edn. Morgan Kaufmann, Burlington, MA
2. Ross Quinlan J (1993) C 4.5: programs for machine learning. Morgan Kaufmann, San Mateo, CA
3. Blom N, Sicheritz-Pontén T, Gupta R, Gammeltoft S, Brunak S (2004) Prediction of post-translational glycosylation and phosphorylation of proteins from the amino acid sequence. Proteomics 4:1633–1649
4. Ramana J, Gupta D (2010) Machine learning methods for prediction of CDK-inhibitors. PLoS One 5(10):e13357
5. Buchwald F, Richter L, Kramer S (2011) Predicting a small molecule- kinase interaction map: a machine learning approach. J Cheminform 3:22
6. Fürnkranz J (1999) Separate-and-conquer rule learning. Artif Intell Rev 13(1):3–54
7. Friedman N, Geiger D, Goldszmidt M (1997) Bayesian network classifiers. Mach Learn 29(2-3):131–163
8. Domingos P, Pazzani M (1997) On the optimality of the simple Bayesian classifier under zero-one loss. Mach Learn 29(2–3):103–130
9. Cortes C, Vapnik V (1995) Support-vector networks. Mach Learn 20(3):273–297
10. Rumelhart DE, Hinton GE, Williams RJ (1986) Learning representations by back-propagating errors. Nature 323(9):533–536
11. Cover TM, Hart PE (1967) Nearest neighbor pattern classification. IEEE Trans Inform Theory 13(1):21–27
12. Friedman JH, Bentley JL, Finkel RA (1977) An algorithm for finding best matches in logarithmic expected time. ACM Trans Math Softw 3 (3):209–226
13. Breiman L (1996) Bagging predictors. Mach Learn 24(2):123–140
14. Freund Y, Schapire RE (1996) Experiments with a new boosting algorithm. International conference on machine learning. Morgan Kaufmann, Bari, Italy
15. Dietterich TG (2000) Ensemble methods in machine learning. Multiple classifier systems. Springer, Berlin, pp 1–15
16. Wolpert DH (1992) Stacked generalization. Neural Netw 5(2):241–259
17. Ting KM (1998) Inducing cost-sensitive trees via instance weighting. Principles of data mining and knowledge discovery. Springer, Berlin, pp 139–147
18. Duda RO, Hart PE (1973) Pattern classification and scene analysis, vol 3. Wiley, New York
19. Kohavi R, John GH (1997) Wrappers for feature subset selection. Artif Intell 97 (1):273–324
20. Hartigan JA (1975) Clustering algorithms. Wiley, New York
21. Johnson SC (1967) Hierarchical clustering schemes. Psychometrika 32(3):241–254
22. McLachlan GJ, Basford KE (1987) Mixture models: inference and applications to clustering. CRC, New York
23. Rakesh A, Srikant R (1994) Fast algorithms for mining association rules. International conference on very large databases. Morgan Kaufmann, Santiago de Chile, Chile
24. Ihaka R, Gentleman R (1996) R: a language for data analysis and graphics. J Comput Graph Stat 5(3):299–314
25. Pedregosa F, Varoquaux G, Gramfort A, Michel V, Thirion B, Grisel O, Blondel M, Prettenhofer P, Weiss R, Dubourg V, Vanderplas J, Passos A, Cournapeau D, Brucher M, Perrot M, Duchesnay E (2011) Scikit-learn: machine learning in Python. J Mach Learn Res 12:2825–2830

Chapter 18

Experimental Design and Power Calculation for RNA-seq Experiments

Zhijin Wu and Hao Wu

Abstract

Power calculation is a critical component of RNA-seq experimental design. The flexibility of RNA-seq experiment and the wide dynamic range of transcription it measures make it an attractive technology for whole transcriptome analysis. These features, in addition to the high dimensionality of RNA-seq data, bring complexity in experimental design, making an analytical power calculation no longer realistic. In this chapter we review the major factors that influence the statistical power of detecting differential expression, and give examples of power assessment using the R package *PROPER*.

Key words RNA-Seq, Gene expression, Sample size, Statistical power, Experimental design

1 Introduction

RNA-sequencing (RNA-seq) has become a routine technique in transcriptome analysis, where identifying differential expression (DE) remains a major task. As expression is a stochastic process in nature, the technological improvement in RNA-seq cannot bypass the presence of biological variability. It has been well recognized that replication is still necessary in making reliable statistical inference of DE [1]. The determination, or choice, of the number of replicates becomes a natural question in experimental design. In this section, we review the major factors affecting statistical power calculation in DE detection using RNA-seq. Specific examples of power assessment using the R package *PROPER*, released by Bioconductor [2], will be given in Subheading 3.

Classical power calculation that deals with a single hypothesis takes a few simple assumptions. These include (1) the effect size, representing the minimum difference that is scientifically meaningful between groups in comparison; (2) within-group variation, representing natural variation in observations regardless of between-group difference; (3) an acceptable type I error rate, usually in the form of *p*-value; and (4) the sample size. With these

Ewy Mathé and Sean Davis (eds.), *Statistical Genomics: Methods and Protocols*, Methods in Molecular Biology, vol. 1418, DOI 10.1007/978-1-4939-3578-9_18, © Springer Science+Business Media New York 2016

values set, one can calculate statistical power, the probability of rejecting the null hypothesis when the effect is as large as assumed. If a certain power level is desired, one can also do a reverse calculation to determine the minimum sample size to achieve the desired power while controlling the type I error rate.

In DE analysis for RNA-seq experiments, we consider similar factors with more complexity since it is a high throughput experiment querying all transcripts simultaneously, and these transcripts are not exchangeable. Below we discuss the main factors affecting the power calculation in RNA-seq:

1. **Average sequencing count levels**. Sequencing can be considered as a counting process. For a gene, the total variation we observe in its read counts from biological replicates reflects both the fluctuation of gene expression (biological variation) and the counting error from sequencing process (technical variation). The technical variation can be well approximated by a Poisson distribution. When counts are low, the variation due to Poisson counting error can easily shadow true DE we are interested in detecting. The average count level for a gene depends on its expression level, the sequencing efficiency (GC content and gene length [3] affect the number of reads yielded from a gene, for example), and the sequencing depth for the entire sample. Sequencing depth can be a factor of choice, at least to some extent. The distribution of expression levels includes which genes are being transcribed as well as the quantity of their transcription. This depends on the transcriptome of study and should not be arbitrarily specified.
2. **Natural variation of gene expression**. Even housekeeping genes do not have constant expression levels, so there is a natural variation of expression level between biological replicates. Gene expression varies differently for each gene, thus the ability to detect DE at a certain fold change varies between genes. In RNA-seq data this is often parameterized as dispersion parameter in a negative binomial model, which has close relationship to the variance parameter in log transformed expression [3].
3. **Type I error control**. The number of true positives identified is closely related to the amount of false positive (type I error) one can tolerate. In high throughput experiments like RNA-seq, the most widely used choice of type I error is false discovery rate (FDR), representing the expected proportion of false discoveries among all declared discoveries (positives). Controlling FDR takes into account of multiple testing, and yet is not overly conservative as controlling the family-wise error rate (the probability of making *any* false discovery). However, we emphasize the difference between nominal FDR and actual

FDR. Most DE detection methods provide only an estimate of FDR, which does not always provide adequate control of the FDR.

4. **Distribution of DE**. Abundant and strong signals are easier to detect than sparse and weak signals. What fraction of genes have DE between the conditions we plan to compare and contrast? For those with DE, how much do they differ? The magnitude of DE is often different between genes, with some genes undergoing dramatic changes and others subtle differences. Thus the distribution of DE size in addition to the fraction of genes with DE affects statistical power.
5. **Goal in DE detection**. The probability of detecting all genes with DE is often very low, especially considering that some DE genes experience only subtle differences. Are we interested in all genes that have any degree of DE, or are we only interested in DE beyond certain magnitude? Are we interested in power in terms of the proportion of true DE genes identified, or the actual number of true DE genes identified? When a large number of genes have DE, even a small fraction is a considerable number.

Making explicit assumptions on all of these factors is challenging, if not unreasonable. One more issue that complicates the sample size determination is that most DE detection methods, adequately, take advantage of the high throughput nature of RNA-seq data and apply some empirical Bayes techniques [3–6], creating dependency among the genes. On the other hand, to make the computation tractable, many sample size determination methods rely on strong and overly simplified assumptions. For example, some simplify by considering single gene expression [7, 8], some simplify by using Poisson model [9], or by setting the effect size, dispersion, and average count the same for all genes [10]. We argue, however, these overly simplified assumptions do not faithfully reflect the complexity of RNA-seq data. We advocate power evaluation in a comprehensive manner, under various scenarios of sample size and sequencing depth, rather than fixing too many high-dimensional factors and producing a single power curve.

2 Materials

As discussed in Subheading 1, multiple factors influence the statistical power in an RNA-seq experiment, including the transcriptome baseline profile (the number of genes, baseline expression levels, natural biological variation within group), target signal (proportion of genes with DE, the magnitude of DE), and technical choice (the number of samples, sequencing depth). Rather than making specific strong assumptions on all of these, it makes much more sense to relate the overall profile, such as the distributions of baseline expression and dispersion, in the planned experiment to an existing real data set. Thus we encourage the use of actual RNA-seq data sets as the basis for simulation. The *PROPER* package provides information extracted from several RNA-seq data sets, including

- Cheung data: lymphoblastoid cell lines from 41 CEU individuals in International HapMap Project. The individuals represent a random sample from a population, thus the expressions show relatively large biological variations.
- Gilad data: human liver sample comparisons between male and female. The biological variations are smaller than those from Cheung data.
- Bottomly data: samples from two strains of inbred mice. Since the data are from inbred animal models, the biological variations among replicates are much smaller.
- MAQC data: benchmark data sets generated for the quality control of the sequencing technology. The replicates are technical replicates so there is no biological variation. These samples represent the lower bound of dispersion observed in RNA-seq data.

These data sets are chosen to represent different levels of dispersion distributions, which DE detection is sensitive to. Most of the real RNA-seq data are expected to have dispersions in between these. Another useful source of RNA-seq data summarized as count tables is ReCount [11]. In addition, pilot data set with count tables generated by investigators, if available, are probably the best to be used for establishing simulation scenarios.

3 Method

In this section we provide detailed examples of using the *PROPER* package to perform in silico experiments and evaluate power in realistic settings.

3.1 Obtaining PROPER

PROPER is a free and open-source R package released via the Bioconductor project. To install *PROPER*, start R and use the installation script provided by Bioconductor by entering

```
source("http://bioconductor.org/biocLite.R")
biocLite("PROPER")
```

The installation only needs to be run once. Packages that PROPER depends on will be automatically installed as well. Once installed, load the package with

```
library(PROPER)
```

Below we give examples for the most common situation of two group comparison. To give the users more flexibility, every function has options beyond we can include in the examples below. We encourage the readers to explore those with the function helps distributed with the package.

3.2 Setting Up Simulation Scenario

3.2.1 Using PROPER Provided Simulation Parameters

- **Simple Setting** Suppose we are considering a transcriptome with 20,000 genes (this depends on the species) and we think the baseline expression level (*lBaselineExpr*) and gene expression variation (*lOD*) are similar to those seen in the Cheung data (a random population of individuals). For signal we expect 5 % of all genes are DE, but we are not sure about the fold changes of each DE gene, thus we leave it at the default setting (normal with mean 0 and standard deviation 1.5). The following commands set up simulation option with these assumptions (note that we omit the magnitude parameter *lfc* since we choose the default setting).

```
sim.opts1 = RNAseq.SimOptions.2grp(ngenes = 20000, p.DE=0.05,
  lOD="cheung", lBaselineExpr="cheung")
```

One does not have to assume that the baseline expression and the variation of expression are based on the same source. For example, if our experiment involves more homogeneous population, we may expect the variation to be closer to that observed in inbred animals. We can simply change the *lOD* option alone to "*bottomly*" by

```
sim.opts2 = RNAseq.SimOptions.2grp(ngenes = 20000, p.DE=0.05,
  lOD="bottomly", lBaselineExpr="cheung")
```

- **Change the DE distribution** As mentioned above, the default setting for the magnitude of DE is a random variable from normal distribution with mean 0 and standard deviation 1.5. A normal random variable with mean 0 is a common choice in simulating the amount of differential expression. This means that among genes that do have differential expression, most of the effect sizes are near zero. Depending on the biological system under study, the user may choose other distributions of effect sizes that are more reasonable for their situation. This is done by passing an argument of *lfc*. For details see *?RNAseq.SimOptions.2grp*. Here we provide two examples below.

  To assume that all DE genes have differential expression of twofold change, while assuming the baseline expression and dispersion resemble those in the *Bottomly* data, we can set

```
sim.opts3 = RNAseq.SimOptions.2grp(ngenes = 20000, p.DE=0.05,
  lOD="bottomly", lBaselineExpr="bottomly", lfc=log (2))
```

A user can also specify a distribution for the magnitude of log fold change (*lfc*). This is done by passing a user-specified function as the *lfc* option. For example, to assume that the magnitude of log fold change (*lfc*) is centered around 1 with standard deviation 0.2, and that the up- and down-regulation is balanced, one may define the following function ***mylfc***.

```
mylfc=function(n){
  rnorm(n,1,.2)*sign(rbinom(n,1,.5)-0.5)
}
# to visualize what data the function mylfc generates you may try
# hist(mylfc(1000))
sim.opts4 = RNAseq.SimOptions.2grp(ngenes = 20000, p.DE=0.05,
  lOD="bottomly", lBaselineExpr="bottomly", lfc=mylfc)
```

3.2.2 Using User Identified Data Source to Set Up Simulation Basis

If we want to use an independent source as pilot data to establish the simulation basis instead of using one of the data sources provided in PROPER, it can be estimated by the ***estParam*** function. The data input can simply be a matrix of sequencing counts or an *ExpressionSet*.[1] For the pilot data object ***myCountMatrix***, do

```
my.param = estParam(myCountMatrix)
sim.opts5 = RNAseq.SimOptions.2grp(ngenes = 20000, p.DE=0.05,
  lOD=my.param$lOD, lBaselineExpr=my.param$lmean)
```

3.3 Running Simulation and DE Detection

Once the simulation scenario is prepared, we are ready to perform experiments in silico and evaluate how well the signals can be detected. This is done with the ***runSims*** function, which generates RNA-seq count data sets at the user-specified sample sizes and simulation settings, analyzes the data sets, and reports with a DE detection method the user chooses. The process will be repeated for a number of (user-specified) times. Since many RNA-seq data sets will be analyzed, this is computationally the most intensive part. However this only needs to be run once under a particular setting, and then different types of power-related quantities can be calculated. *PROPER* currently offers options to use *edgeR*, *DSS*, and *DESeq* as DE detection method. To install *DSS*, run

```
source("http://bioconductor.org/biocLite.R")
biocLite("DSS")
```

In the following example, we simulate data sets with the number of replicates in each group varying at 3, 5, 7, or 10, with the first simulation option we created in Subheading 3.2.

```
simres1 = runSims(Nreps = c(3, 5, 7, 10), sim.opts=sim.opts1,
  DEmethod="edgeR", nsims=20)
```

[1] ExpressionSet is a basic class of object used in Bioconductor. See http://www.bioconductor.org/packages/release/bioc/vignettes/Biobase/inst/doc/ExpressionSetIntroduction.pdf for more details

As a quick example, the above command used only 20 simulations (*nsims=20*). This is the minimum recommended number, which can be used for exploratory analysis or fast comparison between several *simOptions* settings. For a final determined simulation scenario, we recommend *nsims* 100 or above.

```
simres1b = runSims(Nreps = c(3, 5, 7, 10), sim.opts=sim.opts1,
  DEmethod="edgeR", nsims=100)
```

The rest of the chapter will use results from simulation output in *simres1b* for illustration.

3.4 Power Assessment and Comparison

Now that the simulation is completed and we are ready to evaluate power: the ability to identify true DE genes of interest. As in any analysis, there is some probability to produce false positives—genes that are not DE but falsely identified as DE genes, i.e. the type I errors. In order to evaluate the ability to detect true positives, we have to specify the level of false positives we are willing to put up with. The user decides both the type (FDR or raw *p*-value, by setting *alpha.type="fdr"* or *alpha.type="pval"*) and level of type I error control (*alpha.nominal*). Since we are dealing with high throughput experiments that query thousands of genes simultaneously, and in RNA-seq data the power is not uniform for every gene, there are often true but trivial DE that we do not find relevant. So the user also defines the DE signal of interest by setting the minimum magnitude of differential expression (via the option *delta*). The power to detect these signals of interest is referred to as the "*targeted power.*"

The main function for power evaluation is *comparePower*, with default settings *alpha.type="fdr", alpha.nominal=0.1,delta=0.5.* For example, the statistical power assessment using the previous simulation result *simres1b* is obtained by

```
power1b = comparePower(simres1b)
summaryPower(power1b)
   Sample size Nominal FDR Actual FDR Marginal power Avg # of TD Avg # of FD FDC
[1,]        3        0.1       0.40           0.27          43          29 0.67
[2,]        5        0.1       0.22           0.41          68          20 0.29
[3,]        7        0.1       0.15           0.50          83          15 0.18
[4,]       10        0.1       0.10           0.58          98          12 0.12
```

The summary results above give marginal values of errors and sensitivity.

Interpretation of the Marginal Power Summary Table We use the above table as example to illustrate the interpretation of marginal power and error rates. With replicate number 3, we see that FDR is not well controlled by the edgeR method in this simulation. Though we aim to control FDR at 10 % level (nominal FDR), the actual FDR is as high as 40 %. Even with this high level of false discovery, only 27 % of true DE genes with log fold change over 0.5 will be detected on average. We emphasize though, due to the large number total genes (20,000 in this example), even a small proportion may be a considerable number of genes. On average we can identify about 43 true DE genes with the cost of 29 false positives, make the false discovery cost (FDC) 0.67 for each true discovery. With the increase of samples size, as expected, power improves and cost of false discovery drops to 0.12 at 10 replicates each group.

The above example only shows power computed at default settings. In practice, the user may choose different nominal error rate and/or definition for DE of interest. For example, if one wants to re-evaluate power using raw p-values at 0.00001 level (this is lower than classical alpha level 0.05 since we have massive multiple testing), and target DE with at least twofold change, one can simply do

```
power1b.update = comparePower(simres1b,alpha.nominal=1e-5, alpha.type="pval",
  delta=log(2))
summaryPower(power1b.update)
Sample size     Nominal     Actual Marginal  Avg # of TD Avg # of FD  FDC
          type I error type I error   power
[1,]    3        1e-05       0.00150    0.12          16         6.00 0.3700
[2,]    5        1e-05       0.00062    0.26          37         2.60 0.0680
[3,]    7        1e-05       0.00036    0.36          53         1.50 0.0280
[4,]   10        1e-05       0.00012    0.47          69         0.52 0.0075
```

Stratified Targeted Power The marginal power table is a concise summary, but we urge the users to visualize the stratified power. This is important for RNA-seq experiments because the coverage on genes varies a great deal, and coverage is a crucial factor affecting statistical power. Even if one is not interested in genes with very low copies of transcripts, the variation in sequencing efficiency means that low coverage is not necessarily a sign for low expression. To visualize stratified power for the analysis output *power1b*, simply do

```
plotPower(power1b)
```

The *plotPower* function shows stratified power for each strata of expression level, as shown in Fig. 1a. Not surprisingly, the power is low in the first stratum, when average counts are below 10. When counts are as low as such, meaningful biological variation is hard to distinguish from Poisson counting error. However, we see that when counts are above 10, the power is much more acceptable.

Depending on the purpose of the RNA-seq experiment, we may hope to identify most of the true DE genes of interest (thus aiming for a high percentage, i.e., average power), or we may be interested in identifying a number of leads in a hypothesis generating project. In the latter case, the crude number of DE genes rather

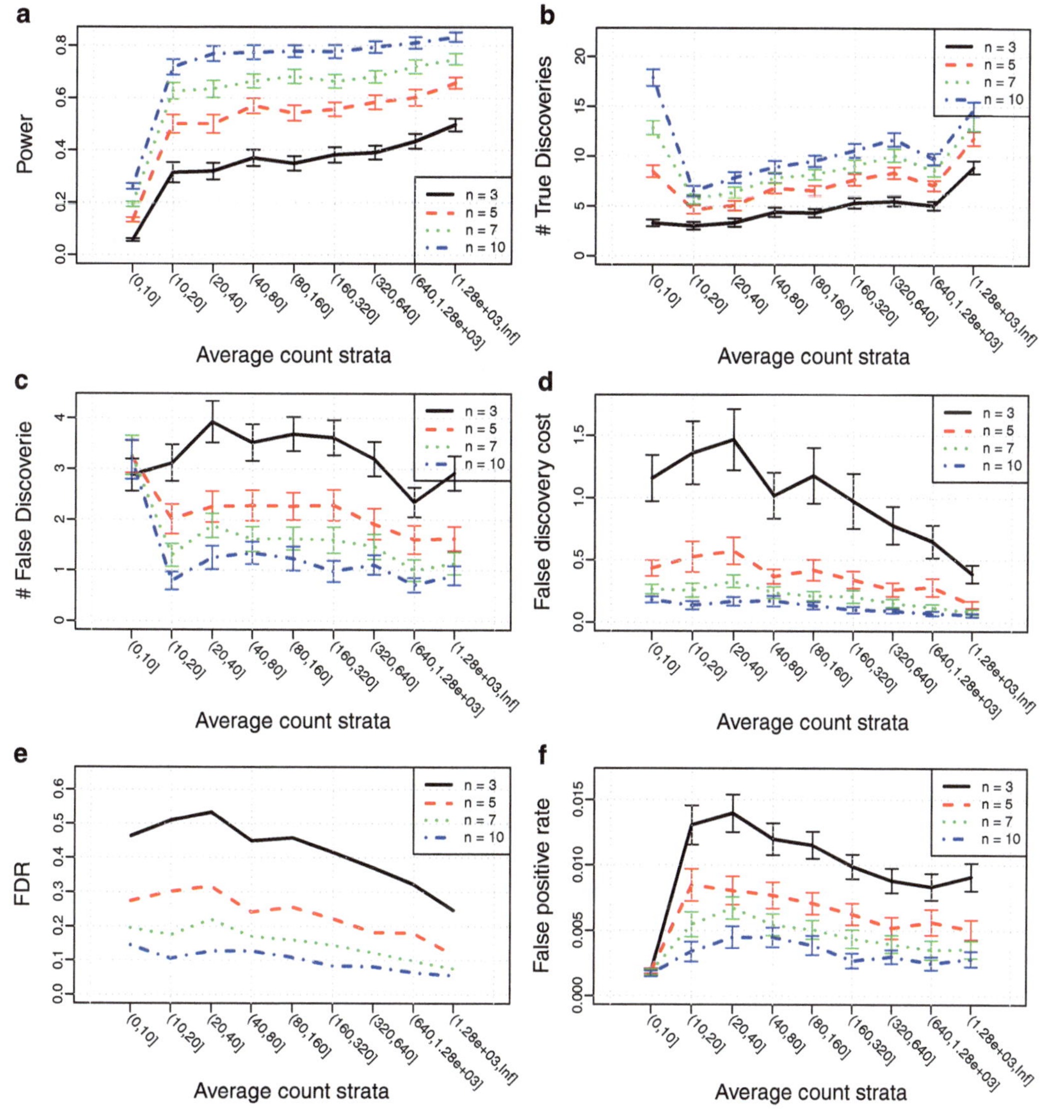

Fig. 1 Comprehensive visualization of stratified power, generated by the function plotAll

than the proportion is of interest. We can visualize this with the *plotPowerTD* function, which is shown in Fig. 1b.

```
plotPowerTD(power1b)
```

Based on the average power in Fig. 1a, we may consider filtering out genes with counts lower than 10 since the statistical power is very low. However, based on Fig. 1b, we realize that though the proportion of true DE genes identified in the first stratum is low, the actual number of DE genes is often higher than other strata, because a large fraction of genes fall in this range. A natural question arises: is it worthwhile to keep this stratum? The function *plotPowerFD* shows the number of false discoveries and *plotFDcost* puts the true and false discoveries together in terms of FDC, which represents the number of false discoveries accompanying each true discovery at the current cutoff.

```
plotFDcost(power1b)
```

For a complete set of figures including the average power, power as crude number of true discoveries, false discoveries, FDC, actual FDR, and actual type I error rates for each strata, one can simply call the function

```
plotAll(power1b)
```

This produces the combined figure as shown in Fig. 1.

3.5 More Samples or Deeper Sequencing

One striking observation in the power curves is that genes with low count numbers are associated with low statistical power even for as many as 10 replicates. To get around this and increase power, we may consider using more samples or increasing the sequencing depth—but which choice provides more benefit? We provide a table that allows the user to compare both types of choices in a simple matrix form. To see power changes associated with various sequencing depths and sample sizes, we use the function *power.seqDepth* and provide two inputs: the simulation result and the initial power assessment output.

```
power.seqDepth(simres1b,power1b)
  3 reps 5 reps 7 reps 10 reps
0.2  0.21  0.34  0.42   0.50
0.5  0.24  0.38  0.46   0.55
1    0.27  0.41  0.50   0.58
2    0.30  0.44  0.53   0.61
5    0.34  0.48  0.57   0.65
10   0.36  0.51  0.60   0.68
```

Each row of the table shows the results if the experiment was done at a sequencing depth relative to the one used in initial power assessment. For example, the third row (with relative depth 1) represents power at the same sequencing depth used in the original simulation setting, and we have 41 % marginal power using 5 replicates in each group. If we use half the sequencing depth (relative depth 0.5) but double the sample size (10 reps), we can get an increased power at 55 % for the same amount of sequencing. Whether to increase the sample size or sequencing depth depends on the availability of biological samples and the sequencing cost, which changes rather rapidly, thus we leave the decision to the user. The table enables the user to make informed decision taking statistical power into account.

3.6 To Filter or Not to Filter

As we realize the particular challenge in detecting true DE genes from noise among genes with very low counts, we often face the difficult choice of whether to filter out these genes. Filtering decreases the number of total tests and thus reduces the burden for multiple testing adjustment, a necessary component in high throughput data analysis. Though we forego the possibility of detecting DE in those genes, we do not generate false discoveries from these either. This increases marginal power, and may reduce overall FDC. The decision to filter should be made prior to actual data analysis, as repeated analysis of the same data set (trying several filters, for example) is another source of increased error. Using the simulation result we can assess whether filtering is likely to help in data generated in similar settings, thus make an informed decision before analyzing the actual experimental data.

To compare how power would have been if we filtered out the first strata (*strata.filtered=1*) in the expression level (*filter.by="expr"*), we use the following

```
powers1b.update = comparePower(simres1b, alpha.type="fdr", alpha.nominal=0.1,
 strata = c(0, 10, 2^(1:7)*10, Inf), filter.by="expr", strata.filtered=1, delta=0.5)
summaryPower(powers1b.update)
Sample size Nominal FDR Actual FDR Marginal power Avg # of TD  Avg # of FD  FDC
[1,]       3            0.1        0.42           0.43          44           31 0.72
[2,]       5            0.1        0.24           0.61          62           20 0.32
[3,]       7            0.1        0.16           0.71          72           14 0.20
[4,]      10            0.1        0.11           0.81          82           11 0.13
```

The option *strata* defines how the average counts are stratified. The values for *strata* define eight strata for average gene counts: (0, 10], (10, 20], (20, 40], (40, 80], (80, 160], (160, 320], (320, 640], (640, 1280], and $(1280, \infty)$. One can choose to filter out more than one strata by changing *strata.filtered*. Compared with the earlier results without filtering, the marginal powers are greatly improved. For example, with 5 replicates in each group, the marginal power increased from 0.41 to 0.61. However, the average number of true discovery also reduced from 68 to 62. Whether filtering is beneficial depends on the balance between the reduction of false positive and the reduction of true positives, which varies from data set to data set, and depends on the definition of target size.

PROPER provides flexible ways for gene filtering. For example, if we only want to filter out genes with counts up to 8, we can re-define the *strata* as shown below:

```
power1b.update = comparePower(simres1b, alpha.type="fdr", alpha.nominal=0.1,
 strata = c(0, 8, 2^(1:7)*10, Inf), filter.by="expr",
 strata.filtered=1, stratify.by="expr", delta=0.5)
summaryPower(power1b.update) # the output is omitted below
```

4 Notes

Results from the above examples are generated for the purpose of illustrating the use of *PROPER*. We do not argue that these necessarily reflect the typical situation in RNA-seq experiments that a user will run into. The users are encouraged to choose simulation scenarios that are closest to their study. One may consider the population heterogeneity of samples (ecological samples from random populations, or controlled experiments on cell lines), tissues types, and species in selecting a data source as simulation basis. One may consult reported differential expression in published results, especially those validated in other platforms, as expected DE distribution.

The DE detection considered here is at the gene or transcript level. Examples of methods using read counts at this level include edgeR [5], DESeq [4], and DSS [3]. Another class of methods such as Cufflinks [12] considers alternative splicing and different isoforms of the same transcript. These first estimate isoform expressions and then perform DE analysis on the estimates. PROPER does no apply to these methods.

The results shown in the examples are generated using R version 3.2.0 and Bioconductor version 3.1.0, PROPER version 1.2.0.

References

1. Hansen KD, Wu Z, Irizarry RA, Leek JT (2011) Sequencing technology does not eliminate biological variability. Nat Biotechnol 29 (7):572–573
2. Gentleman RC, Carey VJ, Bates DM, Bolstad B, Dettling M, Dudoit S, Ellis B, Gautier L, Ge Y, Gentry J et al (2004) Bioconductor: open software development for computational biology and bioinformatics. Genome Biol 5(10):R80
3. Wu H, Wang C, Wu Z (2013) A new shrinkage estimator for dispersion improves differential expression detection in RNA-seq data. Biostatistics 14(2):232–243
4. Anders S, Huber W (2010) Differential expression analysis for sequence count data. Genome Biol 11(10):R106
5. Robinson MD, McCarthy DJ, Smyth GK (2010) edgeR: a Bioconductor package for differential expression analysis of digital gene expression data. Bioinformatics 26 (1):139–140
6. McCarthy DJ, Chen Y, Smyth GK (2012) Differential expression analysis of multifactor RNA-Seq experiments with respect to biological variation. Nucleic Acids Res 40 (10):4288–4297
7. Fang Z, Cui X (2011) Design and validation issues in RNA-seq experiments. Brief Bioinform 12(3):280–287
8. Hart SN, Therneau TM, Zhang Y, Poland GA, Kocher J-P (2013) Calculating sample size estimates for RNA sequencing data. J Comput Biol 20(12):970–978
9. Li C-I, Su P-F, Shyr Y (2013) Sample size calculation based on exact test for assessing differential expression analysis in RNA-seq data. BMC Bioinformatics 14(1):357
10. Li C-I, Su P-F, Guo Y, Shyr Y (2013) Sample size calculation for differential expression analysis of RNA-seq data under Poisson distribution. Int J Comput Biol Drug Des 6 (4):358–375
11. Frazee AC, Langmead B, Leek JT (2011) ReCount: a multi-experiment resource of analysis-ready RNA-seq gene count datasets. BMC Bioinformatics 12(1):449
12. Trapnell C, Roberts A, Goff L, Pertea G, Kim D, Kelley DR, Pimentel H, Salzberg SL, Rinn JL, Pachter L (2012) Differential gene and transcript expression analysis of RNA-seq experiments with TopHat and Cufflinks. Nat Protoc 7(3):562–578

Chapter 19

It's DE-licious: A Recipe for Differential Expression Analyses of RNA-seq Experiments Using Quasi-Likelihood Methods in edgeR

Aaron T.L. Lun, Yunshun Chen, and Gordon K. Smyth

Abstract

RNA sequencing (RNA-seq) is widely used to profile transcriptional activity in biological systems. Here we present an analysis pipeline for differential expression analysis of RNA-seq experiments using the Rsubread and edgeR software packages. The basic pipeline includes read alignment and counting, filtering and normalization, modelling of biological variability and hypothesis testing. For hypothesis testing, we describe particularly the quasi-likelihood features of edgeR. Some more advanced downstream analysis steps are also covered, including complex comparisons, gene ontology enrichment analyses and gene set testing. The code required to run each step is described, along with an outline of the underlying theory. The chapter includes a case study in which the pipeline is used to study the expression profiles of mammary gland cells in virgin, pregnant and lactating mice.

Key words RNA-seq, Differential expression, Generalized linear models, Quasi-likelihood, Variability, Read alignment, Read counts

1 Introduction

RNA sequencing (RNA-seq) is widely used to profile transcriptional activity in biological systems, superseding microarrays as the technique of choice for profiling gene expression [1–3]. One of the most common aims of RNA-seq profiling is to identify genes or molecular pathways that are differentially expressed (DE) between two or more biological conditions. Changes in expression can then be associated with differences in biology, providing avenues for further investigation into potential mechanisms of action.

This article describes an analysis pipeline for the detection of DE genes and pathways from RNA-seq data using the Rsubread and edgeR software packages [4, 5]. We will assume throughout that RNA samples have been extracted from cells of interest under two or more treatment conditions, and that there are independent biological replicates for at least one of the treatment conditions.

Ewy Mathé and Sean Davis (eds.), *Statistical Genomics: Methods and Protocols*, Methods in Molecular Biology, vol. 1418, DOI 10.1007/978-1-4939-3578-9_19, © Springer Science+Business Media New York 2016

We assume that RNA-seq profiling has been applied to each RNA sample and that the pipeline takes the raw sequence reads as input. The analysis strategy we describe can be applied to any RNA-seq gene expression experiment, but we give particular attention to experiments with multiple treatment factors and with small numbers of biological replicates.

The pipeline uses the Rsubread package for mapping reads and assigning them to genes, and the edgeR package for statistical analyses. Generalized linear models (GLMs) are used to accommodate complex experimental designs [6]. Quasi-likelihood *F*-tests are used to conduct hypothesis tests [7]. edgeR provides a range of capabilities. We focus here on the quasi-likelihood features of edgeR rather than exact tests [8, 9] or likelihood ratio tests (LRTs) [6], because the latter have been described previously [10] and because the quasi-likelihood functions provide more robust and reliable error rate control when the number of replicates is small.

The use of the pipeline is demonstrated here with a published RNA-seq data set involving the mouse mammary gland [11]. edgeR, Rsubread and other packages used in this article are publicly available as part of the Bioconductor project [12]. See http://www.bioconductor.org/install for installation instructions.

2 Basic Theory of the Differential Expression Analysis

2.1 Overview

Before we start on the computational aspects of the analysis, it is useful to give an outline of the underlying theory. The next few sections describe the approach taken to process the reads into gene counts and to normalize the counts for library-specific biases. We also give a brief introduction to the statistical models used to represent complex designs, to estimate technical and biological variability and to conduct hypothesis tests.

2.2 From Reads to Genewise Counts

Massively parallel sequencing generates millions of short read sequences from each RNA sample. We will refer to the set of reads produced from a particular sample as a library. The first step in processing the data is to align each read to the reference genome for the organism being studied. This can be done efficiently with a splice-aware aligner such as TopHat [13] or subread [4]. We use subread in this tutorial because it is especially fast and convenient, as well as being available as an R package.

The next step is to assign reads to genes. The number of reads overlapping the exons of a gene can be used as a measure of the expression level of that gene. featureCounts [14] and HTSeq [15] are popular tools for counting the reads assigned to each gene. We use featureCounts in this tutorial because it is fast and available in the Rsubread package.

Table 1
Table of read counts for a simple RNA-seq experiment with four samples

	Wild-type		Mutant	
	Sample 1	Sample 2	Sample 3	Sample 4
Gene 1	24	31	76	59
Gene 2	0	3	7	2
Gene 3	1988	1125	3052	2450
Gene 4	5	0	0	1
...	...	...	...	...

Each column corresponds to a sample from a mouse with a wild-type or mutant genotype. Each row corresponds to a gene in the mouse genome. Each entry represents the read count for a gene in a sample

This process produces a read count for each gene in each sample. An example of the read counts for a simple RNA-seq experiment is shown in Table 1. Two groups are present in the data set (wild-type and mutant), each of which contains samples from two mice, i.e., two biological replicates. After sequencing, reads for each sample are mapped to the mouse genome and summarized into gene-level counts. The final expression profile is represented by a matrix of read counts, with one row for each of the thousands of genes and one column per sample.

Counting reads by genes is far from the only method for RNA-seq data summarization that is available. An alternative would be to obtain a read count for each exon in each sample, in order to investigate isoform levels. Another approach might be to assemble transcripts de novo from the read sequences, and then count the number of reads aligned to each new transcript. We recommend the gene counting approach, however, as it is simple and effective when identification of DE genes is desired, especially when the sequencing depths are not very high.

2.3 Modelling Variability in a Quasi-Negative Binomial Framework

edgeR uses the negative binomial (NB) distribution to model the read counts for each gene in each sample [8, 9]. The NB distribution is ideal as it is a discrete count distribution that provides accurate modelling at low counts. It can account for variability between biological replicates, through the NB dispersion parameter [6]. The NB model can also be extended with quasi-likelihood (QL) methods to account for gene-specific variability from both biological and technical sources [7]. Write y_{gi} for the read count for gene g in sample i, and let $E(y_{gi}) = \mu_{gi}$ be the expected count for this gene in this sample given the sequencing depth and treatment conditions applied to sample i. The variance of the count is a quadratic function of the mean,

$$\mathrm{var}(y_{gi}) = \sigma_g^2(\mu_{gi} + \mu_{gi}^2\phi)$$

where ϕ is the NB dispersion parameter and σ_g^2 is the QL dispersion parameter.

Any increase in the observed variance of y_{gi} will be modelled by an increase in the estimates for ϕ and/or σ_g^2. In this model, the NB dispersion ϕ is a global parameter whereas the QL is gene-specific, so the two dispersion parameters have different roles. The NB dispersion describes the overall biological variability across all genes. The square-root of the NB dispersion is known as the biological coefficient of variation [6]. It represents the observed variation that is attributable to inherent variability in the biological system, in contrast to the Poisson variation from sequencing. The QL dispersion picks up any gene-specific variability above and below the overall level.

A common NB dispersion for the entire data set can be stably estimated by using information across all genes. In practice, a more flexible approach is to fit a mean-dispersion trend across genes. In this approach, ϕ is replaced by a function $\phi(A)$, where A is a measure of the overall expression level of gene g. Then $\phi(A)$ is referred to as the trended NB dispersion for that gene [6]. This approach accounts for empirical mean–variance relationships, e.g., from distributions other than the NB. Trend fitting is also stabilized by sharing information between genes with similar abundances.

Estimation of the gene-specific QL dispersion is difficult as most RNA-seq data sets have limited numbers of replicates. This means that there is often little information to stably estimate the dispersion for each gene. To overcome this, an empirical Bayes (EB) approach is used whereby information is shared between genes [7, 16, 17]. Briefly, a mean-dependent trend is fitted to the raw QL dispersion estimates. The raw estimates are then squeezed towards this trend to obtain moderated EB estimates, which can be used in place of the raw values for downstream hypothesis testing. This EB strategy reduces the uncertainty of the estimates and improves testing power.

2.4 Introducing the Generalized Linear Model

edgeR uses the GLM framework to account for complex experimental designs [6]. The mean count for gene g in sample i is modelled as

$$\log(\mu_{gi}) = x_{i1}\beta_{g1} + x_{i2}\beta_{g2} + \ldots + x_{in}\beta_{gn} + o_i$$

where β_{gj} is the gene-specific value of coefficient j, x_{ij} is the sample-specific predictor for j and o_i is the sample-specific offset. There are n coefficients in total, where each coefficient should describe some factor of the experimental design, e.g., different biological conditions, batch effects. The predictors can be binary or continuous, though we will focus on binary values for simplicity. Setting $x_{ij} = 1$

Table 2
Group-mean design matrix

		Coefficients		
Group	**Sample**	$j = 1$	$j = 2$	$j = 3$
1	$i = 1$	1	0	0
1	$i = 2$	1	0	0
2	$i = 3$	0	1	0
2	$i = 4$	0	1	0
3	$i = 5$	0	0	1
3	$i = 6$	0	0	1

This is displayed as a table of predictor values x_{ij} for each coefficient j in each sample i, for an experimental design containing groups 1, 2 and 3. Each group has two replicate samples. This choice of predictor variables is called the group-mean parametrization. Each of the three estimated coefficients represents the log-expression level (i.e. log-count-per-million) of the gene in the corresponding treatment group

indicates that coefficient j contributes to the expression of sample i, i.e., the factor of the design corresponding to j affects this sample. Finally, the value of β_{gj} describes the impact of that design factor on the expression of gene g in affected samples.

For example, consider a design with three groups 1, 2 and 3. One way to choose the predictor variables x_{ij} is as indicator variables for each treatment group (Table 2). With this choice of design matrix, the coefficient β_{gj} represents the log-average expression level of gene g across the libraries in group j.

Another popular way to choose the predictor variables is to make the first coefficient β_{g1} an intercept term (Table 3). With this choice of predictor variables, the coefficients β_{g2} and β_{g3} represent log-fold expression changes for gene g between group 1 and the other two groups. The different choices for the predictor variables are a matter of convenience, as they lead to equivalent fitted models and to the same downstream differential expression results.

Finally, the offset term is defined as the log-transformed library size. The library size refers to the total number of reads across all genes in each library, and depends on the amount of sequencing resources spent in characterizing each sample. Consider two samples $i = 1$ and 2, where the library size for $i = 1$ is twice that of 2. This means that the expected count μ_{g1} will be twice as large as μ_{g2} for some gene g with the same expression in both samples. The offsets ensure that the differences in library size do not contribute to spurious differences in expression, i.e.,

$$\log(\mu_{g1}) - o_1 = \log(\mu_{g2}) - o_2 .$$

Table 3
Design matrix with a reference level

		Coefficients		
Group	Sample	$j = 1$	$j = 2$	$j = 3$
1	$i = 1$	1	0	0
1	$i = 2$	1	0	0
2	$i = 3$	1	1	0
2	$i = 4$	1	1	0
3	$i = 5$	1	0	1
3	$i = 6$	1	0	1

This is displayed as a table of predictor values x_{ij} for each coefficient j in each sample i, for an experimental design containing groups 1, 2 and 3. Each group has two replicate samples. This choice of predictor variables treats the first treatment group as a reference. The first coefficient β_{g1} represents log-expression of the gene in the first treatment group, while β_{g2} and β_{g3} represent the log-fold-changes in expression for the second and third treatment groups, respectively, compared to the first

Note that it is possible to construct more complex parametrizations involving gene- and sample-specific offsets. For simplicity, these will not be described here.

2.5 Normalizing for Composition Biases

Consider two samples A and B that are sequenced to the same "depth", i.e., the resulting libraries have the same number of reads. Assume that all genes are expressed at the same level in the two samples, except for one DE gene that increases in B. The DE gene will use up more sequencing resources in B, such that fewer reads will be available for all other non-DE genes. This results in "composition bias" where the read counts for the non-DE genes are suppressed in library B [18]. Spurious differences will then be observed upon comparison to the counts in library A.

Composition bias can be eliminated with trimmed mean of M-values (TMM) normalization. Briefly, most genes are assumed to be non-DE between libraries A and B. If this is true, any systematic difference in the gene counts between the two libraries must represent composition bias. The log-fold difference of the counts in B over A is estimated from the data and used to compute a scaling factor, also known as the TMM normalization factor. This factor is used to scale the size of library B to obtain an "effective" size, to be used to compute the GLM offset for B. More intuitively, this is equivalent to scaling the counts in library B upwards by the reciprocal of the estimated factor. This reverses the effect of suppression in B.

3 Tutorial with Real Data

3.1 Description of the Data Set

To demonstrate how the differential expression analysis works, we will use RNA-seq data from the Fu et al. study [11]. The sequence and count data are publicly available from the Gene Expression Omnibus (GEO) at the series accession number GSE60450. This study examines the expression profiles of basal stem-cell enriched cells (B) and committed luminal cells (L) in the mammary gland of virgin, pregnant and lactating mice. Six groups are present, with one for each combination of cell type and mouse status. Each group contains two biological replicates. This is summarized in the table below, where the basal and luminal cell types are abbreviated with `B` and `L`, respectively.

```
> targets
           FileName GEOAccession CellType   Status
1  SRR1552450.fastq   GSM1480297        B   virgin
2  SRR1552451.fastq   GSM1480298        B   virgin
3  SRR1552452.fastq   GSM1480299        B pregnant
4  SRR1552453.fastq   GSM1480300        B pregnant
5  SRR1552454.fastq   GSM1480301        B  lactate
6  SRR1552455.fastq   GSM1480302        B  lactate
7  SRR1552444.fastq   GSM1480291        L   virgin
8  SRR1552445.fastq   GSM1480292        L   virgin
9  SRR1552446.fastq   GSM1480293        L pregnant
10 SRR1552447.fastq   GSM1480294        L pregnant
11 SRR1552448.fastq   GSM1480295        L  lactate
12 SRR1552449.fastq   GSM1480296        L  lactate
```

The name of the file containing the read sequences for each library is also shown. Each file is downloaded from the Sequence Read Archive and has an accession number starting with SRR, e.g., SRR1552450 for the first library in `targets`. All files have been converted to the FASTQ format—*see* Subheading 3.2 for more details.

The experimental design for this study can be parametrized with a one-way layout, whereby one coefficient is assigned to each group. The design matrix contains the predictors for each sample and is constructed using the code below. Each row of this design matrix corresponds to a sample in `targets`. Each column represents a coefficient that corresponds to the group after which it is named. This is effectively an extension of the three-group example described in Table 2.

```
> group <- factor(paste0(targets$CellType, ".", targets$Status))
> design <- model.matrix(~ 0 + group)
> colnames(design) <- levels(group)
> design
```

```
  B.lactate B.pregnant B.virgin L.lactate L.pregnant L.virgin
1         0          0        1         0          0        0
2         0          0        1         0          0        0
3         0          1        0         0          0        0
4         0          1        0         0          0        0
5         1          0        0         0          0        0
6         1          0        0         0          0        0
7         0          0        0         0          0        1
8         0          0        0         0          0        1
9         0          0        0         0          1        0
10        0          0        0         0          1        0
11        0          0        0         1          0        0
12        0          0        0         1          0        0
attr(,"assign")
[1] 1 1 1 1 1 1
attr(,"contrasts")
attr(,"contrasts")$group
[1] "contr.treatment"
```

3.2 Read Alignment and Processing

Read sequences are stored in FASTQ files. Before the differential expression analysis can proceed, these reads must be aligned to the mouse genome and counted into annotated genes. This can be achieved with functions in the Rsubread package. We assume that an index of the mouse genome is already available—if not, this can be constructed from a FASTQ file of the genome sequence with the `buildindex` command. In this example, we assume that the prefix for the index files is `mm10`. The reads in each FASTQ file are then aligned to the mouse genome, as shown below.

```
> library(Rsubread)
> output.files <- sub(".fastq", ".bam", targets$FileName)
> align("mm10", readfile1=targets$FileName, phredOffset=33,
+   input_format="FASTQ", output_file=output.files)
```

This produces a set of BAM files, where each file contains the read alignments for each library. The mapped reads can be counted into mouse genes by using the `featureCounts` function. The code below uses the exon intervals defined in the NCBI annotation of the mm10 genome. Recall that counts for all exons of a gene are added together to obtain the count for each gene. Repeating this for every sample generates a matrix of read counts for each gene in each sample, similar to Table 1.

```
> fc <- featureCounts(output.files, annot.inbuilt="mm10")
> colnames(fc$counts) <- 1:12
> head(fc$counts)
            1   2   3   4   5   6  7  8  9 10 11 12
497097    438 300  65 237 354 287  0  0  0  0  0  0
100503874   1   0   1   1   0   4  0  0  0  0  0  0
```

```
100038431 0   0   0   0   0   0   0   0   0   0   0   0
19888       1   1   0   0   0   0  10   3  10   2   0   0
20671     106 182  82 105  43  82  16  25  18   8   3  10
27395     309 234 337 300 290 270 560 464 489 328 307 342
```

The row names of the matrix represent the Entrez gene identifiers for each gene. In the output from `featureCounts`, the column names of `fc$counts` are the output file names from `align`. Here, we simplify them for brevity.

Notes

1. Sequence data from GEO is normally obtained in the Sequence Read Archive (SRA) format. Prior to read alignment, these files should be converted into the FASTQ format using the `fastq-dump` utility from the SRA Toolkit. See http://www.ncbi.nlm.nih.gov/books/NBK158900 for how to download and use the SRA Toolkit.
2. By default, alignment is performed with `unique` set to `TRUE`. If a read can be aligned to two or more locations, Rsubread will attempt to select the best location based on a number of criteria. Only reads that have a unique best location are reported as being aligned. Keeping this default is strongly recommended, as it avoids spurious signal from non-uniquely mapped reads derived from, e.g., repeat regions.
3. The Phred offset determines the encoding for the base-calling quality string in the FASTQ file. For the Illumina 1.8 format onwards, this encoding is set at +33. However, older formats may use a +64 encoding. Users should ensure that the correct encoding is specified during alignment. If unsure, one can examine the first several quality strings in the FASTQ file. A good rule of thumb is to check whether lower-case letters are present (+64 encoding) or absent (+33).
4. `featureCounts` requires gene annotation specifying the genomic start and end position of each exon of each gene. Rsubread contains built-in gene annotation for mouse and human. For other species, users will need to read in a data frame in GTF format to define the genes and exons.

3.3 Count Loading and Annotation

The count matrix is used to construct a `DGEList` class object. This is the main data class in the edgeR package. The `DGEList` object is used to store all the information required to fit a GLM to the data, including library sizes and dispersion estimates as well as counts for each gene.

```
> library(edgeR)
> options(digits=3)
> y <- DGEList(fc$counts, group=group)
> colnames(y) <- targets$GEO
```

Human-readable gene symbols can also be added to complement the Entrez identifiers for each gene, using the annotation in the `org.Mm.eg.db` package.

```
> require(org.Mm.eg.db)
> y$genes <- select(org.Mm.eg.db,keys=rownames(y),+ columns="SYMBOL")
> head(y$genes)
   ENTREZID  SYMBOL
1    497097    Xkr4
2 100503874 Gm19938
3 100038431 Gm10568
4     19888     Rp1
5     20671   Sox17
6     27395  Mrpl15
```

Notes

1. In the GLM framework, specifying `group` is not strictly necessary for the construction of the `DGEList`. It will not be used in the analysis when a design matrix is supplied. Setting `group` is only done here for the sake of completeness.
2. Users should pick an organism package corresponding to their biological system. For example, the appropriate package for human data would be `org.Hs.eg.db`. More organism packages can be found on the Bioconductor website.

3.4 Filtering to Remove Low Counts

Genes with very low counts across all libraries provide little evidence for differential expression. In addition, the pronounced discreteness of these counts interferes with some of the statistical approximations that are used later in the pipeline. These genes should be filtered out prior to further analysis. Here, a gene is only retained if it is expressed at a count-per-million (CPM) above 0.5 in at least two samples.

```
> keep <- rowSums(cpm(y) > 0.5) >= 2
> y <- y[keep,]
> summary(keep)
   Mode FALSE  TRUE  NA's
logical 11375 15804     0
```

Note that the whole `DGEList` object, including annotation as well as counts, subsets by rows as if it was a matrix. This ensures the annotation and counts continue to be aligned correctly in the subsetted object.

A CPM of 0.5 is used as it corresponds to a count of 10–15 for the library sizes in this data set. If the count is any smaller, it is considered to be very low, indicating that the associated gene is not expressed in that sample. A requirement for expression in two or more libraries is used as each group contains two replicates. This ensures that a gene will be retained if it is only expressed in one group.

Notes

1. Smaller CPM thresholds can be used for larger libraries. As a general rule, a near ideal threshold can be chosen by identifying the CPM at a count of 10, which in this case is near 0.5:

```
> cpm(10, mean(y$samples$lib.size))
      [,1]
[1,] 0.446
```

Users should filter with CPMs rather than filtering on the counts directly, as the latter does not account for differences in library sizes between samples.

3.5 Normalization for Composition Bias

TMM normalization is performed to eliminate composition biases between libraries. This generates a set of normalization factors, where the product of these factors and the library sizes defines the effective library size. The `calcNormFactors` function returns the `DGEList` argument with only the `norm.factors` changed.

```
> y <- calcNormFactors(y)
> y$samples
                 group  lib.size  norm.factors
GSM1480297    B.virgin  23227641         1.237
GSM1480298    B.virgin  21777891         1.214
GSM1480299 B.pregnant   24100765         1.126
GSM1480300 B.pregnant   22665371         1.070
GSM1480301   B.lactate  21529331         1.036
GSM1480302   B.lactate  20015386         1.087
GSM1480291    L.virgin  20392113         1.368
GSM1480292    L.virgin  21708152         1.365
GSM1480293 L.pregnant   22241607         1.005
GSM1480294 L.pregnant   21988240         0.923
GSM1480295   L.lactate  24723827         0.529
GSM1480296   L.lactate  24657293         0.536
```

The normalization factors multiply to unity across all libraries. A normalization factor below unity indicates that the library size will be scaled down, as there is more suppression (i.e. composition bias) in that library relative to the other libraries. This is also equivalent to scaling the counts upwards in that sample. Conversely, a factor above unity scales up the library size and is equivalent to downscaling the counts.

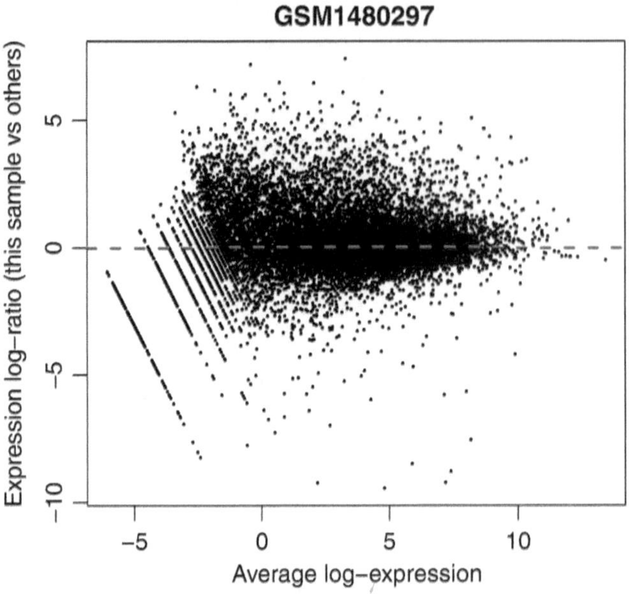

Fig. 1 A mean-difference plot of expression in sample 1 against the average expression across all other samples. Each point represents a gene, and the *red line* corresponds to a log-ratio of zero

The performance of the TMM normalization procedure can be examined using mean-difference (MD) plots. This visualizes the library size-adjusted log-fold change between two libraries (the difference) against the average log-expression across those libraries (the mean). In Fig. 1, an MD plot is generated by comparing sample 1 against an artificial library constructed from the average of all other samples.

```
> plotMD(cpm(y, log=TRUE), column=1)
> abline(h=0, col="red", lty=2, lwd=2)
```

Ideally, the bulk of genes should be centred at a log-fold change of zero. This indicates that any composition bias between libraries has been successfully removed. This quality check should be repeated by constructing a MD plot for each sample.

Notes

1. The MD plot shown above compares each library to all others. An alternative is to compare each library to a single reference library. The code below uses sample 6 as a reference, and compares sample 1 to this reference.

   ```
   > plotMD(cpm(y[,c(1,6)], log=TRUE))
   ```

   This may be easier to interpret as any problems with other libraries will not affect the MD plot for this pair. As a rule of thumb, the reference library should not be very large or small, e.g., use the sample with the median library size.

2. In extreme cases, trends may be observed in the MD plot. These trended biases cannot be removed with scaling normalization methods like TMM. Rather, non-linear methods are required. These were traditionally developed for normally distributed microarray intensities, but can be applied here on log-counts.

```
> logy <- log(y$counts + 0.5)
> normy <- normalizeBetweenArrays(logy, + method="cyclicloess")
> y$offset <- normy - logy
```

3.6 Exploring Differences Between Libraries

The data can be explored by generating multi-dimensional scaling (MDS) plots. This visualizes the differences between the expression profiles of different samples in two dimensions. Figure 2 shows the MDS plot for the mammary gland data.

```
> points <- c(0,1,2,15,16,17)
> colors <- rep(c("blue", "darkgreen", "red"), 2)
> plotMDS(y, col=colors[group], pch=points[group])
> legend("topleft", legend=levels(group),
+   pch=points, col=colors, ncol=2)
```

The distance between each pair of samples in the MDS plot is calculated as the *leading fold change*, defined as the root-mean-square of the largest 500 $\log_2$-fold changes between that pair of

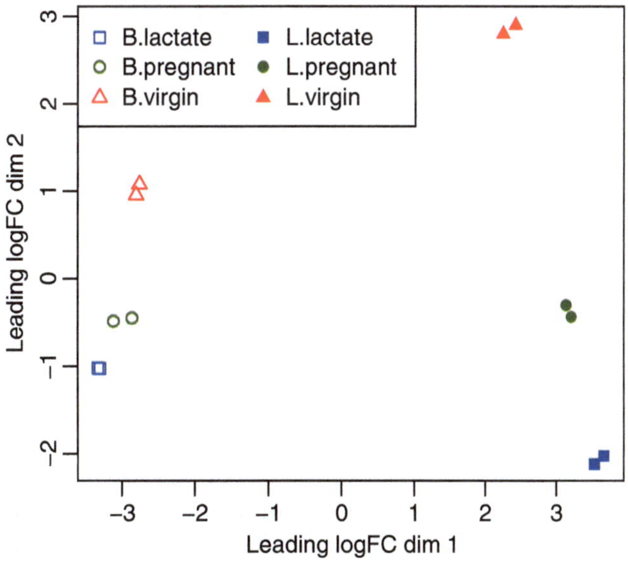

Fig. 2 The MDS plot of the mammary gland RNA-seq data set. Samples are separated by the cell type in the first dimension, and by the mouse status in the second dimension

samples. Replicate samples from the same group cluster together in the plot, while samples from different groups form separate clusters. This indicates that the differences between groups are larger than those within groups, i.e., differential expression is greater than the variance and can be detected. In Fig. 2, the distance between basal samples on the left and luminal cells on the right is about 6 units, corresponding to a leading fold change of about 64-fold ($2^6 = 64$) between basal and luminal. The expression differences between virgin, pregnant and lactating are greater for luminal cells than for basal.

Notes

1. The MDS plot can be simply generated with `plotMDS(y)`. The additional code is purely for aesthetics, to improve the visualization of the groups.
2. Clustering in the MDS plot can be used to motivate changes to the design matrix in light of potential batch effects. For example, imagine that the first replicate of each group was prepared at a separate time from the second replicate. If the MDS plot showed separation of samples by time, it might be worthwhile adding an additional column to `design` to represent this time-based effect.

3.7 Dispersion Estimation

The trended NB dispersion is estimated using the `estimateDisp` function. This returns the `DGEList` object with additional entries for the estimated NB dispersions for all gene. These estimates can be visualized with `plotBCV`, which shows the root-estimate, i.e., the biological coefficient of variation for each gene (Fig. 3).

```
> y <- estimateDisp(y, design, robust=TRUE)
> plotBCV(y)
```

In general, the trend in the NB dispersions should decrease smoothly with increasing abundance. This is because the expression of high-abundance genes is expected to be more stable than that of low-abundance genes. Any substantial increase at high abundances may be indicative of batch effects or trended biases. The value of the trended NB dispersions should range between 0.005 and 0.05 for laboratory-controlled biological systems like mice or cell lines, though larger values will be observed for patient-derived data (> 0.1) or single-cell data (> 1). Note that tagwise and common estimates are also shown here but will not be used further.

For the QL dispersions, estimation can be performed using the `glmQLFit` function. This returns a `DGEGLM` object containing the estimated values of the GLM coefficients for each gene. It also contains a number of EB statistics, e.g., the fitted mean-QL

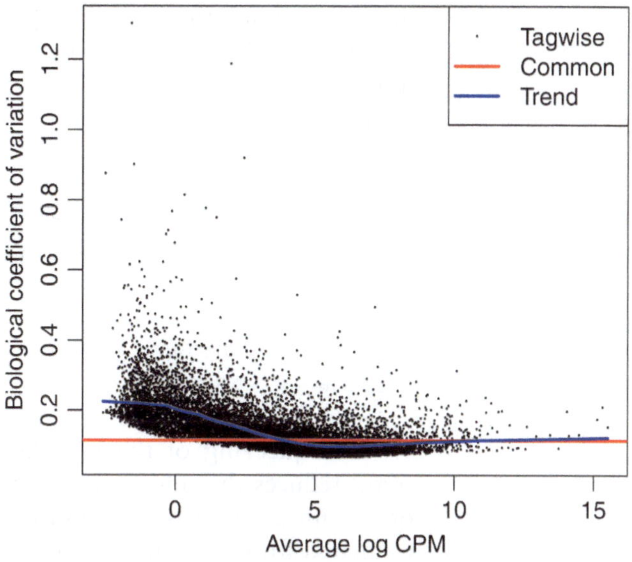

Fig. 3 A plot of the biological coefficient of variation against the average abundance of each gene. Coefficients are shown corresponding to the estimates for the common, trended and tagwise NB dispersions

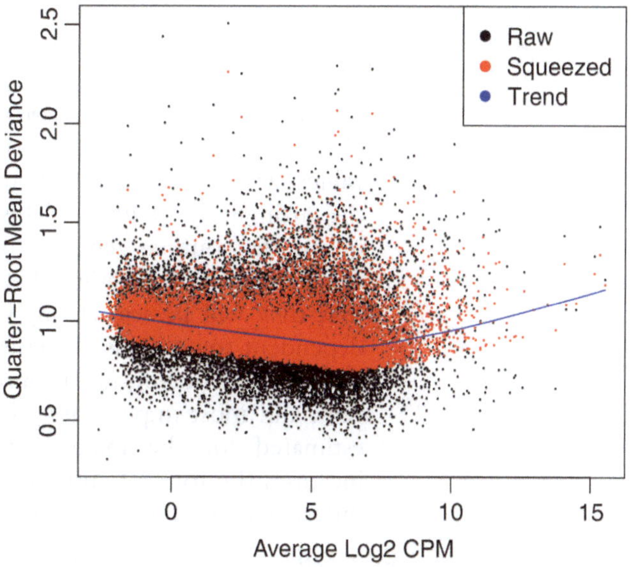

Fig. 4 A plot of the quarter-root QL dispersion against the average abundance of each gene. Estimates are shown for the raw (before EB moderation), trended and squeezed dispersions

dispersion trend, the squeezed QL estimates and the prior degrees of freedom (df). These can be visualized with the `plotQLDisp` function (Fig. 4).

```
> fit <- glmQLFit(y, design, robust=TRUE)
> head(fit$coefficients)
       B.lactate B.pregnant B.virgin L.lactate L.pregnant L.virgin
497097    -11.14     -12.02   -11.23     -19.0     -19.03    -19.0
20671     -12.77     -12.51   -12.15     -14.5     -14.31    -14.1
27395     -11.27     -11.30   -11.53     -10.6     -10.87    -10.9
18777     -10.15     -10.21   -10.77     -10.1     -10.39    -10.4
21399      -9.89      -9.74    -9.79     -10.2      -9.97    -10.0
58175     -16.16     -14.86   -15.99     -13.3     -12.29    -12.1
> plotQLDisp(fit)
```

EB squeezing of the raw dispersion estimators towards the trend reduces the uncertainty of the final estimators. The extent of this moderation is determined by the value of the prior df, as estimated from the data. Large estimates for the prior df indicate that the QL dispersions are less variable between genes, meaning that stronger EB moderation can be performed. Smaller prior df indicate that the dispersions are highly variable, such that strong moderation would be inappropriate.

```
> summary(fit$df.prior)
 Min. 1st Qu. Median  Mean 3rd Qu.  Max.
 3.15  6.76     6.76  6.63   6.76  6.76
```

Notes

1. Setting `robust=TRUE` in `glmQLFit` is strongly recommended [17]. This causes `glmQLFit` to estimate a vector of `df.prior` values, with lower values for outlier genes and larger values for the main body of genes. This has two advantages. First, it means that outlier genes with unusually large or small QL dispersions will be assigned lower prior df values, meaning that they will be less strongly squeezed towards the mean-dependent trend. This prevents genes with extremely low dispersions from being inappropriately called as significant. Second, and most importantly, it allows a higher prior df to be estimated for the main body of non-outlier genes. This increases the total df and increases statistical power to detect differential expression for most genes.
2. Setting `robust=TRUE` in `estimateDisp` has no effect on the downstream analysis, but is nevertheless very useful as it identifies genes that are outliers from the mean-NB dispersion trend. Outliers are marked by small `prior.df` values:

```
> o <- order(y$prior.df)
> y$genes[o[1:6],]
      ENTREZID     SYMBOL
 615     12835     Col6a3
1616    215866  LOC215866
2745    140703      Emid1
```

```
7070     20390     Sftpd
7789     21943   Tnfsf11
16048    11828      Aqp3
```

In mouse data sets, this set of outliers may be enriched for sex-linked genes, when replicates are from different sexes; ribosomal genes, when there are technical issues with cDNA preparation; or immunoglobulins, when the cell population is contaminated with plasma cells. If an obvious pattern can be identified among the outlier genes, the associated set of genes can be removed beforehand in order to avoid distorting the trended NB dispersion estimates. In this case, there is no apparent enrichment of genes from any particular group, so no action is required.

3. Estimating the QL dispersion requires special care for genes that have all counts exactly zero within a treatment group. `glmQLFit` handles this situation by reducing the effective residual degrees of freedom for such genes. The output vector `df.residual.zeros` contains the effective residual degrees of freedom used to estimate the raw QL dispersions.

3.8 Testing for Differential Expression

The final step is to actually test for significant differential expression in each gene, using the QL *F*-test. The contrast of interest can be specified using the `makeContrasts` function. Here, genes are detected that are DE between the basal pregnant and lactating groups. This is done by defining the null hypothesis as `B.pregnant - B.lactate = 0`.

```
> con <- makeContrasts(B.pregnant - B.lactate, + levels=design)
> res <- glmQLFTest(fit, contrast=con)
```

The top set of most significant genes can be examined with `topTags`. Here, a positive log-fold change represents genes that are up in `B.pregnant` over `B.lactate`. Multiplicity correction is performed by applying the Benjamini–Hochberg method on the *p*-values, to control the false discovery rate (FDR). The total number of DE genes in each direction at a FDR of 5 % can be examined with `decideTestsDGE`.

```
> topTags(res)
Coefficient: -1*B.lactate 1*B.pregnant
      ENTREZID  SYMBOL  logFC  logCPM   F     PValue       FDR
18071    12992 Csn1s2b  -6.09   10.18 421  4.71e-11  7.45e-07
22881   211577  Mrgprf  -5.15    2.74 345  1.30e-10  8.06e-07
12177   226101    Myof  -2.32    6.44 322  1.97e-10  8.06e-07
  851   381290  Atp2b4  -2.14    6.14 320  2.04e-10  8.06e-07
 9279   140474    Muc4   7.17    6.05 308  2.63e-10  8.31e-07
18829   231830 Micall2   2.25    5.18 282  4.49e-10  1.18e-06
 2491    24117    Wif1   1.82    6.76 260  7.28e-10  1.58e-06
```

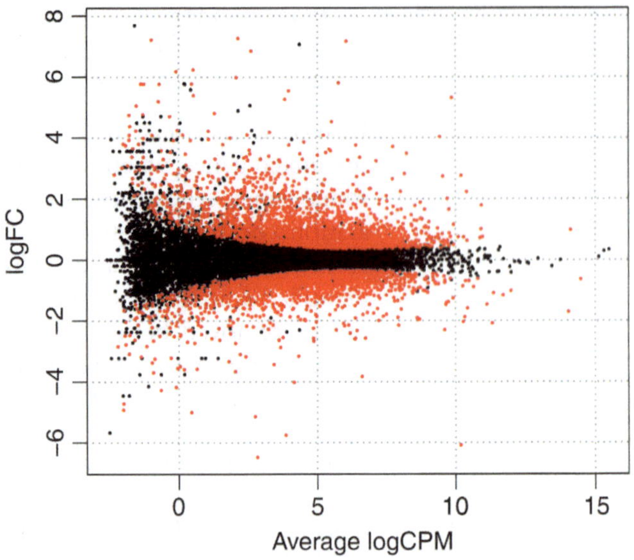

Fig. 5 A smear plot showing the log-fold change and average abundance of each gene. DE genes are marked in *red*

```
18684    12740 Cldn4   5.32  9.87 299  8.35e-10  1.58e-06
22829    21953 Tnni2  -5.75  3.86 314  9.02e-10  1.58e-06
19483   231991 Creb5  -2.57  4.86 241  1.17e-09  1.85e-06
```

The top gene Csn1s2b has a large negative log2-fold-change, showing that it is far more highly expressed in lactating than pregnant mice. This gene is known to be a major source of protein in milk.

There are in fact nearly 2500 DE genes in this comparison:

```
> is.de <- decideTestsDGE(res, p.value=0.05)
> summary(is.de)
   [,1]
-1  2094
0  11307
1   2403
```

The differential expression test results can be visualized using a smear plot (Fig. 5). The log-fold change for each gene is plotted against the average abundance, i.e., `logCPM` in the result table above. Significantly DE genes at a FDR of 5 % are highlighted in red.

```
> plotSmear(res, de.tags=rownames(res)[is.de!=0])
```

Notes

1. While the LRT is a more obvious choice for inferences with GLMs, the QL *F*-test is preferred as it reflects the uncertainty in estimating the dispersion for each gene.

2. The expression supplied to `makeContrasts` is assumed to equate to zero, in order to define the null hypothesis for the contrast. The sign of the terms in the expression determine how the log-fold change is to be interpreted. For example, setting `B.lactate - B.pregnant` in `makeContrasts` would return positive log-fold changes for genes that are upregulated in the basal lactating group.

4 Advanced Usage

4.1 Analysis of Variance

The differential expression analysis of two-group comparison can be easily extended to comparisons between three or more groups. This is done by creating a matrix of contrasts, where each column represents a contrast between two groups of interest. In this manner, users can perform a one-way analysis of variance (ANOVA) for each gene.

As an example, suppose we want to compare the three groups in the luminal population, i.e., virgin, pregnant and lactating. An appropriate contrast matrix can be created as shown below, to make pairwise comparisons between all three groups.

```
> con <- makeContrasts(
+   L.PvsL = L.pregnant - L.lactate,
+   L.VvsL = L.virgin - L.lactate,
+   L.VvsP = L.virgin - L.pregnant, levels=design)
```

The QL *F*-test is then applied to identify genes that are DE among the three groups. This combines the three pairwise comparisons into a single *F*-statistic and *p*-value. The top set of significant genes can be displayed with `topTags`.

```
> res <- glmQLFTest(fit, contrast=con)
> topTags(res)
Coefficient: LR test of 2 contrasts
      ENTREZID  SYMBOL  logFC.L.PvsL  logFC.L.VvsL  logCPM     F
19230  19242       Ptn         -1.54          7.26    7.96  2390
15679  13645       Egf         -5.36         -7.22    3.79  1164
21207  52150     Kcnk6         -2.42         -7.00    5.94  1021
18071  12992  Csn1s2b         -8.55        -11.36   10.18  1051
23907  15439        Hp          1.08          5.42    4.90   988
22626  14183     Fgfr2         -1.15          3.95    7.37   953
20062  11941    Atp2b2         -7.37        -10.56    6.62  1135
2901   20856      Stc2         -1.81          3.19    6.09   918
9083   13358   Slc25a1         -4.13         -4.91    7.50   888
8278   17068      Ly6d          3.42          9.24    4.65   884
         PValue       FDR
19230 3.78e-17 5.97e-13
15679 3.66e-15 2.89e-11
```

```
21207 8.39e-15 3.25e-11
18071 9.87e-15 3.25e-11
23907 1.03e-14 3.25e-11
22626 1.30e-14 3.25e-11
20062 1.61e-14 3.25e-11
2901  1.65e-14 3.25e-11
9083  2.03e-14 3.30e-11
8278  2.09e-14 3.30e-11
```

Notes

1. Note that the three contrasts of pairwise comparisons are linearly dependent. Constructing the contrast matrix with any two of the contrasts would be sufficient to specify an ANOVA test. For instance, the contrast matrix shown below produces the same test results but with a different column of log-fold changes.

```
> con <- makeContrasts(
+    L.PvsL = L.pregnant - L.lactate,
+    L.VvsP = L.virgin - L.pregnant, levels=design)
```

If all three contrasts are present in the contrast matrix, then only the log-fold changes of the first two contrasts are shown in the output of `topTags`.

4.2 Complicated Contrasts

The GLM framework is highly flexible in that arbitrary contrasts can be specified by the user. Suppose we are interested in the differences in the time effect from late pregnancy to early lactation between the two cell populations, i.e., whether the change in expression between pregnant and lactating groups for basal cells is the same as that for luminal cells. An appropriate contrast can be made as shown below.

```
> con <- makeContrasts(
+    (L.pregnant-L.lactate)-(B.pregnant-B.lactate),
+    levels=design)
```

The contrast is passed to `glmQLFTest` to test for genes that are DE under this comparison. The top set of DE genes can be viewed with `topTags`. A positive log-fold change represents a stronger time effect in the luminal over the basal population.

```
> res <- glmQLFTest(fit, contrast=con)
> topTags(res)
Coefficient: 1*B.lactate -1*B.pregnant -1*L.lactate 1*L.pregnant
      ENTREZID    SYMBOL logFC  logCPM   F    PValue      FDR
8947     19041       Ppl  4.62    6.96 525 9.60e-12 1.52e-07
19483   231991     Creb5  5.61    4.86 439 2.93e-11 2.07e-07
7967     20512    Slc1a3 -5.03    3.69 418 3.93e-11 2.07e-07
4354    217294  BC006965  3.88    4.67 372 8.14e-11 2.92e-07
1929     14598      Ggt1 -3.17    6.38 357 1.04e-10 2.92e-07
```

```
9083     13358  Slc25a1 -3.47   7.50 354 1.11e-10 2.92e-07
12763   192166    Sardh -2.92   5.11 342 1.36e-10 3.01e-07
25207    19659     Rbp1  4.40   6.81 336 1.65e-10 3.01e-07
15727    67547  Slc39a8 -6.19   5.11 378 1.72e-10 3.01e-07
6536     14063    F2rl1  3.92   5.60 302 2.95e-10 4.34e-07
```

4.3 Gene Ontology Enrichment Analysis

Further analyses are required to interpret the differential expression results in a biological context. One common downstream procedure is a gene ontology (GO) enrichment analysis. Genes are grouped into GO categories, or GO terms, by some common biological property. Then, given a set of genes that are up- or down-regulated under a certain contrast of interest, a GO enrichment analysis will find which GO terms are over-represented (or under-represented) using annotations for the genes in that set.

Suppose we want to identify GO terms that are over-represented in the basal lactating group compared to the basal pregnancy group. This can be achieved by applying the `goana` function to the differential expression results of that comparison. The top set of most enriched GO terms can be viewed with the `topGO` function.

```
> con <- makeContrasts(B.lactate - B.pregnant, levels=design)
> res <- glmQLFTest(fit, contrast=con)
> go <- goana(res, species = "Mm")
> topGO(go, n=10)
                                             Term Ont    N  Up Down
GO:0044822                     poly(A) RNA binding MF 1077  84  324
GO:0003723                             RNA binding MF 1391 118  383
GO:0042254                    ribosome biogenesis BP  190   5  102
GO:0022613 ribonucleoprotein complex biogenesis BP  285  18  127
GO:0022626                     cytosolic ribosome CC   85   0   61
GO:0005730                              nucleolus CC  667  74  213
GO:0030529              ribonucleoprotein complex CC  592  34  195
GO:0006364                        rRNA processing BP  136   2   72
GO:0016072                rRNA metabolic process BP  139   4   72
GO:0003676                  nucleic acid binding MF 2748 329  593
            P.Up   P.Down
GO:0044822 1.000 2.85e-38
GO:0003723 1.000 6.49e-36
GO:0042254 1.000 2.69e-35
GO:0022613 1.000 6.31e-33
GO:0022626 1.000 1.23e-31
GO:0005730 0.989 7.50e-29
GO:0030529 1.000 2.35e-28
GO:0006364 1.000 8.95e-25
GO:0016072 1.000 5.09e-24
GO:0003676 1.000 1.52e-23
```

The row names of the output are the universal identifiers of the GO terms, with one term per row. The `Term` column gives the names of the GO terms. These terms cover three domains - biological process (BP), cellular component (CC) and molecular function (MF), as shown in the `Ont` column. The `N` column represents the total number of genes that are annotated with each GO term. The `Up` and `Down` columns represent the number of genes with the GO term that are significantly up- and down-regulated in this differential expression comparison, respectively. The `P.Up` and `P.Down` columns contain the *p*-values for over-representation of the GO term across the set of up- and down-regulated genes, respectively. The output table is sorted by the minimum of `P.Up` and `P.Down` by default.

Notes

1. Users can specify the domain of the enriched GO terms in `topGO`. For instance, `topGO(go,ontology="BP")` lists the top set of most enriched GO terms that are related to a biological process. This avoids other domains that are not of interest.
2. The `goana` function uses the NCBI RefSeq annotation. Therefore, the Entrez Gene identifier (ID) should be supplied for each gene as the row names of `res`.
3. Obviously, users should set `species` according to the organism being studied.

4.4 Rotation Gene Set Tests

Another downstream step uses the rotation gene set test (ROAST) [19]. Given a set of genes, the aim of this procedure is to test whether the majority of the genes in the set are DE across the contrast of interest. It is useful when the specified set contains all genes involved in some pathway or process, such that systematic differential expression across the set indicates a change in the activity of the entire pathway or process.

In our case study, suppose we are interested in three GO terms related to cytokinesis. Each term will be used to define a set containing all genes that are annotated with that term. The names of these terms can be viewed as shown below.

```
> cyt.go <- c("GO:0032465", "GO:0000281", "GO:0000920")
> term <- select(GO.db, keys=cyt.go, columns="TERM")
> term
        GOID                          TERM
1 GO:0032465    regulation of cytokinesis
2 GO:0000281          mitotic cytokinesis
3 GO:0000920 cytokinetic cell separation
```

The first step is to extract the set of genes for each GO term from the GO database. We construct a list of three components, each of which is a vector of Entrez Gene IDs for all genes annotated with one of the GO terms. We then convert the Gene IDs into row indices of the `fit` object using the function `ids2indices`.

```
> Rkeys(org.Mm.egGO2ALLEGS) <- cyt.go
> ind <- ids2indices(as.list(org.Mm.egGO2ALLEGS),
+     fit$genes$ENTREZID)
```

Finally, we proceed to run ROAST on the defined gene sets for the contrast of interest. Suppose the comparison of interest is between the virgin and lactating groups in the basal population. We use `mroast` to test for multiple gene sets.

```
> con <- makeContrasts(B.virgin-B.lactate, levels=design)
> rst <- mroast(y, index=ind, design=design, nrot=9999,
+    contrast=con)
> rst
           NGenes PropDown PropUp Direction PValue     FDR
GO:0032465    42     0.167 0.476        Up 0.0004 0.00105
GO:0000920    16     0.688 0.188      Down 0.0014 0.00202
GO:0000281    26     0.385 0.423        Up 0.0058 0.00580
           PValue.Mixed FDR.Mixed
GO:0032465       2e-04    2e-04
GO:0000920       1e-04    1e-04
GO:0000281       1e-04    1e-04
```

Each row corresponds to a single gene set, i.e., GO term. The `NGenes` column gives the number of genes in each set. The `PropDown` and `PropUp` columns contain the proportions of genes in the set that are down- and up-regulated, respectively, with absolute fold changes greater than $\sqrt{2}$. The net direction of change is determined from the significance of changes in each direction, and is shown in the `Direction` column. The `PValue` provides evidence for whether the majority of genes in the set are DE in the specified direction, whereas the `PValue.Mixed` tests for differential expression in any direction. FDRs are computed from the corresponding *p*-values across all sets.

A barcode plot can be produced with the `barcodeplot` function to visualize the results for any particular set. In this case, visualization is performed for the gene set defined by GO:0032465 (Fig. 6). Here, genes are represented by bars and are ranked from left to right by decreasing log-fold change. This forms the barcode-like pattern. The line above the barcode shows the relative local enrichment of the vertical bars in each part of the plot. This particular plot suggests that most genes in this set are up-regulated in the virgin group compared to the lactating group.

```
> res <- glmQLFTest(fit, contrast=con)
> barcodeplot(res$table$logFC, ind[[1]], main=names(ind)[1])
```

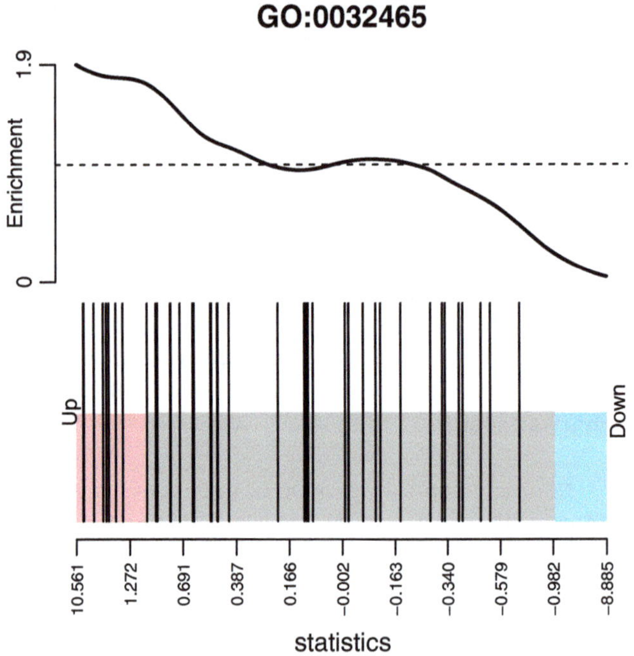

Fig. 6 A barcode plot of genes in one of the GO terms, for the comparison between virgin and lactating groups in the basal population

Notes

1. Unlike `goana`, ROAST is not limited to GO terms. Any predefined gene set can be used, for example, KEGG pathways or MSigDB gene sets. A common application is to use a set of DE genes that was defined from an analysis of an independent data set. ROAST can then determine whether similar changes are observed in the contrast of interest for the current data set.
2. Even for GO-defined gene sets, `goana` and ROAST have different behaviours. In `goana`, the significance of differential expression for a GO term is determined relative to other DE genes that are not annotated with that term. In ROAST, only differential expression for the genes in the set is relevant to the significance of that set and its corresponding term. `goana` depends on a significance cutoff to choose DE genes, whereas ROAST does not require a cutoff and evaluates all genes in the set.
3. The `roast` function can be used to test each gene set individually.
4. ROAST estimates p-values by simulation, so the results may change slightly between runs. More precise p-values can be obtained by increasing the number of rotations, albeit at the cost of increased computational time.

5. The smallest p-value that can be reported is $1/(2 \cdot nrot + 1)$ where nrot is the number of rotations. This lower bound can be decreased by increasing `nrot`.

4.5 Session Information

The session information describes the versions of R and of the packages that were used in the analysis. This is useful for record-keeping purposes, and ensures that an analysis can be reproduced even when the software is updated over time.

```
> sessionInfo()
R version 3.2.0 (2015-04-16)
Platform: x86_64-w64-mingw32/x64 (64-bit)
Running under: Windows 7 x64 (build 7601) Service Pack 1

locale:
[1] LC_COLLATE=English_Australia.1252
[2] LC_CTYPE=English_Australia.1252
[3] LC_MONETARY=English_Australia.1252
[4] LC_NUMERIC=C
[5] LC_TIME=English_Australia.1252

attached base packages:
[1] parallel stats4 stats graphics grDevices utils
[7] datasets methods base

other attached packages:
 [1] GO.db_3.1.2       org.Mm.eg.db_3.1.2 RSQLite_1.0.0
 [4] DBI_0.3.1        AnnotationDbi_1.30.1 GenomeInfoDb_1.4.0
 [7] IRanges_2.2.1     S4Vectors_0.6.0    Biobase_2.28.0
[10] BiocGenerics_0.14.0 edgeR_3.10.0     limma_3.24.5

loaded via a namespace (and not attached):
[1] locfit_1.5-9.1 lattice_0.20-31 grid_3.2.0   splines_3.2.0
[5] statmod_1.4.21
```

Acknowledgements

This worked was funded by the University of Melbourne (Elizabeth and Vernon Puzey Scholarship to Aaron T.L. Lun), by the National Health and Medical Research Council (NHMRC) (Fellowship 1058892 and Program 1054618 to Gordon K. Smyth), by the NHMRC Independent Research Institutes Infrastructure Support (IRIIS) Scheme, and by a Victorian State Government Operational Infrastructure Support (OIS) Grant.

References

1. Mortazavi A et al (2008) Mapping and quantifying mammalian transcriptomes by RNA-Seq. Nat Methods 5.7:621–628
2. Wang Z, Gerstein M, Snyder M (2009) RNA-Seq: a revolutionary tool for transcriptomics. Nat Rev Genet 10.1:57–63

3. Shendure J, Aiden EL (2012) The expanding scope of DNA sequencing. Nat Biotechnol 30.11:1084–1094
4. Liao Y, Smyth GK, Shi W (2013) The Subread aligner: fast, accurate and scalable read mapping by seed-and-vote. Nucleic Acids Res 41.10:e108
5. Robinson MD, McCarthy DJ, Smyth GK (2010) edgeR: a Bioconductor package for differential expression analysis of digital gene expression data. Bioinformatics 26.1:139–140
6. McCarthy DJ, Chen Y, Smyth GK (2012) Differential expression analysis of multifactor RNA-Seq experiments with respect to biological variation. Nucleic Acids Res 40.10:4288–4297
7. Lund et al SP (2012) Detecting differential expression in RNA-sequence data using quasi-likelihood with shrunken dispersion estimates. Stat Appl Genet Mol Biol 11.5:Article 8
8. Robinson MD, Smyth GK (2008) Small-sample estimation of negative binomial dispersion, with applications to SAGE data. Biostatistics 9.2:321–332
9. Robinson MD, Smyth GK (2007) Moderated statistical tests for assessing differences in tag abundance. Bioinformatics 23.21:2881–2887
10. Anders S et al (2013) Count-based differential expression analysis of RNA sequencing data using R and Bioconductor. Nat Protoc 8:1765–1786
11. Fu NY, Rios A, Pal B, Soetanto R, Lun ATL, Liu K, Beck T, Best S, Vaillant F, Bouillet P, Strasser A, Preiss T, Smyth GK, Lindeman G, Visvader J (2015) EGF-mediated induction of Mcl-1 at the switch to lactation is essential for alveolar cell survival. Nat Cell Biol 17.4:365–375
12. Huber W et al (2015) Orchestrating high-throughput genomic analysis with Bioconductor. Nat Methods 12.2:115–121
13. Trapnell C, Pachter L, Salzberg SL (2009) TopHat: discovering splice junctions with RNA-seq. Bioinformatics 25.9:1105–1111
14. Liao Y, Smyth GK, Shi W (2014) featureCounts: an efficient general-purpose read summarization program. Bioinformatics 30:923–930
15. Anders S, Pyl PT, Huber W (2015) HTSeq—a Python framework to work with high-throughput sequencing data. Bioinformatics 31.2:166–169
16. Smyth GK (2004) Linear models and empirical Bayes methods for assessing differential expression in microarray experiments. Stat Appl Genet Mol Biol 3.1:Article 3
17. Phipson B et al (2013) Empirical Bayes in the presence of exceptional cases, with application to microarray data. Tech. rep. Bioinformatics Division, Walter and Eliza Hall Institute of Medical Research, Melbourne, Australia, May 2013. http://www.statsci.org/smyth/pubs/RobustEBayesPreprint.pdf
18. Robinson MD, Oshlack A (2010) A scaling normalization method for differential expression analysis of RNA-seq data. Genome Biol 11.3:R25
19. Wu D et al (2010) ROAST: rotation gene set tests for complex microarray experiments. Bioinformatics 26.17:2176–2182

Index

A

B

C

D

E

F

G

Ewy Mathé and Sean Davis (eds.), *Statistical Genomics: Methods and Protocols*, Methods in Molecular Biology, vol. 1418, DOI 10.1007/978-1-4939-3578-9, © Springer Science+Business Media New York 2016